EAI/Springer Innovations in Communication and Computing

Series Editor

The impact of information technologies is creating a new world yet not fully understood. The extent and speed of economic, life style and social changes already perceived in everyday life is hard to estimate without understanding the technological driving forces behind it. This series presents contributed volumes featuring the latest research and development in the various information engineering technologies that play a key role in this process. The range of topics, focusing primarily on communications and computing engineering include, but are not limited to, wireless networks; mobile communication; design and learning; gaming; interaction; e-health and pervasive healthcare; energy management; smart grids; internet of things; cognitive radio networks; computation; cloud computing; ubiquitous connectivity, and in mode general smart living, smart cities, Internet of Things and more. The series publishes a combination of expanded papers selected from hosted and sponsored European Alliance for Innovation (EAI) conferences that present cutting edge, global research as well as provide new perspectives on traditional related engineering fields. This content, complemented with open calls for contribution of book titles and individual chapters, together maintain Springer's and EAI's high standards of academic excellence. The audience for the books consists of researchers, industry professionals, advanced level students as well as practitioners in related fields of activity include information and communication specialists, security experts, economists, urban planners, doctors, and in general representatives in all those walks of life affected ad contributing to the information revolution.

Indexing: This series is indexed in Scopus, Ei Compendex, and zbMATH.

About EAI - EAI is a grassroots member organization initiated through cooperation between businesses, public, private and government organizations to address the global challenges of Europe's future competitiveness and link the European Research community with its counterparts around the globe. EAI reaches out to hundreds of thousands of individual subscribers on all continents and collaborates with an institutional member base including Fortune 500 companies, government organizations, and educational institutions, provide a free research and innovation platform. Through its open free membership model EAI promotes a new research and innovation culture based on collaboration, connectivity and recognition of excellence by community.

Milan Turčáni
Editor

15th International Scientific Conference on Distance Learning in Applied Informatics

DiVAI 2024

Springer

EAI
RESEARCH MEETS INNOVATION

Editor
Milan Turčáni
Constantine the Philosopher University
Nitra, Slovakia

ISSN 2522-8595 ISSN 2522-8609 (electronic)
EAI/Springer Innovations in Communication and Computing
ISBN 978-3-031-81260-6 ISBN 978-3-031-81261-3 (eBook)
https://doi.org/10.1007/978-3-031-81261-3

This Springer imprint is published by the registered company Springer Nature Switzerland AG
The registered company address is: Gewerbestrasse 11, 6330 Cham, Switzerland

Preface

Dear readers!

At the beginning of the new millennium in 2000, e-learning enthusiasts at the Department of Informatics, Faculty of Natural Sciences, Constantine the Philosopher University in Nitra decided to organize the first meeting of e-learning enthusiasts as part of a conference focused on the application of ICT in the teaching of computer science subjects. In the early years, the conference organizers were very careful about the number of participants. The participants mostly came from computer science departments at universities in the Czech and Slovak Republics, which was mainly related to the considerable reluctance of other departments to use modern information technologies to support teaching at their institutions. In the initial years of the conference, which we called DiVAI (Distance Education in Applied Informatics), a strong community of experts in the field of informatics managed to prove that the supporting education using Internet tools has its justification and a permanent place, especially in the distance form of student education. Departments of informatics in Slovakia and the Czech Republic began using these tools and created an appropriate educational environment for their activities. At the FNS UKF Department of Informatics in Nitra, such a tool was the LMS MOODLE management education system, which is constantly used not only at our institution but also at most universities in Europe.

Later, we expanded participation in the conference to educators and experts from the surrounding countries, especially the Czech Republic, Poland, Slovenia, Lithuania, Latvia, and Hungary. For the 14th year, we will welcome participants from Serbia, Ukraine, Azerbaijan, and other countries. The conference, within the framework of higher education, has one common feature: the use of Internet services and tools to support teaching at various types and levels of education, thereby removing obstacles to ongoing cooperation in this area. After the 9th conference and based on reviews and feedback from participants, we submitted our results in the form of proceedings for indexing in the Thomson Reuters WoS database to enhance the scientific quality of the conference outputs. After a period, we were pleased by the positive evaluation and subsequent indexing of the proceedings in the WoS database. In previous editions of the conference, we published accepted and peer-reviewed papers in printed form. We requested this form of publication from the renowned publishing house Wolters Kluwer, which has a representative office in Prague.

The jubilee 15th year of the DiVAI 2024 conference is significant not only for its milestone but also for the change of editor and publisher. This edition will be managed by the Springer Publishing House, and the outputs will be published in their publication series. We believe that, after a thorough review and selection of the best contributions, you will find the same high-quality professional material in the field of higher education using modern, high-quality information resources.

In conclusion, I wish all participants, as well as readers of the conference results, which come from professional practice, and all those interested in these issues at all levels of education, a valuable experience and the acquisition of new knowledge in the field.

I would also like to express my gratitude to all the members of the program committee and the organizing committee for their willingness and helpfulness in preparing and conducting the DiVAI 2024 conference and for editing the final publication. I believe that the submitted contributions will be published in a high-quality edition by Springer and their subsequent indexing will be successful in both the Scopus and WoS databases.

Nitra, Slovakia Milan Turčáni

Conference Organization

Organizing Committee

Milan Turčáni	Constantine the Philosopher University in Nitra, Nitra, Slovakia
Zoltán Balogh	Constantine the Philosopher University in Nitra, Nitra, Slovakia
Michal Munk	Constantine the Philosopher University in Nitra, Nitra, Slovakia
Martin Magdin	Constantine the Philosopher University in Nitra, Nitra, Slovakia
Jan Francisti	Constantine the Philosopher University in Nitra, Nitra, Slovakia
L'ubomír Benko	Constantine the Philosopher University in Nitra, Nitra, Slovakia
Kristián Fodor	Óbuda University, Budapest, Hungary
Štěpán Hubálovský	University of Hradec Králové, Hradec Králové, Czech Republic
Petra Poulová	University of Hradec Králové, Hradec Králové, Czech Republic
Darina Tóthová	EUNIS Slovakia, Slovakia
Lucie Rohlíková	University of West Bohemia in Pilsen, Plzeň, Czech Republic
Jan Skalka	Constantine the Philosopher University in Nitra, Nitra, Slovakia

Technical Program Committee

Tibor Tóth	Constantine the Philosopher University in Nitra, Nitra, Slovakia
Katarína Látečková	Constantine the Philosopher University in Nitra, Nitra, Slovakia
Mária Burianová	Constantine the Philosopher University in Nitra, Nitra, Slovakia
Matúš Šášik	Constantine the Philosopher University in Nitra, Nitra, Slovakia

Contents

About the Editor

Milan Turčáni obtained his engineering degree (Ing.) in 1978 in the field of Technical Cybernetics—Automation and Regulation at the Faculty of Electrical Engineering of the Slovak Technical University in Bratislava. He received his doctoral degree in 1988 in Technical Cybernetics at the Faculty of Electrical Engineering of the Slovak Technical University in Bratislava. In 1998 he habilitated in the field of Theory of Mathematics Teaching, and in 2005 he inaugurated in the field of Educational Technology at the University of Constantine the Philosopher in Nitra, Slovak Republic. Currently he is a professor at the Department of Informatics of the Constantine the Philosopher University in Nitra. His research interests include: Application of new forms of education using ICT—E-learning and research and development of systems for the Internet of Things (IoT). His current research interests include reading comprehension using eyetracking technologies. Milan Turčáni is a member of the editorial boards of *Acta Pragiensis Informatika ČVUT Praha*, *ERIES ČZU Praha*, and *Problems of Education in the twenty-first Century*. He is a member of several conference program committees and founder of the DiVAI conference.

Current Issues in Effective Learning: Methodological and Technological Challenges and Opportunities Based on Modern ICT and Artificial Intelligence

György Molnár and Enikő Nagy

1 Introduction

The educational environment of the twenty-first century is constantly changing, presenting teachers and students alike with new methodological and technological challenges. Today, the current focus on effective learning is not only on traditional pedagogical methods but increasingly on modern information and communication technologies (ICT) and artificial intelligence (AI)-based solutions. ICT tools and platforms, such as online learning systems, digital textbooks and virtual classrooms, are opening up new dimensions in education, enabling personalised learning and distance learning [22]. At the same time, innovative applications of AI, such as adaptive learning systems and student performance analytics, are revolutionising learning effectiveness and pedagogical practices. However, all these advances also pose major challenges for the education system, as it is necessary to ensure adequate infrastructure, continuous training for teachers and digital access for students. The primary aim of this chapter is to provide an overview of current issues in effective learning, with a special focus on the methodological and technological opportunities and challenges of modern ICT and AI-based learning [9, 18]. Education is undergoing a radical change. Students are using AI tools that go beyond traditional learning

G. Molnár (✉)
Kandó Kálmán Faculty of Electrical Engineering, Óbuda University, Budapest, Hungary

Apáczai Csere János Faculty of Humanities, Education and Social Sciences, Széchenyi University, Győr, Hungary
e-mail: molnar.gyorgy@uni-obuda.hu

E. Nagy
Antal Bejczy Center for Intelligent Robotics, John von Neumann Faculty, Óbuda University, Budapest, Hungary
e-mail: nagy.eniko@nik.uni-obuda.hu

M. Turčáni (ed.), *15th International Scientific Conference on Distance Learning in Applied Informatics*, EAI/Springer Innovations in Communication and Computing,
https://doi.org/10.1007/978-3-031-81261-3_1

aids. These tools personalise learning: AI can create individual learning plans and suggest resources. Interacting with AI can improve learning skills and help students develop critical thinking, problem-solving and technical literacy. Virtual assistants can help them increase their efficiency, manage scheduling, organisation and time management. This growing trend of AI-driven learning forms the basis of our research. Accordingly, it is increasingly important to research the impact of these tools on education by investigating traditional e-learning technologies and their AI-based equivalents [19]. In light of this, the existence and continuous development of appropriate digital competences on the part of educators is now essential for the implementation of effective learning. Digital competence development is therefore of increasing importance in the digital world. It is important to continuously develop and update digital skills in education and industry to keep pace with technological change. To do this, good-quality educational materials and training are available to help us develop our digital competences to an appropriate level. These are supported by digital guides and various national (DOS, DJP) and international plans (Digital Education Action Plan 2021–27) [11]. If we had to define effective learning, we could say that it basically means the success of the learning process in such a way that the learning objectives can be achieved by the individual in a short time, i.e. a rapid learning process that meets the requirements at the output. Effective learning is a process in which the learner is able to acquire new knowledge, skills and competences quickly and thoroughly, while retaining and applying them over the long term. The characteristics of effective learning include a high level of understanding, the development of problem-solving skills and the practical application of new knowledge. Another definition of effective learning is the process of optimising the learner's effort and time to achieve the desired learning outcomes. It involves the use of appropriate learning strategies and techniques, active participation and the use of continuous feedback to improve.

2 Digital Learning–Teaching

Digitalisation processes not only affect the school system but also the people who work within it. Three interpretative frameworks can be distinguished, which have a fundamental impact on schools and their actors. According to the social constructivist view, digital development is transforming our whole way of thinking, making knowledge ubiquitous and, at the same time, more open but also more sensitive. Social pessimists, on the other hand, argue that this transformation is rather negative, that people and their knowledge become more superficial, that process and immediate reward as external motivation become more important than content, and that internal control and time perception are lacking [14]. The third interpretation is associated with the biological optimists, who argue that change does occur, but that we manage it with our existing emotional and cognitive capacities, and that as a consequence, activities do not deteriorate or get lost but simply transform, following changes in the external environment [3, 4, 13, 16, 17]. Each interpretation is, in our view, not exclusive but co-existent between education and social actors. As educators and researchers, we must strive to better understand the effects of digital change, so that we can counteract the negative and amplify the positive effects as far as possible, preparing students and learners to adapt to current processes.

Digital learning is a learning method that allows people in different types of training to study online through an electronic interface. There are many ways of doing this, ranging from delivering the whole training exclusively online to blended learning and occasional solutions, i.e. activities used as occasional supplements to the educational process. Digital learning can offer a number of advantages, including universal accessibility, flexibility and the differentiation and customisation of learning materials according to the level and composition of learners within the institution. For those who find it difficult to participate or meet requirements through the formulae found in the traditional education system, digital learning can provide a way to learn at their own pace through independent or semi-independent learning. However, the use of digital learning needs to be very carefully planned in order to support the learning process as effectively as possible. In each case, it is important to consider whether it can really support a particular session and, if so, how, using what method and what range of tools. It should also not be thought of as 'all-powerful'. A lack of motivation, self-learning ability or will can significantly sabotage the success of digital learning, and there are a number of other factors that may only become apparent in the middle or at the end of the process, such as a lack of personal contact, which can also be reflected in performance. Every person, and therefore every learner, is different, and while digital learning may be the most appropriate for some, the traditional method may be the most effective for others. It is important to find the most appropriate learning method for each learner, with the effective involvement of educators, to help them achieve their learning goals.

In the wake of the coronavirus epidemic in 2020, there was a sudden switch to digital, work-based education at all levels of the school system. Generally speaking, the first experiences were very negative, as neither teachers nor pupils and students were prepared for the new system. Digital learning materials were not available, not everyone had the right tools, and the digital skills deficit made the transition difficult. Based on the results of our research from 31 March 2020, we present the key features of the digital switchover at that time.

An online questionnaire survey was used to gather information on the situation from $N = 902$ students, including primary, secondary, vocational and higher education students [21]. A 5-point Likert scale (where 5 is strongly agree and 1 is strongly disagree) self-report survey showed that students rated their ICT competences on average as 4, with the same level of confidence in their ability to use ICT tools. Despite the positive responses, it was shown that there were already major gaps in the regular use of ICT for learning purposes and in the use of programmes and platforms. Although the inclusion of digital tools in the classroom had become a more noticeable trend in the past, it had not yet reached the level where pupils had a digital learning routine. Numerically, it could be shown that learners were using their digital devices more on a daily basis, with a significant increase in the level of device use for learning purposes. The effectiveness of the digital switchover was also influenced by the initial lack of standardised interfaces and digital classrooms. Typically, e-mail accounts were used to send and receive learning materials and assignments in primary and secondary schools, while Google Classroom became more and more common, where learning materials and assignments could be easily placed, and communication was possible. Later on, Skype, Google Meet, Microsoft Teams and other applications were widely used, which, in addition to real-time conferencing, also provided a single, complex system for managing course materials and assignments. The use of these platforms can still be observed today.

It is felt that the education system during and after the digital education era had to change fundamentally to be successful and sustainable. It became clear to more and more institutions, teachers and trainers that face-to-face education cannot be mapped one-to-one into the online space, so new methods and practices had to be developed.

There have been significant changes in assessment. Since it was difficult to check the clarity of writing in papers online, alternative assessments became increasingly prominent. There was a proliferation of free-word marking systems and an increase in the weighting and number of tasks to be submitted, as these options were easier to implement in an online environment. In higher education and vocational education and training, the use of project assignments and assessment through them has become more common. A new feature of the vocational education and training system has been the introduction of portfolios as learning outcome requirements, replacing the previous item-based assessment of lexical knowledge. What these have in common is that they are becoming more widespread not only because they are delivered online (and in some cases regulated by law), but also because they equip students with different skills and competences that they can use in the labour market. The use of project assignments allows learners to move at their own pace, to determine the focus of learning materials and information and to highlight cause and effect, thus developing critical and analytical thinking, digital literacy and communication skills in presentation more effectively.

During and after digital work-based learning, great progress was observed at the system level. The development of educational actors and the emergence of higher digital skills have allowed a shift in focus from overcoming difficulties and solving new situations to the actors, content and methods. The development of technology and the increase in the sophistication and complexity of virtual learning environments [7] have pushed the system towards efficiency, as communication between teachers and students, the placement of learning materials and the management of tasks have become common on platforms. On reflection, the tests thus designed and used, when properly set up, provided students with immediate feedback, and answers and errors were easily retrievable, thus facilitating student progress. Perhaps it is precisely because of this speed and intensity of development that a new trend is emerging in the way learners are changing their behaviour and consuming content, namely acceleration, as they typically watch, listen to and process video material at double or triple speed, looking for the essential elements. A modern way to do this is offered by the H5P modern technology, following the

SCORM format, which can be used and designed for students embedded in or independent of a learning environment.

Studies by several researchers point out that digitalisation processes will have a significant impact on the labour market and the world of work, indicating that these effects will manifest themselves in differentiated ways (relevant research by Molnár, Buda, Námesztovszki) [10]. In other words, it will shape the different forms of work and their digital support in different ways, depending on the professional sector. The effects of digitalisation will be interwoven with macroeconomic effects, with the stakes being the quality of the jobs created, changed or 'shifted' as a result of the evolution of labour markets, wages and social inequalities. The impact of this global phenomenon on the different aspects and nature of education has already been highlighted by several national researchers [18]. Our digital lifestyle and technological progress are a constant catalyst for the development and enhancement of digital competences, which is in line with the concept of the European Digital Agenda for Education (2021–2027).

3 The Impact of Technological Innovation

The digital economy is growing faster and faster (OECD, 2015), from general commerce (e-commerce) to transport (automated vehicles), healthcare (e-data and personalised medicine) and human interaction (social networks). Telecommunication and information technologies (ICT) play an important role in people's professional and private lives; individuals, businesses and governments are increasingly connected through their various devices at home, at work, in public spaces and when travelling. All of this represents a personal transmission pathway, following millions of individual networks, ranging from residential consumer networks to interconnected networks spanning the globe. The interconnection of fixed, mobile and scattered radius networks, the use of communication between devices, the use of cloud services, the use of sensors and the application of data analytics all combine to provide the opportunity for machine learning, remote control and the development of autonomous intelligent systems [12]. An increasing number of devices are connected to the Internet of Things, which increasingly enables significant interconnections between ICT and the economy. Along with this comes the emergence of 'Big Data', as the continuous growth in the number of computer-driven processes and economic activities on the Internet also leads to the incessant generation of large amounts of data. This data is often used by organisations in highly creative ways to create new products, processes, methods and markets. Big Data can also be a catalyst for outstanding technological innovation. The cost of collecting, storing and analysing data is decreasing, while the proliferation of smart ICT applications is increasing the amount of data that can be a major source of innovation and efficiency gains, provided, of course, that the issues around private data are addressed in a meaningful way. All this requires the right mathematical and IT background and professional analysis of the data obtained [19, 20].

So, we can see that the digital economy is emerging not only in the global economy but also in education. Innovation in the education system contributes to innovation in the teaching–learning process and thus ultimately to the successful acquisition of skills and competences. The successful innovation of economies and societies is undoubtedly based on an appropriate educational structure. If education systems cannot rise to this challenge, they themselves will need to innovate. Education policies to foster innovation have focused primarily on the development of STEM skills (the ability to apply the quadruple of science, technology, engineering and mathematics in a coordinated way). While STEM professionals are undoubtedly important for certain types of innovation, particularly technological ones, a broader perspective has recently emerged that requires the presence of competences beyond these skills. The international REFLEX survey, an interview conducted 5 years after graduation, shows that innovative workers, i.e. those who work in innovative organisations and are involved in the introduction of innovations, use more of these skills in their jobs than their non-innovative counterparts.

The need for innovation in education has been demonstrated over the past decades, and as soon as possible, given the circumstances. Several initiatives have been taken to try to develop general frameworks, principles and guidelines for innovation strategies. Already in 2009, the European Union recognised the importance of education- and training-driven innovation strategies as an essential component of overall innovation strategies. Figure 1 shows the performance of innovation systems in EU Member States. They support several approaches at the same time, stimulating competitive actors, whether it is a new technological tool or the testing of a new school model. To promote quality, education systems need to provide accurate information on how effective new innovations and technologies are—do they work, for how long, and against what criteria? Schools and education systems should also carry out qualitative performance and impact evaluations of new innovations and then share their results with the wider community. This will hopefully ensure that, through appropriate regulation, the educational establishment is increasingly able to meet the challenges of the times.

Hungary is strong in the number of foreign doctoral students, in government support for business R&D, in high-technology exports and in the number of joint publications between academia and business. It is particularly strong in cross-sectoral cooperation and in funding and support schemes. Recently, the government has made it a top priority to improve the infrastructure of higher education and research institutions and link them with businesses to create marketable, innovative products and services. The results of this policy are already clear, with an indicator of innovative enterprises and attractive research systems that exceed last year's performance in terms of the share of innovative enterprises [5].

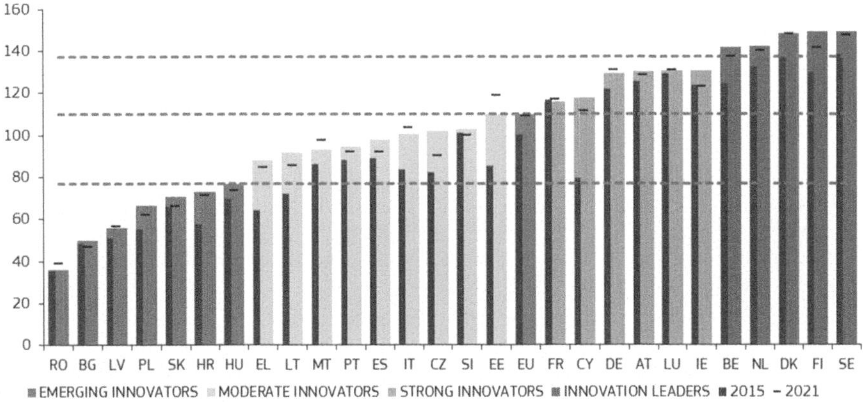

Fig. 1 Performance of EU Member States' innovation systems according to 'EU innovation performance continues to improve'

4 Future Impact of Robotics and Artificial Intelligence

The timeline for the development of artificial intelligence shows an explosion of progress from 1993—the so-called second winter. We are now living in the heyday of artificial intelligence; it is all around us, available in places we would not expect it to be. AI assistants are helping in our everyday lives, and AI systems are providing serious help in areas including medicine, combating online hackers stealing data, streamlining legal research, targeting customers, stock market transactions and preventing crime. It is therefore not surprising that the field of education is not excluded from the list of AI applications. It is not yet clear exactly what role AI will play in this field in the future. Will it transform the learning experience, and if so, at what level and to what extent, making education more interactive and authentic? As the old Scandinavian saying goes: 'Prediction is a difficult business, especially when it comes to the future' (Farvel, 1948). It is particularly difficult to draw any preliminary conclusions, given the fundamental contradiction between machines and humans. Computers will follow instructions exactly, will not deviate from them and will never break the rules and routines they are supposed to follow. This is essentially the basis of Artificial Intelligence (AI) systems [22]. AIs are insensitive, unemotional and 'loyal' to the codes that govern their operation. Humans, on the other hand, are intuitive, emotional creatures and are willing to deviate from or even break the rules when they see fit! The technologies to support education are already available to us; just think of the various e-learning systems. The key issue now is efficiency and taking personal attitudes into account.

According to Rubens, machine learning and data mining techniques may be the technologies that will enable the next version of Web 3.0 and e-learning (Fig. 2). The use of e-learning systems, in most cases, means nothing more than the transfer of face-to-face classroom work to an online interface [15]. Teaching–

Fig. 2 WEB evolution: self-made figure

learning processes are designed to equip learners with as many skills, abilities and motivation as possible, which they can develop in a self-directed way [6], as this is what guarantees them personalised success both in the classroom and in the world of work. Making the learning process more efficient is the 'learner-centred' knowledge transfer, where students are no longer passive recipients and repeaters of information [8]. AI as an intelligent system using semantic standards—together with even virtual space and 3D visualisation—can be the solution to achieve learner-centred knowledge transfer in e-learning. It is predicted that AI will have a major impact on e-Learning 3.0 development, with Web 3.0 becoming the catalyst for the next generation of e-learning.

According to Wheeler, key technologies for e-learning 3.0, the intelligent learning environment, are: remote computing; augmented smart mobile technology [23]; Artificial Intelligence (AI); semantic technologies; collaborative intelligent filtering; Internet of Things (IoT) [1] and wireless communications; and 3D visualisation (Wheeler, 2009). The question is how these technologies can work together. To envision the next generation of e-learning, we need an AI-enabled model that moves away from the Web-based model that currently 'anchors' the learner to the user interface of a computer, laptop, tablet or smartphone [2]. The idea is to transform this environment into a so-called knowledge-valued interface for the learner and to reformulate the concept of Ambient Intelligence for education. Ambient Intelligence is the possibility of integrating AI into our immediate environment, be it a building, a vehicle or anything else. In addition to the wide range of services it provides, AI systems also offer the potential for personalisation, adaptability and predictive behaviour thanks to deep learning technology. Elements that will support education in the near future:

- 3D visualisations, realistic images that further enhance the immersion level and the effectiveness of the e-learning environment. Different interaction modes are possible, such as virtual reality, augmented reality and mixed reality.
- Personalised techniques to create a personalised learning pathway.

- Biometric identification for easier, faster and simpler identification of the person. On the one hand, it eliminates the need for human intervention and makes the creation and management of the learner's profile more efficient.
- Gamification is another e-learning methodology that is in itself an effective and participatory activity. Interactive and challenging elements allow learners–players to acquire different levels, encouraging them to immerse themselves in the subject matter and develop different strategies through play.

In addition to these, there are many more future trends that will emerge in the coming years, which can be integrated into education to achieve a number of additional teaching–learning goals, complemented by catching up and talent management, or even the development of other skills and competences, such as STEM.

5 Conclusion and Outlook

Based on our many years of experience and the research we have conducted, we believe that the effectiveness—and goodness, if you can call it that—of teaching also depends on teachers knowing when to use digital technology and when to stick to traditional methods. This knowledge will apply not only to the present but, we argue, to the entire twenty-first century. AI-enabled applications are mostly intended to assist or replace human work, so the most important aspect is that they can 'own' human intelligence as deeply as possible. This is often done so well that, for example, in communication, it is often impossible to tell whether a human or a machine is responding in the background. Nevertheless, it is important to bear in mind that a skilled, talented and inspiring teacher cannot be fully replaced or imitated by any machine. Blended learning, in the right forms, can provide a healthy balance between electronic and human learning processes, but only if the e-learning system supporting the training is designed with this in mind. The optimisation and efficiency issues of fully electronic learning solutions require a more complex infrastructure, constantly looking for ways to make the human-machine interface more intimate and personal. Digital guides can also help to develop tacit knowledge. Tacit knowledge is knowledge that is difficult to acquire in a formalised, transferable form and is based on experience and practical work. Digital technologies allow guides to share their experience and practical knowledge more easily in online communities or video tutorials. Today, the current issues of effective learning are increasingly centred around modern information and communication technologies (ICT) and artificial intelligence (AI)-based methodologies and technologies. ICT tools, such as online learning platforms and virtual classrooms, and AI-based applications, such as adaptive learning systems, offer new opportunities for personalised and effective learning. At the same time, these innovations bring with them a number of challenges, such as digital inequality, developing the digital competences of educators, and addressing data protection and ethical issues. The integration of new technologies and methods into education can bring significant benefits, provided that adequate infrastructural support is provided, continuous training for teachers is ensured and digital access for students is guaranteed. The future of effective learning depends to a large extent on the successful use of these technologies and methodologies and on addressing the challenges they pose.

References

1. Balogh Z., Fodor K., Magdin M., Francisti J., Koprda S., and Kővári A.: "Development of Available IoT Data Collection Devices to Evaluate Basic Emotions," Acta Polytech. Hungarica, vol. 19, Dec. 2022.
2. Balogh Z., Munk M., Turcáni, M.: 2011. Analysis of Students' Behaviour in the Web-based Distance Learning Environment, Recent Researches in Circuits, Systems, Communications and Computers, pp. 339–344.
3. Csaba Pléh - Anna Rácz - Péter Soltész - Péter Kardos - Péter Kardos - Eszter Berán - Zsolt Unoka: The soul in the WEB world: relationships and learning in the new ICT environment. Hungarian Psychological Review 2014, no. 4, pp. 679–705.
4. Csaba Pléh: The cognitive consequences of the web world, or does the internet make you dumb or bright. Korunk 2011, no. 8, 9–19.
5. European Innovation Scoreboard 2022. https://ec.europa.eu/assets/rtd/eis/2022/ec_rtd_eis-country-profile-hu.pdf
6. Gogh E., Kovari A.: Experiences of Self-regulated Learning in a Vocational Secondary School. Journal of Applied Technical and Educational Sciences, 2019, 9(2), pp. 72–86.
7. Holik, I., Sanda, I. D.: Values and assessment in today's changing educational environment, In: Fehér, Ágota; Mészáros, László (eds.) " .purify it to bring forth more fruit" (Jn 15,2) VIII Christian Educational Science Conference, 2022, pp. 135–146.
8. Karl, É., Molnár, Gy.: ICT-enhanced STEM skills development opportunities for students. Óbuda University, Kandó Kálmán Faculty of Electrical Engineering, 2022, pp. 144–155.
9. Molnár, G., & Szűts, Z.: Modern ICT and network-based learning support in non-formal and informal learning. Létünk, 2019, 49(1), pp. 189–198. source: https://epa.oszk.hu/00900/00997/00050/pdf/EPA000997_letunk_2019_1_189-198.pdf
10. Molnár, G.: Pedagogy, innovation, technology, digital culture – new directions of digitalisation, Budapest, Hungary, Typotex Kiadó 2022, 126 p., ISBN: 9789634932499
11. Mónika, Rajcsányi-Molnár; Krisztina, Sitku; István, András: INTERSECTORAL COLLABORATION FOR A REGIONAL SKILLS ECOSYSTEM: A HUNGARIAN CASE EURASHE 32nd Annual Conference "Skills for Europe" Bucharest. 2023.
12. Mustafa Zengin, Zafer Albayrak: Designing a New Data Encryption Algorithm Using a Genetic Code Method, Acta Polytechnica Hungarica, 19(2), 2022, pp. 238–252.
13. Pléh, Csaba: The school of the future and the people of the past. Korunk, 2021, (02), 38–46.
14. Rajcsányi-Molnár, M.; Bacsa-Bán, A..: On the road to digitalisation: student experiences of digital education in a higher education institution, JOURNAL OF APPLIED TECHNICAL AND EDUCATIONAL SCIENCES, 2021, 11: 1 pp. 88–110.
15. Rubens, N., Kaplan, D., Okamoto, T.: E-Learning 3.0: anyone, anywhere, anytime, and AI., 2011, https://asset-pdf.scinapse.io/prod/2559362885/2559362885.pdf
16. Susan Greenfield: Identity in the 21st century. HVG Books, Budapest, 2010
17. Susan Greenfield, Mind Change: How Digital Technologies Are Leaving Their Mark On Our Brains, Random House, New York, 2015
18. Szűts, Z.: Digital pedagogy methodologies in the VUCA (rapidly changing, unpredictable, complex, controversial) world. School Culture, 2022, 30(7). source: http://misc.bibl.u-szeged.hu/57260/1/iskolakultura_2020_007_076-090.pdf

19. Kővári, A.; Rajcsányi-Molnár, M.: Mathability and Creative Problem Solving in the MaTech Math Competition, Acta Polytechnica Hungarica 17: 2 2020, pp. 147–161.
20. I. Uvaliyeva, Z. Rakhmetullina, O. Baklanova, and G. Györök: "The development of the staking ensemble of methods for analyzing academic data," Acta Polytech. Hung., vol. 19, no. 11, 2022.
21. Molnár, György: "Contributions to the present and future of digital pedagogy, results and perspectives", MTA-BME Open curriculum development research group publications iv., 2018
22. Katona J., Kovari A.: Examining the learning efficiency by a brain-computer interface system. Acta Polytechnica Hungarica, 2018, 15(3), 251–280.
23. Andrea, Tick; Judit, Beke: Online, Digital or Distance? – Spread of Narratives in ICT-supported Education, Journal of Higher Education Theory and Practice 21: 6 pp. 15–31., 17 p. 2021

Stress vs. Calmness: Machine Learning Classification Using Publicly Available Dataset

Zoltán Balogh and Kristián Fodor

1 Introduction

Today's modern era, with significant and regular advances in technology, provides many possibilities in data studies and analysis. One of them is the current development in wearable devices that provide us with rich biological data. Wearable devices include technologies worn by the general population, such as wristwatches, glasses, chest belts, and the like.

There are various machine learning approaches that can be used with smart wearables, including deep learning, which is used for activity and pattern recognition. Recent machine learning approaches use classification and feature engineering to achieve higher accuracy [1].

Machine learning approaches can be used to overcome the shortcomings of existing technologies. For example, a known shortcoming of fall detection sensors is false alarms and maintenance costs. A machine learning approach integrated with the Internet of Things (IoT) has, according to authors [2], a significant impact on solving these problems with the help of wearable fall detection sensors. Other examples of the use of machine learning can be for the prediction of pulse frequency and skin temperature.

Z. Balogh (✉)
Kandó Kálmán Faculty of Electrical Engineering, Óbuda University, Budapest, Hungary

Faculty of Natural Sciences and Informatics, Constantine the Philosopher University, Nitra, Slovakia
e-mail: zbalogh@ukf.sk

K. Fodor
Kandó Kálmán Faculty of Electrical Engineering, Óbuda University, Budapest, Hungary
e-mail: fodor.kristian@kvk.uni-obuda.hu

M. Turčáni (ed.), *15th International Scientific Conference on Distance Learning in Applied Informatics*, EAI/Springer Innovations in Communication and Computing,
https://doi.org/10.1007/978-3-031-81261-3_2

Data obtained from non-wearable devices must be reliable. Inaccuracies in the data that can be caused by the sensors themselves should be minimized for the sake of the accuracy of the analyses. Overall, although consumer wearable devices that collect heart rate (HR) are probably not yet advanced enough to replace medical-grade equipment anytime soon, these devices can be used to collect HR data in research [3].

2 Related Work

Today's physiological monitoring devices enable high-quality data collection [4]. Nevertheless, various factors can affect data collection and therefore need to be considered.

Incorrect placement of sensors results in meaningless measurements. To avoid such problems, internationally used standards for the correct measurement of physiological data have been established. For sensors that have no defined standards, tests must be performed to verify the best placement. The authors [5] verified that different body placements of sensors result in different signal patterns and classification accuracy.

Another important factor is the sampling rate, which must be appropriate to the signal to achieve a compromise between the amount of data to be processed and its quality. It has been confirmed that a sampling rate of 20 Hz is sufficient for monitoring daily life activities, while audio, speech, and biomedical signals must be sampled at a higher frequency of 40 kHz [6].

Signals can also be degraded by instrument noise, random noise, electrical and magnetic noise, etc., as well as poor sensor contact with the skin and body movements, resulting in noise and incomplete data. Signal processing techniques are needed to remove all unwanted effects from signals. Noise can be reduced using filters such as Kalman filters, Butterworth filters, Median filters, Wiener filters, etc. The choice of the best filter in each case depends on the type of signal features to be extracted and the type of noise [6]. In most experiments, the integration of data obtained from different sensors and devices creates problems. The first challenge is determining the data integration point. For calculations to be accurate, data synchronization must be achieved using timestamps.

Eliminating errors such as incomplete data, duplicates, inconsistencies, or data processing errors is possible using algorithms such as least squares, regression analysis, independent component analysis (ICA), and principal component analysis (PCA) [4].

A study by [7] tested stress detection based on signals such as HR, BVP (blood volume pulse), IBI (interbeat interval), GSR, and temperature data. The authors of the study used smartwatches to obtain this data. Their research demonstrated the effectiveness of the combination of electrodermal activity (EDA) and heart rate (HR) signals for stress detection, achieving high classification accuracy.

The effectiveness of combining Electrodermal Activity (EDA) and Heart Rate (HR) signals for stress detection is underscored by the growing body of research focusing on the utilization of physiological signals for mental health monitoring. EDA, known for its sensitivity to sympathetic nervous system activity, is a prominent indicator of stress, revealing changes in skin conductivity due to perspiration levels [8, 9]. This sensitivity makes EDA a reliable measure for stress detection, as demonstrated by the high accuracy rates achieved in various studies [10]. Heart Rate Variability (HRV), derived from ECG (Electrocardiogram) data, is another critical physiological signal associated with stress, although it is not directly mentioned in the provided context. However, the mention of ECG signals in the context of stress detection suggests the importance of cardiac-related measures [11]. The combination of EDA and HR signals could potentially enhance the accuracy of stress detection models by providing a more comprehensive physiological profile of an individual's stress response. Research has shown that wearable devices equipped with sensors to measure these signals can effectively detect stress in real time [11]. The integration of machine learning algorithms further improves the classification and analysis of stress levels, with techniques such as Support Vector Machine (SVM) and Stacking Ensemble Learning (SEL) showing promising results in accurately classifying stress conditions [10, 12]. Moreover, the development of low-cost and low-power sensors that can collect EDA and ECG signals underscores the feasibility of continuous stress monitoring in everyday settings [13]. The potential for wearable technology to provide insights into an individual's mental health through the analysis of EDA and HR signals is significant, offering a pathway to more personalized and immediate mental health support [14–17]. The combination of EDA and HR signals for stress detection is highly effective, leveraging the strengths of each physiological measure to provide a nuanced understanding of stress responses. This approach, supported by advancements in wearable technology and machine learning, holds great promise for enhancing mental health monitoring and intervention [18–20].

The detection of stress in the university campus environment would help students to increase their academic success and quality of life. Authors [21] conducted extensive research on university students. They developed the studentLife application, which collects data on activity, sleep, conversations, and questionnaires filled out by participants. Their aim was to evaluate mental well-being, stress, loneliness, and their correlation with academic performance. They showed that students' sociability, placement changes, and activity decline as middle school approaches. They found a correlation between GPA, conversations, and students' internal mobility.

3 Materials and Methods

Machine learning approaches in general, and in certain healthcare applications in particular, require the availability of sufficient training data to be able to generalize well on unseen data.

A number of datasets are publicly available that contain data recorded by various sensors that fit the criteria. PhysioNet is a large database that offers large collections of physiological and clinical data for research purposes in many areas, such as arrhythmia recognition, stress detection, and activity recognition. It was founded by the National Institutes of Health (NIH) and is managed by the MIT Laboratory for Computational Physiology.

The WESAD dataset is one of many publicly available datasets collected in a laboratory environment using wearable devices by [22]. The study involved 15 participants who were monitored using the Empatica E4 device, which recorded data such as heart rate, skin conductance, and body temperature.

It contains physiological and movement data recorded from chest and wrist sensors from 15 users during a laboratory study. In our research, we will focus on the data obtained from a sensor on the wrist (Empatica E4). Data such as body temperature, electrodermal activity, blood volume pulse, and movement information were recorded. The WESAD dataset is a valuable resource for wearable stress and affect detection, featuring physiological and motion data collected from wrist- and chest-worn devices. It includes sensor modalities such as blood volume pulse, electrocardiogram, electrodermal activity, electromyogram, respiration, body temperature, and three-axis acceleration, providing a comprehensive view of the subjects' physiological responses. Additionally, the dataset covers three affective states (neutral, stress, and amusement) and includes self-reports from subjects obtained through established questionnaires. Machine learning methods applied to the dataset have shown promising results, with accuracies of up to 80% in three-class classification and up to 93% in binary stress classification [1, 23]. However, the dataset's effectiveness in different contexts, such as user experience evaluation, may vary, highlighting the need for further exploration and adaptation [2]. Based on observations and questionnaires, the data were labeled and divided into categories such as stress, fun, meditation, etc. A neutral state was also recorded. The fun condition included watching fun videos. The stress condition includes cases such as public speaking and mental arithmetic tasks. Finally, the meditation condition was guided meditation after stressful and fun conditions.

3.1 Data Preprocessing

The first step before analysis was to process and format the data for further proper editing. The data was saved in pickle format, which is a Python library used to store data or machine learning models. To efficiently work with the dataset during future editing and analysis, we saved all measurements in one file and added an ID variable that indicates the specific user to whom the measurement belongs. Before creating a uniform file, it is necessary to pay attention to the sampling frequency, which was defined by the authors of the dataset. The frequencies for individual sensors were as follows:

- ACC: 32 Hz
- BVP: 64 Hz
- EDA: 4 Hz
- TEMP: 4 Hz

Since in our research we focused on data from smart bracelets, it was also important to properly filter only those data that were important to us, i.e., ACC (motion sensor), BVP (blood volume pulse sensor), EDA (electrodermal sensor), and TEMP (temperature sensor).

Another variable was a label that served as an identifier or a target variable, which also needed some modifications. At the current stage, they were divided into several groups, such as stress, entertainment, and meditation. In our work, we are focused on the classification of data into two groups (stress and stress-free or calmness). Based on this, we created two possible data distributions. The first solution was to divide the data into two categories, and therefore, we unified the data that were labeled as normal state and meditation into the category of stress-free data, while stress data could remain unedited. The second solution was to divide the data into four categories, such as fun, calm, normal state, and stress, which could represent the level of emotions.

As already mentioned, the data were in pickle format. This format is not suitable for analysis or the training of the machine learning model itself; therefore, the next modification was the transformation of the pickle format into csv, which is compatible with statistical libraries, with which we will then be able to perform further analysis. For this modification, we used the pandas Python library, which is suitable for data analysis and manipulation.

Analyzing the resulting file, we noted problems that had to be modified subsequently. The data table showed many blanks and outliers that were not relevant to model training and therefore had to be removed. The reason for the missing data was a large difference in the sampling frequencies of the sensors. At the same time, we noticed an imbalance in the data.

Solving the problem of empty values, we compared three possible solutions:

- Removal of empty values
- Interpolation of empty values
- Addition of values based on the average

Among the solutions tested was the removal of values because the resulting data was correctly labeled. The drawback of this method was the high amount of empty data, which meant that after removing them, the resulting amount of data was 16 times smaller than in the original table. Supplementing the values based on the average was not appropriate. The data was supplemented randomly, and with a high number of empty values, the resulting solution was too unreliable.

The data imbalance could have led to problems in training the model, as the model could have tended to learn to prefer categories with more examples. The balance of the dataset is important, so we compared oversampling and undersampling algorithms to achieve better model classification success results. We used the unbalanced-learn library and its SMOTE method.

3.2 Algorithm Selection

According to the analysis, we found that tree-based algorithms such as Decision Tree (DT) and Random Forest (RF) perform well. We will focus on these algorithms, but due to uncertainty, since the minimum difference in the data can have a different result for each algorithm, we will include more algorithms. This will give us the opportunity to compare the results on our data and then choose the best one when deploying the model.

Dividing data into training and test sets is crucial for evaluating model performance. While the common practice involves a random train/test split, this may not always be optimal, especially when dealing with imbalanced data or rare examples. To enhance model evaluation, strategies based on predictors' variability can be employed to ensure the inclusion of rare instances without losing representativeness. Additionally, the choice of evaluation strategy significantly impacts model interpretations and performance, particularly in the context of spatiotemporal datasets. Understanding the nonlinear correlation between in-distribution and out-of-distribution performance under subpopulation shifts is essential for developing reliable and fair machine learning models. By considering these factors and employing appropriate evaluation strategies, one can enhance model performance and ensure robustness against random influences. An important step was to divide the data into training and test sets to correctly evaluate the performance of the model on new data. This split consisted of splitting the data into x and y, where x was all the data used to train the model and the y variable contained the target variables. 80% of the data was used for training, and the remaining 20% was used for testing the model. At the same time, we used a random state to be sure that our results were not influenced by random factors, but they were really the result of the algorithm.

4 Results

In the research, we worked with the WESAD dataset. We used machine learning technology to process, analyze, train, and evaluate a stress detection machine learning model. We compared the performance of the models and optimized their accuracy based on conditions such as the size of the data or the number of variables used in training.

We tested different algorithms and their ability to classify new data. For each algorithm, we considered the training efficiency based on the size and balance of the data. In addition, we also investigated the effect of the number of target variables, comparing models trained on binary and multivariate classes. We validated our models using cross-validation and analyzed their performance using metrics such as accuracy.

The results in Table 1 show the accuracies of each of the tested algorithms with respect to the balance of the data. In this test, we evaluated the ability of the

Table 1 Accuracy of models on data with each variable

Machine learning algorithm	Unbalanced data (%)	Undersamped data (%)	Oversampled data (%)
Logistic regression	82,2	60,9	62,2
Decision tree	99,8	99,4	99,7
Random Forest	99,8	99,7	99,8
SVM	82,2	67,1	–
k-NN	98,0	96,9	98,6
SGD	92,2	61,3	–
Naive Bayes	91,6	67,1	–
GBT	91,9	93,1	–
XGBoost	99,6	99,5	99,2
ANN	92,5	84,7	–

algorithms to classify the data into two target variables and their ability to assign the data to the correct category based on the balance of the data. The results in this table indicate that the algorithms worked well on unbalanced data when compared to data that was balanced. We've seen a high success rate trend, which points to a few issues we've had to deal with.

When training the models, we had to minimize the risk of overfitting that occurred with tree-based algorithms. At the same time, we noticed a trend in the classifications, which indicated that the error in the classifications occurred at the borderline values. At the same time, all the problems encountered in this test indicate how important it is to set the hyperparameters correctly. For some algorithms, they can increase the accuracy and overall efficiency of the algorithm, but setting them incorrectly can increase the risk of overfitting or the time needed to optimize them.

Table 2 shows the results of training the models after significant changes in the data as well as hyperparameter settings. We tested the data, which was cleaned from outliers and duplicate values, removed empty values, balanced using the oversampling method, and divided into four target categories. Models like Random Forest, XGBoost, and k-NN achieved high levels of accuracy, while models like Logistic Regression and SGD had low levels of accuracy.

One of the important factors that influenced the results is the ability of the models to capture the complexity of the data. High-accuracy models such as Random Forest are better able to capture complex relationships in the data, which may have led to their higher accuracy. Conversely, models such as Logistic Regression have a limited ability to capture this complexity, resulting in a low level of accuracy.

In addition to differences in model accuracy, we also observed significant differences in training speed. Models such as Logistic Regression were relatively quick to train, due in part to their simplicity and small number of hyperparameters. Conversely, models based on trees and complex algorithms such as XGBoost and Random Forest had longer training speeds.

To identify suitable strategies to control overfitting, it was appropriate to analyze the visualization of performance metrics to better understand the functioning of the

Table 2 Accuracy of models in percentage at four target variables

Machine learning algorithm	Without setting hyperparameters	Optimal setting of hyperparameters
Logistic regression	38,8	39,3
Decision tree	98,0	94,4
Random Forest	98,2	97,5
SVM	48,6	–
k-NN	95,6	97,3
SGD	31,7	40,6
Naive Bayes	37,9	–
GBT	86,7	98,0
XGBoost	96,0	98,2
ANN	81,1	–

models and find ways to improve them. This analysis included the study of training and validation curves, which gave us important information about how the model behaves depending on different hyperparameter settings.

Figure 1 shows the accuracies of the training and validation sets when training the Random Forest model. The blue curve represents the training accuracy with a decreasing tendency. A gradual decline may signal that the model is starting to overfit. The optimal state occurs when the result is consistent within one level, indicating that further addition of training data does not significantly affect its performance. A rising curve indicates the learning of the model and points to the lack of a training set.

The green curve shows the accuracy on the test set. The gradual increase in accuracy indicates that the model was gradually improving accuracy on new data. The gradual convergence of the training and testing curves could indicate that the model is fitting the data correctly and there is no overtraining.

When training a machine learning model, it is important to analyze the importance of the variables we use to train the model. Figure 2 shows a graph of the importance of variables in model training. The high importance of TEMP and EDA values indicated that they had a strong influence on the model predictions. The very low importance of the GDP variable indicated that the model did not consider this feature relevant for prediction. In this case, it was appropriate to remove it.

Figure 3 shows a visualization of the ROC curve. A value of 0.95 indicates a high level of model accuracy, meaning that the model was able to classify 95% of positive cases as positive and 95% of negative cases as negative. This high value indicates the robustness and accuracy of the model in distinguishing between classes.

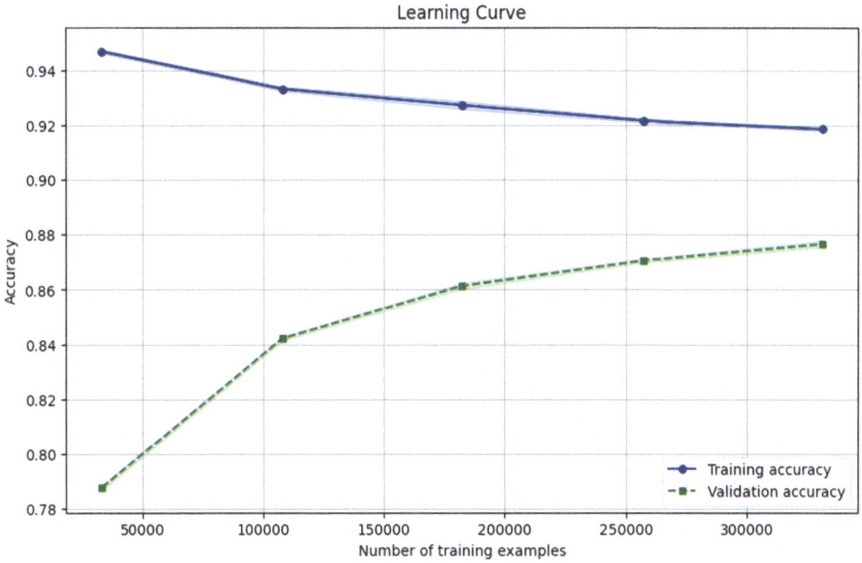

Fig. 1 Training and validation curve of the Random Forest model, where the training accuracy is at 92% at the end of the learning process and validation is at 88%

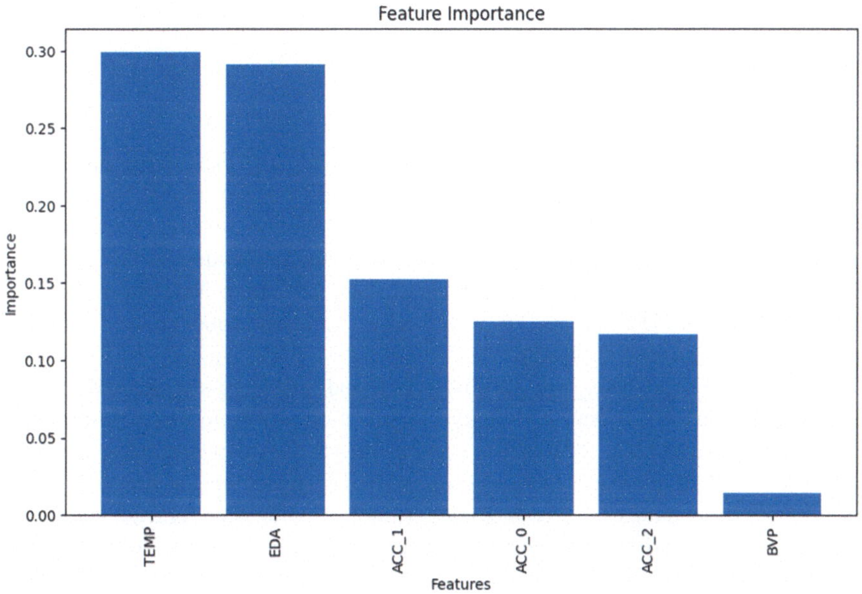

Fig. 2 Feature importance plot highlighting the importance of each feature in the model

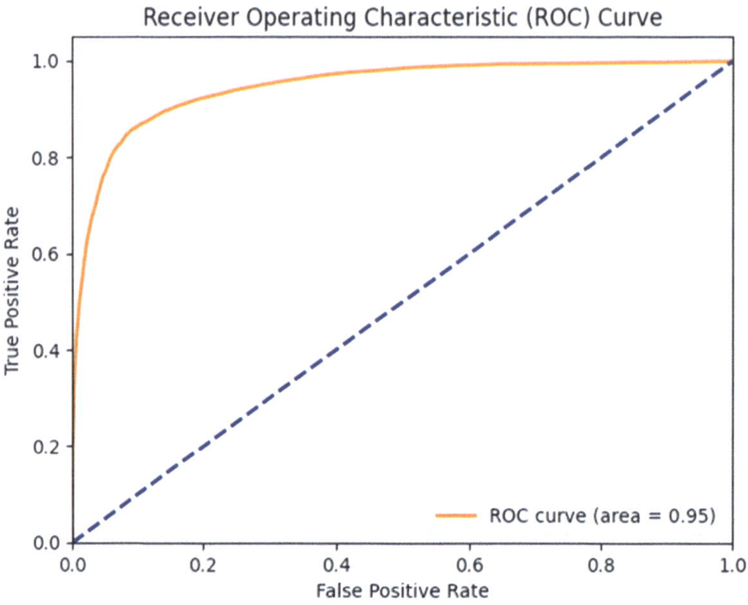

Fig. 3 The ROC curve for the Random Forest model, which stops at 95%, indicates a high classification accuracy

5 Conclusion

In this research, we dealt with the issue of processing and visualizing data from sensory networks with the aim of training a machine learning model for emotion detection. During the work, we examined and compared several algorithms to find the best solution. The analysis included Logistic Regression, Decision Tree, Random Forest, Support Vector Machine (SVM), k-Nearest Neighbors (k-NN), Stochastic Gradient Descent (SGD), Naive Bayes, Gradient Boosting Trees (GBT), and XGBoost algorithms [24, 25].

Based on our analysis and results, we found that Logistic Regression or Naive Bayes performed best for binary classification with minimal requirements for hyperparameter optimization and training. When it was necessary to capture complex relationships, the Random Forest and GBT algorithms achieved the best results. The disadvantages of tree-based algorithms were the need for a large training sample, optimal setting of hyperparameters, and slower training. An important factor for achieving the best results was appropriate preprocessing of the data, including normalization, balancing, and removal of outliers and missing values.

The visualizations we used during the development of the machine learning models helped us a lot in the implementation. They helped us better understand the behavior of individual algorithms and identify areas that required optimization. They also helped us present the results in a simple and clear way.

Overall, we have achieved the goal of creating a machine learning model for emotion detection using data from wearable devices and providing a basis for further research and implementation in this area. One possibility for implementation is the creation of an application that would allow monitoring the emotional state of the user using wearable devices in real-time, which could bring significant benefits in the field of health and wellness.

As mentioned previously, machine learning approaches in healthcare applications require the availability of sufficient training data to be able to generalize well on unseen data. To replicate our study on different data, it is necessary to collect a sufficient amount. In our previous studies, where we have worked with different biodata, we also concluded that Random Forest was the best classifier. However, it is important to address that this might not be the case for all types of data.

Acknowledgments This work was supported by the Scientific Grant Agency of the Ministry of Education of the Slovak Republic (ME SR) and the Slovak Academy of Sciences (SAS) under contract No. VEGA 1/0385/23.

References

1. Z. Lu, A. Narayan, and H. Yu, "A Deep Learning Based End-to-End Locomotion Mode Detection Method for Lower Limb Wearable Robot Control," in *2020 IEEE/RSJ International Conference on Intelligent Robots and Systems (IROS)*, 2020, pp. 4091–4097, https://doi.org/10.1109/IROS45743.2020.9341183.
2. T. Chaudhuri, Y. C. Soh, H. Li, and L. Xie, "Machine learning driven personal comfort prediction by wearable sensing of pulse rate and skin temperature," *Build. Environ.*, vol. 170, p. 106615, 2020, https://doi.org/10.1016/j.buildenv.2019.106615.
3. B. W. Nelson, C. A. Low, N. Jacobson, P. Areán, J. Torous, and N. B. Allen, "Guidelines for wrist-worn consumer wearable assessment of heart rate in biobehavioral research," *npj Digit. Med.*, vol. 3, no. 1, p. 90, 2020, https://doi.org/10.1038/s41746-020-0297-4.
4. A. Alberdi, A. Aztiria, and A. Basarab, "Towards an automatic early stress recognition system for office environments based on multimodal measurements: A review," *J. Biomed. Inform.*, vol. 59, pp. 49–75, 2016, https://doi.org/10.1016/j.jbi.2015.11.007.
5. U. Lindemann, A. Hock, M. Stuber, W. Keck, and C. Becker, "Evaluation of a fall detector based on accelerometers: A pilot study," *Med. Biol. Eng. Comput.*, vol. 43, no. 5, pp. 548–551, 2005, https://doi.org/10.1007/BF02351026.
6. R. Khusainov, D. Azzi, I. E. Achumba, and S. D. Bersch, "Real-Time Human Ambulation, Activity, and Physiological Monitoring: Taxonomy of Issues, Techniques, Applications, Challenges and Limitations," *Sensors*, vol. 13, no. 10, pp. 12852–12902, 2013, https://doi.org/10.3390/s131012852.
7. Y. S. Can, B. Arnrich, and C. Ersoy, "Stress detection in daily life scenarios using smart phones and wearable sensors: A survey," *J. Biomed. Inform.*, vol. 92, p. 103139, 2019, https://doi.org/10.1016/j.jbi.2019.103139.
8. E. Hosseini *et al.*, "A Low Cost EDA-based Stress Detection Using Machine Learning," in *2022 IEEE International Conference on Bioinformatics and Biomedicine (BIBM)*, 2022, pp. 2619–2623, https://doi.org/10.1109/BIBM55620.2022.9995093.
9. J. Francisti, Z. Balogh, J. Reichel, M. Magdin, Š. Koprda, and G. Molnár, "Application Experiences Using IoT Devices in Education," *Appl. Sci.*, vol. 10, no. 20, pp. 1–14, 2020, https://doi.org/10.3390/app10207286.

10. L. Zhu *et al.*, "Stress Detection Through Wrist-Based Electrodermal Activity Monitoring and Machine Learning," *IEEE J. Biomed. Heal. Informatics*, vol. 27, no. 5, pp. 2155–2165, 2023, https://doi.org/10.1109/JBHI.2023.3239305.
11. L. Zhu, P. Spachos, and S. Gregori, "Multimodal Physiological Signals and Machine Learning for Stress Detection by Wearable Devices," in *2022 IEEE International Symposium on Medical Measurements and Applications (MeMeA)*, 2022, pp. 1–6, https://doi.org/10.1109/MeMeA54994.2022.9856558.
12. T. Ninh, S. Smyth, M.-T. Tran, and C. Gurrin, "Analysing the Performance of Stress Detection Models on Consumer-Grade Wearable Devices," 2021.
13. A. Mohammadi, M. Fakharzadeh, and B. Baraeinejad, "An Integrated Human Stress Detection Sensor Using Supervised Algorithms," *IEEE Sens. J.*, vol. 22, no. 8, pp. 8216–8223, 2022, https://doi.org/10.1109/JSEN.2022.3157795.
14. Ž. Stržinar, A. Sanchis, A. Ledezma, O. Sipele, B. Pregelj, and I. Škrjanc, "Stress Detection Using Frequency Spectrum Analysis of Wrist-Measured Electrodermal Activity," *Sensors*, vol. 23, no. 2, 2023, https://doi.org/10.3390/s23020963.
15. Z. Balogh, K. Fodor, M. Magdin, J. Francisti, S. Koprda, and A. Kővári, "Development of Available IoT Data Collection Devices to Evaluate Basic Emotions," *Acta Polytech. Hungarica*, vol. 19, Dec. 2022.
16. M. Magdin, Z. Balogh, J. Reichel, J. Francisti, S. Koprda, and G. Molnár, "Automatic detection and classification of emotional states in virtual reality and standard environments (LCD): comparing valence and arousal of induced emotions," *Virtual Real.*, Mar. 2021, https://doi.org/10.1007/s10055-021-00506-5.
17. K. Fodor, Z. Balogh, and J. Francisti, "Visualization of Physiological Response in the Context of Emotion Recognition," in *Progress in Artificial Intelligence*, 2022, pp. 387–398.
18. J. Francisti, Z. Balogh, J. Reichel, Ľ. Benko, K. Fodor, and M. Turčáni, "Identification of heart rate change during the teaching process," *Sci. Rep.*, vol. 13, no. 1, p. 16674, 2023, https://doi.org/10.1038/s41598-023-43763-x.
19. S. Ishaque, N. Khan, and S. Krishnan, "Detecting stress through 2D ECG images using pretrained models, transfer learning and model compression techniques," *Mach. Learn. with Appl.*, vol. 10, p. 100395, 2022, https://doi.org/10.1016/j.mlwa.2022.100395.
20. R. K and V. R. Murthy Oruganti, "Transfer Learning for Subject-Independent Stress Detection using Physiological Signals," in *2020 IEEE 17th India Council International Conference (INDICON)*, 2020, pp. 1–6, https://doi.org/10.1109/INDICON49873.2020.9342505.
21. R. Wang *et al.*, "StudentLife: assessing mental health, academic performance and behavioral trends of college students using smartphones," in *Proceedings of the 2014 ACM International Joint Conference on Pervasive and Ubiquitous Computing*, 2014, pp. 3–14, https://doi.org/10.1145/2632048.2632054.
22. P. Schmidt, A. Reiss, R. Duerichen, C. Marberger, and K. Van Laerhoven, "Introducing WESAD, a Multimodal Dataset for Wearable Stress and Affect Detection," in *Proceedings of the 20th ACM International Conference on Multimodal Interaction*, 2018, pp. 400–408, https://doi.org/10.1145/3242969.3242985.
23. L. Malviya *et al.*, "Mental Stress Level Detection Using LSTM for WESAD Dataset," in *Proceedings of Data Analytics and Management*, 2023, pp. 243–250.
24. M. Zengin, Z. Albayrak, Designing a New Data Encryption Algorithm Using a Genetic Code Method, *Acta Polytechnica Hungarica*, 19(2), 2022, pp. 238–252.
25. Kővári, A.; Rajcsányi-Molnár, M., Mathability and Creative Problem Solving in the MaTech Math Competition, *Acta Polytechnica Hungarica* 17: 2 2020, pp. 147–161.

Possibilities and Limits of the Use of Extended Reality in Chemistry Education at Lower Secondary School

Kristýna Pospíšilová and Martin Bílek

1 Introduction

Digital technologies are increasingly influencing our lives, placing a great deal of emphasis on education to respond to these changes. Existing teaching methods and organizational forms of learning are undergoing innovation, and it is necessary to understand the new role of the school as an educational institution focused on restructuring the knowledge, attitudes, habits and values of students and teachers [1]. In 2014, the government of the Czech Republic approved the Strategy for Digital Education from 2020 [2], thanks to which digital technologies are gradually being incorporated into teaching, as its integration into education is an inevitable part of the ever-changing technological world. Despite the fact that the use of technology is becoming more common in schools, there is a constant search for ways to streamline its impact on student learning [3].

Don Tapscott, a Canadian marketing strategy specialist, refers to the current generation of learners as the "networked generation" [4]. The "networked generation" is used to using digital technology from a young age and is accustomed to a constant flow of information. It is a generation that learns very quickly in areas that they enjoy. Pupils then find it harder to accept the traditional way of teaching, and it is necessary to respond to this [5]. Therefore, funding is allocated to schools to prevent the digital divide and for digital learning technologies. Primary and secondary schools can use advanced digital technologies across all subjects. Such technologies include robotic kits, 3D printers, electronic building blocks, measurement sensors and a host of other devices and applications, including augmented, mixed and virtual reality [6, 7].

K. Pospíšilová · M. Bílek (✉)
Faculty of Education, Charles University, Prague, Czech Republic
e-mail: martin.bilek@pedf.cuni.cz

M. Turčáni (ed.), *15th International Scientific Conference on Distance Learning in Applied Informatics*, EAI/Springer Innovations in Communication and Computing,
https://doi.org/10.1007/978-3-031-81261-3_3

This is also the case in chemistry teaching, which in its early stages struggles with the higher abstractness of the curriculum. This concerns, in particular, the microworld, which is essential for explaining the changes in substances on which chemistry, as a science, is based. In this area, a variety of graphical materials are used as supporting digital technologies, ranging from static and dynamic images, animations and simulations to applications based on extended and augmented reality (ER and AR). And it is the use of augmented reality (AR) that was the focus of our investigation, the main objective of which was to identify the possibilities and limits of using selected ER applications produced by VividBooks [8] aimed at explaining the context of the topic "Particle composition of substances" in the teaching of chemistry in the eighth grade of primary school.

2 Theoretical Background

Augmented or mixed reality (ER/AR) are interactive representations of the world by combining real and virtual environments [9]. Here, reality is augmented with computer-generated elements. The resulting environment provides both real and augmented information about the surrounding world. Unlike virtual reality (VR, hereafter abbreviated as VR), which leads to complete immersion in an artificially created world, augmented reality enhances what we see by adding additional digital content to the real world [10]. Initially, entry into the virtual world required impractical headsets that drew users to the computer. Over time, ER began to work through various devices (smartphone, tablet, glasses, etc.) that capture the real world and create live animations, texts or sounds that are then viewed by the user in the real-world environment [11].

A typical VR system presents the user in a virtual space that may have nothing to do with their current space. In contrast, a typical ER/AR system adds extra information to the user's current space. The ER/AR user is much more sensitive to subtle variations between added visual stimuli set in the real world. Therefore, the visual stimuli should be strictly aligned with real-world motion [12].

Azuma [12] defines ER/AR as systems that satisfy the following three conditions:

1. They combine reality and virtual reality.
2. They are interactive in real time.
3. They display elements in 3D.

These basic conditions define ER/AR in relation to other technologies, such as 3D films or the fusion of 2D graphics and video [12]. ER/AR complements the real world with virtual objects that relatively coexist with the environment. Over the years, there has been an expanding group of areas that can benefit from so-called extension or augmentation. Apart from the original focus of use for military, medical and industrial applications, ER/AR systems are mainly used commercially or for entertainment. However, they have many other uses, e.g. in education, navigation, design, tourism, sports broadcasting, etc. [13]. ER/ARs have been adopted as an effective support for teaching methods. This new, interactive approach to teaching and learning better meets the needs of the twenty-first-century learner who wants fun, interactivity, participation and manipulation of objects [9]. There are many benefits to be gained from using ER/AR, such as reduced cognitive load, increased motivation and interest in learning, new opportunities for individual learning, concretization of abstract concepts and increased student interaction [10]. It is the concretization of abstract concepts that helps to present things to multiple senses, and thus to remember them better.

Another advantage of using ER/AR is that most students already have the equipment that these technologies require (tablet, smartphone, computer with webcam). Traditional projectors, interactive whiteboards or classrooms equipped with computers or tablets, as well as mobile phones, can be used to utilize ER/AR elements in schools. However, tablets generally have a broader range of applications due to the size of the display [14]. According to the results of various studies [15], however, in addition to the above advantages of using ER/AR, several disadvantages can be mentioned. One of the main disadvantages is the possibility of attention loss, where the learner is so fascinated by the ER/AR that there is a loss of attention over the actual content of the message. The same disadvantage can also be caused by the difficulty of controlling the technology. Another disadvantage is the display of text, which may be harder to read or distract from the context of the message. A final disadvantage is the a priori rejection of the educational role of technology. Nevertheless, ER/AR is used in a number of educational areas. They make it possible to display information about the real world that we could not detect with our own senses.

Despite the great popularity of technology and the Internet, people still prefer reading books instead of reading from a computer screen, tablet or other device [16]. The same applies to textbooks, the printed versions of which are still widely used in practice. Combining textbooks with technology results in ER/AR-enabled textbooks. Such a textbook may be printed, but pointing a webcam at a segment of the book makes it “move”. This technology allows any existing book to be later developed into an ER/AR edition. By using ER/AR, textbooks become a dynamic source of information without the need to carry a computer, just a mobile device [16].

Teaching abstract concepts in chemistry with concrete examples is not always easy for a teacher, even if he or she has a chemistry classroom or laboratory. In laboratory exercises or demonstration experiments, although the phenomenon can be observed visually, pictures, diagrams or animations are needed for further, concrete explanations. These tools, which were previously used mainly to visualize compounds at the molecular level, are now used as teaching media [17]. ER/ARs give students the opportunity to understand more abstract chemical concepts or to visualize the arrangement of atoms in a molecule. Chemistry is not a popular subject due to its difficulty, so motivation to learn and explore is necessary. Using ER/AR, the structure of matter can be explored that would be impossible to visualize in a 2D environment. The technology allows students to see, for example, a molecule from all angles or to visualize the organization of atoms in a molecule. These possibilities increase students' interest, and there is greater engagement in the learning process [18].

3 Educational Extended Reality VividBooks

The ER application of Czech provenance, VividBooks, was created in 2020 by V. Škop and F. Cáb to support the teaching of physics, and later the application was extended to support the teaching of chemistry and other subjects [19]. The chemistry textbooks in VividBooks allow students to show the composition and behaviour of substances in an illustrative and comprehensible way. Each lesson includes introductory text, an animation (ER), questions on the topic ordered from lower cognitive learning objectives to higher ones, as well as a summary, a worksheet for students and a teacher's guide [19].

When opening the Vivid Chemistry app, even just the animations (ER) without the accompanying text and questions can be shared. In the app, the relevant book has to be opened, and the pupil is presented with the option to use the ER. By pointing the device at the drawing, the animation will move. Unlike viewing the Vivid Chemistry animations, which students can use without an Internet connection, when using ER it is necessary to be online [19].

Figure 1 shows on the left the basic worksheet for the topic "Solubility of vitamin C in water and oil". The application allows students to view the animation offline or online, using the ER version of the drawing (Fig. 1, right). At the bottom is the accompanying text on the topic.

4 Research Design

In our research investigation, we focused on monitoring the longer-term and repeated use of ER applications in the thematic unit Particle Composition of Matter in the teaching of chemistry in Year 8 of primary school. A case study of three classes of Year 8 primary school pupils (labelled A, B, C) was selected ($n = 65$, 29 girls and 36 boys). All pupils had a basic understanding of VividBooks applications from previous chemistry lessons prior to the observed lessons. To investigate the effect of incorporating ER features into chemistry lessons on students' knowledge and attitudes, we planned to implement a total of six ER apps over the course of two months of instruction. These were lessons with the following topics (see Table 1).

Prior to the first chemistry lesson of the study period, we conducted initial group interviews with students in each class, focusing on their opinions and previous experiences with VividBooks AR, which they had already encountered occasionally during the course of their lessons. This was followed by the implementation of the lessons taught. After the implementation of all the planned lessons, group interviews

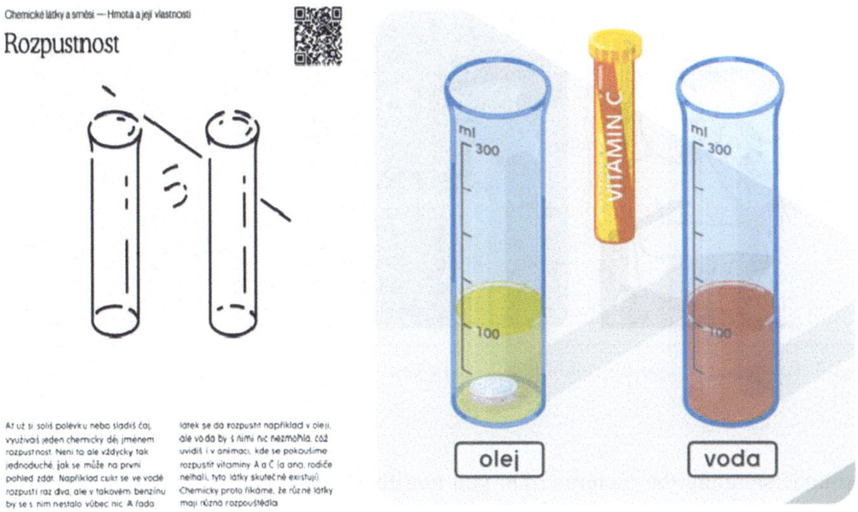

Fig. 1 Basic worksheet for the topic "Solubility" [19]

Table 1 Lessons with extended reality applications

Topic	ER application(s)	Goals
Atomic structure	Atoms and elements—Atomic theory—Atomic structure	To recognize atomic structure (atomic nucleus and electron shell)
Atom models	Atoms and elements—Atomic theory—Atom models	To describe the development of atom models
Electron shell	Atoms and elements—Atomic theory—Atomic Shell	To identify the valence layer and valence electrons
Ion formation	Atoms and elements—Atomic theory—Ions	To identify cations and anions
Electronegativity	Chemical bond—Electronegativity	To distinguish non-polar, polar and ionic bonding
Chemical bond	Chemical bond—Chemical bond	To distinguish single, double and triple chemical bonding

were again conducted to identify students' opinions and attitudes towards ER and the way it was implemented in the realized lessons.

5 Research Realization and Results

For the initial interviews, the following questions were formulated and answered spontaneously by students in each class: How do you like VividBooks? Do you enjoy working with the app? Have you ever opened the VividBooks app at home? And if so, have you ever used the app to learn? Do you find using the app better

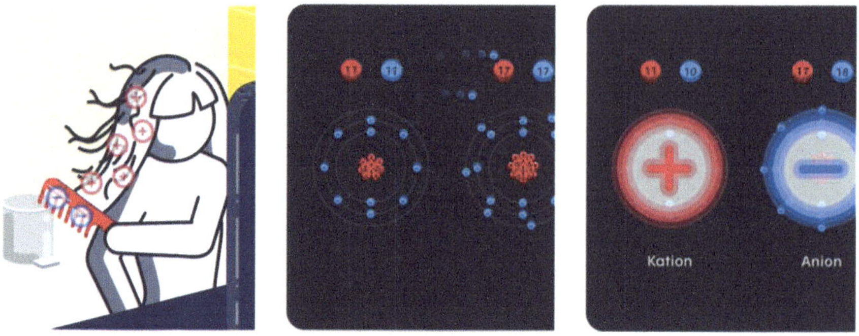

Fig. 2 Formation of ions by combing hair and electron transfer, formation of anion and cation [19]

for understanding the material? Do you find teaching without the app or teaching with the app better? From the students' responses, it can be seen that they mostly treat the app as something new and fun ("It's fun." "It's good for explaining the material." "Awesome for me, I understand it more when I see it on video." "I enjoy it a lot because we can see the process and then I understand it better." Only a small proportion of pupils have a neutral or negative view of the app: "Something is good." "Yeah, it was pretty good." "I don't enjoy it, I don't understand it." "I don't like it, it doesn't explain much." When asked if the Vividbooks apps help them to understand the material, students mostly answered that the apps help them to understand. The majority of pupils were inclined to think that teaching with the app or partial teaching with the app combined with traditional teaching is better. "Teaching with the app. It's fun and I understand more." "With the app because it's easier to understand." "I find it better with an app because it's better to understand and it's easier for me." "Teaching with an app. I'm tired of looking at a whiteboard for 45 minutes and sitting like a potato. It's best when we have VividBooks in group work. I learn more." Responses preferring teaching without the app were rare: "Probably a traditional class, the teacher is very good at explaining it so I understand." "I think classical teaching because I can see it live."

As an example of recording the course of teaching using ER Vividbooks, we can mention the fourth lesson of the whole series with the topic of the formation of ions, where the animation with AR elements from the book *Atoms and Elements—Atomic Theory—Ions* was used (Fig. 2). Ions contains a total of two animations. The first animation shows the formation of ions when combing hair and its consequence of hair being attracted to the comb. The second animation shows the formation of an anion and cation using Bohr's model of the sodium and chlorine atoms.

The structure of the lesson consisted of a common repetition of the characteristics of the atom—electrically neutral, proton number, number of protons and electrons in the atom, their charge, etc. After the repetition, the first animation was presented, and the pupils were asked to try to explain what was happening during the animation. Pupils were then given a worksheet in pairs with a tablet, and their task was to look at the second animation—moving an electron, learning about the concepts of anion and cation and their formation. The pupils' work was accompanied by questions: Which element atoms are in the animation? What does the animation describe? In what ways can we disturb an atom? What happens when this happens? After summarizing the formation of the ions, we made a short notation and used the equation to write the formation of the ions from the animation. Finally, using the electron cards and a sketch of the different layers of the shell of an atom, the pupils were asked to piece together the electron shell of oxygen, and then sulphur, to form the ions O^{2-} and S^{6+}. Next, they determined the number of their protons and electrons, wrote the equation and named the ions.

The pupils spontaneously commented, "We did that last year in physics, but with a balloon." "Yeah, that's right, because plus and minus attract." Questions were formulated for the second animation: Atoms of which elements are in the animation? What does the animation describe? In what ways can we distort the atom? What happens when this happens? Each group then commented on what they observed in the animation, "The first atom is sodium and the second atom is chlorine." "The first one passes the electron to the second one." "Sodium is positively charged and chlorine is negatively charged." "So either I add a proton and you get a cation or you add an electron and you get an anion." "It wouldn't be sodium anymore because it would have a different proton number."

The aim of the lesson was therefore to understand the formation of ions and to introduce the new concepts of anion and cation. The pupils had no difficulty with the animations and were able to describe them well. They had a little difficulty specifying the formation of a cation, with some pupils thinking that a positive ion is formed by adding a positive charge. However, when they replayed the animation, they corrected their mistake. In the next activity without the app, students were asked to create an oxygen anion and a sulphur cation. The students correctly formed atoms of each element, but about half of the pairs were taking electrons away when forming the anion because there is O^{2-}, and vice versa when forming the cation.

The following questions were formulated for the exit interviews with the students: how did you enjoy using VividBooks in your last chemistry lessons? Which apps do you remember? (+ Can you remember what the app explains?) Did anything bother you about the apps you used? What didn't you like about them? Are there any that you found too complicated or incomprehensible? Did you enjoy working with the apps? Can you remember which one you worked with the most? Why did you enjoy working with it? What did it help you understand? Have you opened any of the apps we used in recent chemistry classes at home? Why? Have you shown the app to anyone else? And did you use it to explain a principle or process? Did you find using the app helpful in understanding the material? Which applications can you think of that apply more, and which less? Did you feel that any of the apps tended to obscure what they were meant to help explain? Which one was it and why? Did you find teaching better with or without the apps? Why do you think so? Is it just because they are new and fun for you, or do you see other reasons there?

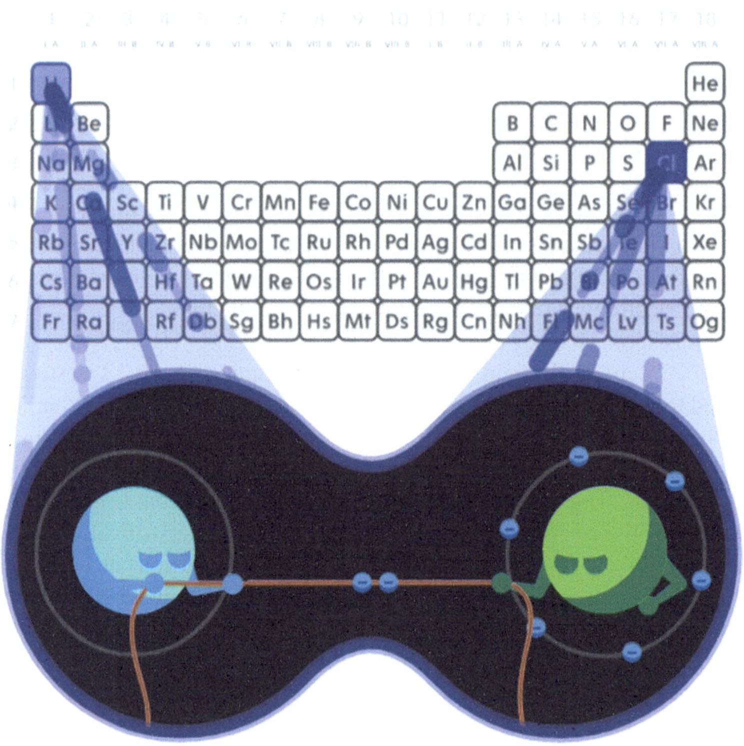

Fig. 3 Electronegativity of hydrogen and chlorine—animation [19]

Pupils again, in the majority, evaluated positively the use of ER applications in lessons with the topic "Particle composition of substances". "Yeah, I liked it." "We could easily do more with it." "The valence electrons, I liked that." "I mean, it was easy, the valence electrons were always in that last layer, and they filled in the layers gradually. I don't quite remember, the first one was definitely two, then eight, and then I don't know." When asked if they remembered any types of bonds they were familiar with using the app, and if they remembered anything more specific from the bonds, there were also interesting responses: "Well, and then the single, double and triple bonds." "The polar, non-polar and ionic, those were from the electronegativity one. I don't remember the values, but every time it was subtracted, that was the bond according to that result." "Well they were sort of tugging on those electrons, and the bigger the difference between the elements, the stronger the electrons took." (Fig. 3).

Of the negative comments, they mainly recalled the texts embedded in the ER applications. "What bothered me the most was the text. It's really awfully long and cluttered, and again, without that, most of the time the animations couldn't be fully understood. Such a vicious circle." "Also the text. It's too long." Other pupils nodded that the texts for the animations did not suit them either.

The pupils' evaluation of how ER applications help understanding the curriculum also yielded interesting answers. "It's fine for understanding, but it's better with commentary." "I liked probably all of them, I just don't know the atom models, I either missed those or don't remember them at all." "I think if they'd put some audio that talks instead of the long text it would be totally cool." Pupils also universally agreed that they found learning better with the use of the app than without. "Definitely with the app. All the lessons are terribly the same. Notes, textbooks, presentations, at most a picture. So this is at least some change." "I don't think it's just because it's new. It's just interesting and it brings it more in line with reality when it's moving." "I can definitely remember it better when I see it on the animation, and it's better to understand."

6 Conclusion

In the initial interviews, students had only a few isolated experiences with VividBooks. Their reactions to working with the apps were mostly positive. Pupils commented positively on the fun form of animations with AR elements, the departure from traditional teaching, the simplification of the curriculum or the better understanding of the storyline. However, a few pupils were rather negative, as they found the animations more difficult to understand or even unnecessary. Pupils may have considered the animations unnecessary because they were mostly presented with animations that were confronted with a real experiment, such as changing the state of matter or separating components from mixtures, when they were first introduced to VividBooks. The pupils, therefore, appreciated the real experiment more than the animation. In the exit interviews, conducted after the systematic inclusion of animations in topics related to the microworld, no pupils identified animations (ER) as useless. The clarity and better memorization and understanding of the material due to the animations used were repeated in the positive evaluation. Pupils in all classes agreed that they liked the animations, but the accompanying text was too long and incomprehensible. Some pupils found it difficult to understand the animations without help and described them as good for revision but not for understanding the material itself. A frequent request from pupils for changes to the animations was to accompany them with spoken words, as some pupils found that it was not only the visuals that helped their understanding. This does not mean that pupils wanted interpretation with the animations, but the important information that was in the accompanying text would be a spoken part of the animations. The pupils worked very willingly with the AR VividBooks applications. The two topics that the pupils had the most difficulty with were "Models of the atom" and "Chemical bonding". In the exit interviews, pupils did not recall the topic "Models of the Atom" themselves in either class, and approximately half of the pupils in each class claimed to have missed the lesson when this was not actually the case. Even after the clues that were on the animations, many students did not remember the topic. Maybe it was because the other topics were more interrelated (atom–atom shell–ion formation–chemical bonding), so e.g. the terms atom, electrons, valence electrons were often repeated. The second topic, which the students identified as more challenging, was "Chemical bonding". Specifically, the formation of double and triple bonds was described by pupils as confusing, but after a joint explanation, everything became clearer. All classes recalled this topic during the exit interviews. Unlike the atom models, the chemical bond was presented to pupils last, so pupils mentioned it as one of the first. Pupils recalled the other animations in some detail or complemented each other. Some pupils knew what was happening in the animation but forgot how to name it. Pupils were not given any help with the animations, and the principles in the animations were not explained to them by the teacher or assistant. However, it often happened that pupils had to be "pushed" to do the right thing. This was usually only ensured by stopping the animation and asking them what had just happened. Overall, we can summarize that the regular inclusion of animations with AR elements in the teaching of chemistry in primary school on topics related to the microworld has had a positive effect. We have observed improved attention, concentration and greater independence of the pupils when teaching in this way. In the exit interviews, pupils were mostly able to recall what the individual animations represented and were able to describe and explain them.

References

1. Šafránková, D.: Pedagogika. Praha: Grada, (2019)
2. MŠMT. Strategie digitálního vzdělávání od roku 2020, (2014). https://www.msmt.cz/vzdelavani/skolstvi-v-cr/strategie-digitalniho-vzdelavani-do-roku-2020, last accessed 2023/04/04
3. Sarker, M. N. I., Wu, M., Cao, Q., Alam, M., Li, D.: Leveraging Digital Technology for Better Learning and Education: A Systematic Literature Review. International Journal of Information and Education Technology, (2019). https://pdfs.semanticscholar.org/fcb1/d0d75bf9146a7b61fabb24a38758bd31dbb7.pdf, last accessed 2023/04/04
4. Brdička, B.: Síťová generace podle Tapscotta. Metodický portál RVP.CZ, (2008). https://clanky.rvp.cz/clanek/s/P/11753/SITOVA-GENERACE-PODLE-TAPSCOTTA.html, last accessed 2023/04/07
5. Brdička, B.: Vliv technologií na děti prudce roste. Metodický portál: Inspirace a zkušenosti učitelů, (2011). https://spomocnik.rvp.cz/clanek/11103/, last accessed 2023/04/04
6. Balogh, Z., Turčáni, M., Magdin, M.: Design and Creation of a Universal Model of Educational Process with the Support of Petri Nets. In: Li, S., Jin, Q., Jiang, X., Park, J. J. (eds.) Frontier and Future Development of Information Technology in Medicine and Education, Springer Netherlands, Dordrecht. 1049–1060 (2014)
7. EDU.CZ. Inspirace k pořízení digitálních učebních pomůcek. (2022), https://www.edu.cz/inspirace-k-nakupu-digitalnich-pomucek/, last accessed 2023/04/04
8. Aplikace VividBooks. VividBooks, Praha, (2022). https://app.vividbooks.com/, last accessed 2022/10/02
9. Chen, M. P., Wang, L. Ch., Zou, D., Lin, S. Y., Xie, H., Tsai, Ch. Ch.: Effects of captions and English proficiency on learning effectiveness, motivation and attitude in augmented-realityenhanced theme-based contextualized EFL learning. Computer Assisted Language Learning, (2020). https://doi.org/10.1080/09588221.2019.1704787, last accessed 2022/09/18

10. Elmqaddem, N.: Augmented reality and virtual reality in education. Myth or reality?. International journal of emerging technologies in learning. (2019). https://online-journals.org/index.php/i-jet/article/view/9289, last accessed 2024/04/07
11. Johnson, L., Smith, R., Levine, A., Haywood, K. The 2010 Horizon Report: Australia– New Zealand Edition. Austin, Texas: The New Media Consortium. (2010). https://files.eric.ed.gov/fulltext/ED513479.pdf, last accessed 2024/04/22
12. Azuma, R. T.: A Survey of Augmented Reality. Hughes Research Laboratories, Malibu, (1997). https://www.cs.unc.edu/~azuma/ARpresence.pdf, last accessed 2023/04/04
13. Van Krevelen, D.W.F., Poelman, R. A.: Survey of Augmented Reality: Technologies, Applications and Limitations. The International Journal of Virtual Reality, (2010). https://www.researchgate.net/publication/279867852_A_Survey_of_Augmented_Reality_Technologies_Applications_and_Limitations, last accessed 2022/08/28
14. Warthová, V.: Rozšířená realita ve školství. Metodický portál: Spomocník, (2013). https://spomocnik.rvp.cz/clanek/17151/ROZSIRENA-REALITA-VE-SKOLSTVI.html, last accessed 2022/07/29
15. Radu, I. Augmented reality in education: a meta-review and cross-media analysis. : Personal and Ubiquitous Computing, 1533–1543 (2014). https://link.springer.com/article/10.1007/s00779-013-0747-y, last accessed 2022/08/27
16. Kesim, M., Ozarslan, Y. Augmented reality in education: current technologies and the potential for education. Procedia-social and behavioral sciences, 47, 297-302 (2012).
17. Irwansyah, S. F., Yusuf, Y. M., Farida, I., Ramdhani, M. Augmented Reality (AR) Technology on The Android Operating System in Chemistry Learning. IOP Conference Series: Materials Science and Engineering, (2018). https://iopscience.iop.org/article/10.1088/1757-899X/288/1/012068, last accessed 2024/04/07
18. Urzúa Reyes, M. D.: Augmented Reality for Learning Chemistry. Institute for the Future Education, (2021). https://observatory.tec.mx/edu-bits-2/augmented-reality-for-learning-chemistry, last accessed 2022/08/04
19. VividBooks pro školy. VividBooks, Praha: Vividbooks, (2022). https://www.vividbooks.com/, last accessed 2022/08/04

Google Applications in the Position of Learning Management System 2.0 in Comparison with Moodle

Jana Burgerová and Vladimír Piskura

1 Introduction

There is a generation of young people studying at universities, referred to by terms such as the Millennial Generation [1], the Net Generation [2], or Digital Natives, who, according to Prensky [3], are considered to be 'native speakers' of the digital language of computers and the Internet. Other authors refer to them as Generation Z, replacing Generation Y [4], the iGeneration [5], or by the term Digital Learners, which offers a more global vision of the twenty-first-century learner [6].

Regardless of the label, the young people of this era are generally considered universally competent in working with digital technologies. However, there is research that speaks against this expectation and does not support the view of this generation's automatic capacity to transfer digital competences and digital skills into the professional or academic environment. In fact, there is no evidence that today's students want to use current technologies for educational purposes. However, the issue of digital skills needs to be viewed comprehensively in view of several indicators such as age, geographical location, or social status. In the conducted study, this is confirmed by Oyedemi [7], whose findings reveal that forms of Internet access, the cost of access, social stratification, and inequalities have implications for the patterns of digital skills, with content creation being a challenge for many youths. Through the Digital Skills Indicator, Helsper, Schneider, van Deursen, and van Laar [8] defined four dimensions that constitute digital skills: (1) technical and operational skills; (2) information navigation and processing skills; (3) communication and interaction skills; and (4) content creation and production

J. Burgerová · V. Piskura (✉)
Faculty of Education, University of Presov, Presov, Slovakia
e-mail: jana.burgerova@unipo.sk; vladimir.piskura@unipo.sk

M. Turčáni (ed.), *15th International Scientific Conference on Distance Learning in Applied Informatics*, EAI/Springer Innovations in Communication and Computing,
https://doi.org/10.1007/978-3-031-81261-3_4

skills. In their research, they declare that many young people did not master a range of skills, including critical information navigation and processing skills. Our experience from the pregraduate training of future teachers of primary and pre-primary education at the Faculty of Education of the University of Presov in Presov also speaks against the generally overrated digital skills of current students. The results of the testing we conducted in 2020, whether the initial testing of basic computer skills, which consisted of editing a document in a word processor using its basic tools, or the complex testing of digital skills through the IT Fitness test on a sample of 532 students, pointed to significant gaps in the digital skills of our students [9]. Similarly, the first-hand experience of teachers of subjects where all or part of the teaching is conducted via computers and tablets, and where it is possible to directly observe students' work, points to the absence of skills for working with spreadsheets, graphic editors, presentation software, the use of cloud computing, and generally moderately advanced user skills when working with a PC.

In the context of the above, we were interested in how learners of the Web 2.0 generation [5] perceive a new e-learning platform built on more Web 2.0-enabled services suitable for them through freely available Google Apps (Google Workspace). As a result of the main features of the LMS Moodle, this platform is based on Web 1.0 principles, as it forms a robust closed system, whereas Google applications form a comprehensive and open system of services and applications supporting modern web principles that today's learners are closely familiar with. However, today's LMSs still represent a place for the student and the teacher to find everything they need for studying and teaching. We cannot say the same about Google applications, but due to their high level of integration with each other, they can act as such. Therefore, we perceive these comparisons as a very interesting challenge.

2 Theoretical Background

Today, e-learning in its many forms is very popular in the higher education environment, and Web 2.0 tools are increasingly coming to the fore. It is a dynamically changing web that is characterized by the ability of users to intervene in the creation of its content. In the information phase that preceded the research, we encountered and familiarized ourselves with a robust information base in the field related to the issue under investigation. However, most of the research has focused on specific tools, their application, their features, and their impact, not on the environment itself. Google and its applications form the prototype of Web 2.0 tools; they are simple, sophisticated, available for free, and provide elements of interaction or sharing. Their implementation in education is being addressed by several authors. For instance, Peacock and Grande [10] developed a collaborative classroom based on a learning management system by integrating freely available Google Apps such as Google Blogger, Forms, Docs, and Sites. This model was then research-validated with a questionnaire among students, with positive results. Google Apps proved to be an effective tool for education, management, and online collaboration of students. Similar research can be found in Karabayeva [11], Sureephong and Singjai [12], Holmes et al. [13], Ilie and Almasan [14], and Ilie [15]. Many authors also address the effectiveness of using individual Web 2.0 tools, or the frequency and popularity of their use by students. Chircu and Neacsu [16] focused on which Web 2.0 tools students prefer and how often they use them. The results show the positive role of their use in education, with Facebook groups proving to be the most widely used tool.

Compared to Google Apps, the Moodle system is much more established in higher education. This has been proven by the amount, but also by the variability of research. Most of the authors focus on the impact of selected LMS tools on different aspects of the educational process. Min et al. [17] investigate the impact of LMS on student learning, Maldonado Berea and Vega Gea [18] the impact of LMS on the attitudes of university students, and Adamkovičová and Burgerová [19], on the other hand, the impact of the communication dimension of e–learning on students' attitudes towards the e–learning course.

In the research presented here, we investigate whether the environment in which e-learning is implemented influences students' knowledge and attitudes. In the literature, this issue has been discussed in articles comparing different LMS platforms such as Moodle, Blackboard, ATutor, and SuccessFactors from several perspectives (Mohd Kasim and Khalid [20]). A comparison of the Moodle LMS with Google Sites, a tool that allows users without programming language skills to create their own website, can be found in Herrera et al. [21]. In general, most articles deal with the combination of 1.0 and 2.0 tools, or the integration of 2.0 tools into the Moodle LMS in order to complement the absent features. Rodríguez Ruibal [22] describes the synchronous communication tool from Google and its implementation into Moodle to support virtual tutorials. Daraghmi et al. [23] address the shortcomings of Moodle by implementing cloud applications such as Google Drive, One Drive, and Dropbox as supplementary storage, and Facebook as a chat tool that offers wider opportunities to support discussions. Interactive work with tools through HTML tags, iFRAME, is described by Ornat and Zallas [24], and similar work can be found in Castro Fernandez [25], Bettoni et al. [26], and Barnard-Ashton [27].

However, none of these studies compared Google Apps as a system that could behave as a fully fledged LMS due to the high degree of integration of the individual applications, which could thus have an impact on selected aspects of the educational process. We are aware that Google already operates a Classroom service, which is a single educational environment with all the aspects of an LMS and is directly predestined for education. However, in the case of Google Classroom, we have encountered problems with the creation of new 'verified' Google accounts. As students currently use a wide variety of different accounts and often have trouble remembering many logins (even though there are different Password Managers), we wanted to avoid creating yet another duplicate Google account.

3 Hypotheses

One of the most important aspects of the educational process is student evaluation, motivation to work, and student attitudes towards study. Therefore, we decided to compare Google Apps with LMS Moodle and investigate them in the environment of the Faculty of Education. We have defined two hypotheses for the research:

- H1: After completing an e-learning course delivered through LMS Moodle, students will achieve better results in the final knowledge tests than students completing a course delivered through Google services.
- H2: Learners will evaluate an e-learning course integrated into an LMS Moodle environment more positively than an e-learning course integrated into a Google Apps environment.

4 Research Methodology

4.1 General Background

The research was carried out between January 2020 and December 2021 at the Faculty of Education of the University of Presov in Presov. We applied a quantitative approach to the research. Within the framework of the research, a knowledge test and a non-standardized questionnaire were used as research instruments of a natural pedagogical experiment. The aim of the presented study was to verify the impact of a specific e-learning environment on the knowledge and evaluation of the e-course by students of the University of Presov through a natural pedagogical experiment.

4.2 Sample

The total sample size was 301 students divided into 3 control and 3 experimental groups in 3 diverse subjects. The sample of the conducted research consisted of all students of the winter semester of the academic year 2020/2021 in the subjects Didactics of ICT in Primary Education: 1 year of Master's studies (Subject 1), Theory and History of Literature for Children and Young Adults: 1 year of Master's studies (Subject 2), and Science Education: 1 year of Bachelor's studies (Subject 3). The choice of subjects was made in accordance with ensuring the greatest possible external validity of the research, in accordance with the possibility of maximizing the generalization of the observed phenomenon. The selected subjects differed in their content, form, teacher, level of higher education, year of study, and many other characteristics. The total number of all students in individual subjects is shown in Table 1.

Table 1 Number of students in individual groups

N			
Subject	CG	EG	∑
1	83	78	161
2	23	21	44
3	48	48	96
∑	154	147	301

Note: *CG* Control group, *EG* Experimental group

4.3 Procedure

In the process of verifying hypotheses H1 and H2, we used statistical analysis to determine whether there is a statistically significant difference between students' course evaluations and students' knowledge levels. Hypothesis H2 is further specified into seven domains in which we observed students' evaluations.

Students were randomly divided into a control group and an experimental group at the beginning of the semester. In both groups, we observed the form of e-learning distribution within the e-learning support of face-to-face teaching through LMS Moodle and Google Apps. We focused on seven specific areas within e-learning distribution in both learning environments: distribution, layout, structure, testing, assignment submission, communication, and collaboration. The control group worked with LMS Moodle, and the individual aspects of LMS Moodle with respect to these areas were as follows:

Distribution: The student logged into LMS Moodle using the same login credentials as for the modular academic information system MAIS (via the Lightweight Directory Access Protocol). Once logged into the LMS system, the student had to browse through the year categories to access the relevant course. To enter the course, the student needed a login key from the teacher. Once all students were logged in, logging into the course was systematically disabled (eliminating contamination of research groups).

Layout and visual aesthetics: We have adapted all the courses in question to have approximately the same layout. The header of the course consisted of an introductory thematic graphic with a welcome note from the course authors. Instructions and organizational guidelines followed, and then individual topics. The layout of each topic was also adapted to be approximately the same in each course. The topics in each course had a colour-coded header with graphics reflecting the particular lesson. Below the banner were the learning objectives, either unzipped (in Subject 1) or zipped into a single file called After studying, you will know (via the Page module).

Structure: Moodle is a modular system that allows the user to customize the learning environment to a certain extent, i.e. to collapse/expand some modules/windows. In the standard display, the student had a hypertext menu on the left (Navigation, Settings window), the course itself in the middle, and various activity overviews and a calendar on the right. The educational content was mediated to the learners by classical content carriers such as page, URL, file, book, and dictionary modules. Their arrangement was cascaded in each course and had a fixed structure. The core learning material was conveyed by the Book file, which allowed us to create structured texts in the form of groups of several web pages linked together by a hypertext menu. The Book file also allowed students to easily print structured hypertext learning texts.

Testing: This took place through the Test activity. During the semester, students had the opportunity to take self-tests in each lesson. The test icon was always placed below the study material in the structure of each topic. After taking the test, the student could see his/her results, but never the correct answers.

Assignment submission: This was carried out in the standard way through the Assignment activity. After clicking on the assignment icon, the student uploaded the required file in the prescribed format. A discussion forum in the format of Standard Forum for Common Use was used to upload short text assignments, where the requirement was established for students to be able to see each other's assignments.

Communication: The Discussion Forum was used for asynchronous communication among students. Students were assigned rights to post topics to the discussion forum. The News Forum was used for general notifications, where the student had limited posting rights. Synchronous communication was conducted through the Chat module, through which virtual office hours were held on a designated day of the week.

Collaboration: Student collaboration was implemented in the creation of collaborative assignments through a Wiki activity that allows multiple students to create/collaborate on the creation of a shared hypertext document.

During the semester, the experimental group worked with selected Google Apps, i.e. all aspects of e-learning were conducted through Google Apps. Distribution, Communication, and Collaboration were conducted through the services of Google Drive, Google Mail, Google Forms, and Google Docs. All materials and activities were duplicated from the LMS Moodle to the above applications. The individual aspects of Google Apps with respect to the above areas were as follows:

Distribution: All experimental courses (learning materials, activities) were in a virtual Google Drive (GD) space. A Google account is required to access the GD. Since Google services are frequently used by the current generation (which became the basis for the conducted research), we used students' private accounts in the research. At the initial briefing, we did not observe any absence of Google accounts in the experimental groups. To access the course, the student had to fill in a registration form, where he/she had to fill in basic information such as first name, last name, email, study group, and the course he/she was enrolling in. The teacher could then share the desired course with the student, i.e. the GD folder without editing rights. The student was prompted by an invitation email to log in to the course for the first time.

Layout, visual aesthetics: The layout of all the research courses on GD was diametrically different from that of the LMS Moodle course, but all the materials were identical. The course was composed of documents located in one folder. Both folders and documents were numbered and arranged chronologically by title. The header with the introductory graphics and welcome note can be found in folder 00 Introductory lesson—instructions and organizational guidelines.

The layout of each lesson on GD was similar. The lesson in each course contained a 01 Introduction file, which contained an image with a colour-coded header with graphics reflecting the particular lesson. Again, the learning objectives were located below the banner.

Structure: A GD course consists of folders in which documents are located. There are no options to customize the learning environment on GD. The learner was only able to change the folder display in the form of a list or a grid, or change the criteria for the arrangement of the files. The educational content was delivered to the students by Google Docs or files in *.docx, *.xlsx, *.pdf, and *.jpeg formats. In total, Google Drive allows working with up to 30 file types. Their arrangement in each folder had the same structure following the name of the document. The arrangement of the study material in each topic was the same as in LMS Moodle, except for the folders, which could not be placed under the document type in the GD item structure; therefore, they were always arranged on top. The main study material was mediated by a pdf file. This file allowed us to create simple study materials with hyperlinks and a hypertext menu (similar to the Book module in LMS Moodle). All materials were available online via a web browser or an app on mobile devices and, once downloaded, could be accessed offline.

Testing: Students were tested through Google Forms in the form of a quiz. Students had the opportunity to take self-tests in each lesson during the semester. The test icon was located below the study material in the structure of each topic. After taking the quiz, the student could see his/her results and the correct answers. The test was not limited by the number of attempts.

Assignment submission: This was conducted only in subject No. 1. Students in the experimental group submitted assignments via a shared folder on GD for which they were granted editing rights. The students found the folder in their GD in the 'Shared with me' section. They could upload documents from their devices (pc, smartphone) to this folder. A shared text Google document with editing rights for them was used to upload short text assignments where the requirement was established for students to be able to see their classmates' assignments.

Communication: The Gmail service was used for asynchronous communication with course users. Bulk email was used for general notifications, as we created a database of email addresses for all students in the course through the registration form.

Collaboration: Student collaboration was implemented in the creation of group assignments. For this purpose, as with traditional assignments, a shared document was created with a list of students' names listed in a spreadsheet, to which students were granted editing rights. In this case, however, multiple students (collaborating to create a single assignment) were grouped in a single cell of the spreadsheet.

In the conducted research, we monitored the level of students' knowledge, which was expressed as the total score achieved in the final knowledge tests, and the level of students' subjective evaluation of the e-learning distribution, which was expressed as a summary index of evaluations using a scale. Students were informed at the beginning of the semester how to complete the course by means of an electronic test. Students in the control group in the first year of the Master's degree were familiar with the e-learning support for face-to-face teaching through LMS Moodle, while students in the first year of the Bachelor's degree were informed by a short briefing on how to use the system. All experimental groups were also briefed on how to work with Google Apps. The experimental group students were also given an information leaflet with instructions on how to work with the Google Apps system.

4.4 Measure

In the research, we monitored two dependent variables:

1. The level of students' knowledge.
2. The extent of students' subjective evaluation of e-learning distribution.

The impact of e-learning distribution on students' knowledge was measured by means of a knowledge test, which had different parameters in each subject and was created by an experienced teacher of the given subject. In all the tests, we always considered only the first attempts. Each of the tests was administered electronically in the LMS Moodle environment during the credit week and examination period.

The test in Subject 1 consisted of a test task bank containing 90 questions, in Subject 2 of 96 questions, and in Subject 3 of 251 questions. In each of the tests, an aliquot amount was randomly administered to the student with respect to the number of questions in each category. Students were always given 15 min to complete the test. Correct answers from all tests were always published after all groups had been tested at the end of the examination period. The statistical method of analysis of covariance was used to test the statistical significance of the differences in achieved knowledge between the control and experimental groups.

To test the impact of e-learning distribution on students' attitudes, we used our own research instruments, questionnaires, which were in a non-standardized form with items arranged in a scale. The basis for constructing the questionnaire was an existing questionnaire by R. Wright [28] called Learner Questionnaire for Online or eLearning Courses. Its validity was assessed by two independent experts. Estimation of the reliability of the scales in terms of their internal consistency was done by calculating Cronbach's α, which was 0.90, indicating the overall internal consistency of the research instruments. Each item contained a six-grade scale, and a summary score was calculated for each category, which is the result of the sum of the scores of the individual items belonging to that category. Questionnaire items were categorized into seven domains: distribution (1), layout (2), structure (3), testing (4), assignment submission (5), communication (6), and collaboration (7). The last three categories (5,6,7) were only monitored in Subject 1. There were a total of 53 scaled items in the questionnaire in each category. The questionnaire was administered electronically in all three subjects through Google Forms (URL link in LMS Moodle); it was completed in class and was a prerequisite for the final knowledge test. The data that support the findings of this study are openly available in osf data repository at https://bit.ly/3Tm1GER.

4.5 Data Analysis

For the purpose of testing the impact of e-learning distribution on students' knowledge, analysis of covariance (ANCOVA) was conducted. The aim was to test whether the differences between the experimental group (ES) and the control group (CS) in the dependent variables would be significant if the two groups were equivalent at the beginning of the experiment. Testing the impact of the e-learning distribution on students' attitudes was implemented using the Welsch test, which is analogous to the t-test for independent samples but does not assume equal variance in the two samples (highly unrealistic in most cases). The above frequentist techniques were used to assess whether the difference between the groups was statistically significant. Therefore, Bayesian statistics were used to assess the relative degree of empirical evidence in favour of the null (H0) or alternative hypothesis (Ha). Thus, for each of the hypotheses tested, a so-called Bayes factor was calculated to indicate the extent to which our beliefs in favour of one of the hypotheses need to be revised in light of the observed data. A Bayes factor greater than 3 can already be considered significant evidence in favour of a given hypothesis.

5 Research Results

Statistical analysis was conducted in the programming language R version 3.3.2. In the first step, data screening was conducted with a focus on examining the distributional characteristics of the variables, as well as the occurrence and influence of outlying values. For participants who had taken part in either the entrance or final measurements in their subject, missing data were absent (due to electronic administration). The availability of data from the entrance or final measurements was a sufficient prerequisite for inclusion of the participant in the analyses (hence, the 'pairwise deletion' method was used). None of the values were assessed as outliers, so there was no reason to consider any of the participants as coming from a different population, mainly because of the clearly definable sampling frame of the target population. Despite the non-parametric nature of several variables, no non-linear transformations of the data were performed due to inefficiency, and parametric statistical tests known to be robust to common violations of distributional assumptions (Welsch test, Analysis of Covariance) were used. Thus, all variables have an original (naturally interpretable) scale. In the attitude questionnaire, it was

Table 2 Means and standard deviations for control and experimental groups in individual subjects for the posttest

Subject	EG	Mean	SD	N
1	0	71.81	13.58	79
	1	71.55	12.21	73
2	0	70.74	11.85	23
	1	63.38	13.67	21
3	0	59.72	13.27	43
	1	61.09	12.86	46

EG Experimental group

the mean of Likert-type scale items, and in the knowledge tests, the percentage of correct responses.

For the purpose of testing H1, analysis of covariance (ANCOVA) was implemented. This linear model tests whether the population means of the dependent variable are the same at all levels of the independent variable (in this case, placement in the EG or CG), while considering possible differences in the observed covariate (in this case, the pretest). Thus, the aim is to test whether the differences between EG and CG in the dependent variables would be significant if the two groups were equivalent at the beginning of the experiment. Testing H2 was conducted using the Welsch test, which is analogous to the t-test for independent samples but does not assume equal variation in the two samples (highly unrealistic in most cases).

The above frequentist techniques were used to assess whether the difference between the groups was statistically significant, i.e. for the purpose of maintaining a controlled type I error rate. However, frequentist statistics cannot formally demonstrate the eventual validity of the null hypothesis (e.g. no difference between groups). Therefore, Bayesian statistics were used to assess the relative degree of empirical evidence in favour of the null (H0) or alternative hypothesis (Ha). Thus, for each of the hypotheses tested, a so-called Bayes factor was calculated to indicate the extent to which our beliefs in favour of one of the hypotheses needed to be revised in the light of the observed data. A Bayes factor greater than 3 can already be considered significant evidence in favour of a given hypothesis.

The impact of e-learning distribution on students' knowledge was measured using a knowledge test. The results of the final tests in each subject can be seen in Table 2, and the observed data distributions in Figs. 1, 2 and 3.

It is apparent from the data visualizations alone that the EG and CG are at the same level across all subjects in terms of average test performance. This indication was also formally confirmed by the non-significant p-values in the analysis of covariance presented in Table 3.

The implemented Bayesian analysis of covariance has shown that, assuming the validity of H0, i.e. p(D|H0), the observed data in Subject 1 and Subject 3 are 5.59 times and 4.31 times more probable than the data that can be expected if the alternative hypothesis, i.e. p(D|Ha), is valid.

In subject No. 2, the threshold for statistical significance ($p = 0.06$) can occasionally be considered questionable. However, the calculated values of the Bayes factor probabilities show ($BF10 = 1.52$) that the data are ambiguous ($BF10 < 3$),

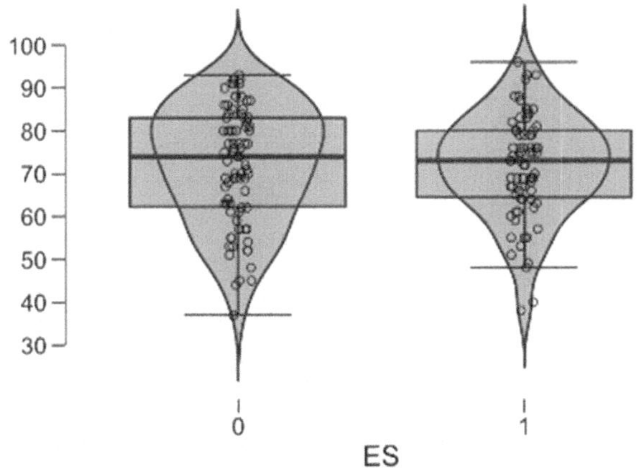

Fig. 1 Observed distribution in subject No. 1

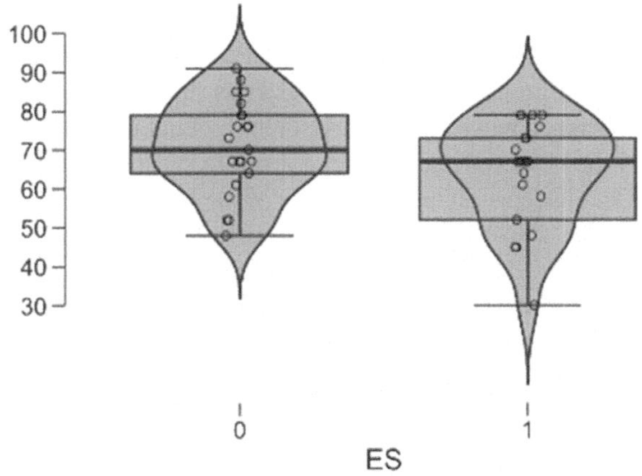

Fig. 2 Observed distribution in subject No. 2

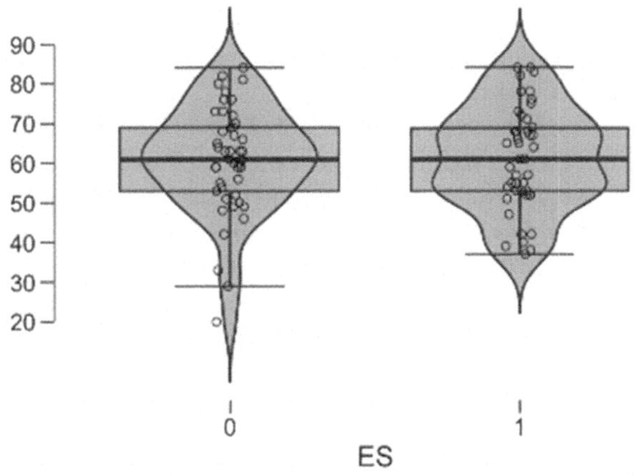

Fig. 3 Observed distribution in subject No. 3

Table 3 Analysis of covariance

Subject	F	p	ω^2
1	F(1, 149) = 0,12	0,74	0,00
2	F(1, 41) = 3,85	0,06	0,06
3	F(1, 86) = 0,11	0,74	0,00

Table 4 Average rating and its variability

Domain	Subject	EG	N	Mean	SD	SE
Summative assessment ($\sum$)	1	0	82	4.827	0.538	0.059
		1	75	4.908	0.478	0.055
	2	0	21	4.695	0.659	0.144
		1	24	4.586	0.579	0.118
	3	0	46	4.582	0.623	0.092
		1	47	4.688	0.728	0.106

and thus, the observed level of evidence in favour of Ha is not sufficient to declare the validity or invalidity of Ha or H0. This is primarily due to the small sample size in both groups (total $N = 41$). Thus, it can be concluded that the hypothesis of a difference between EG and CG in the final test within Subject 2 was not supported, or the observed data did not constitute sufficient empirical evidence to make a more reliable decision on the validity of the hypothesis in question.

The above data provides empirical support for the conclusion that there is no difference in knowledge acquisition between the Moodle and Google platforms in all subjects. Thus, it can be concluded that hypothesis H1 may be rejected.

The impact of e-learning distribution on students' attitudes was investigated using the research instrument questionnaire across seven domains of assessment in Subject 1 and four domains of assessment in Subjects 2 and 3. The basic descriptive statistics under summative assessment across all subjects are presented in Table 4.

As can be seen in Table 4, students rated the courses in general more positively than negatively, averaging 4.71 out of a 6-point scale. Welsch test was used to test the statistical significance of the differences between EG and CG, and the results for each domain of assessment, as well as for the overall course assessment, can be seen in Table 5.

As can be seen from Table 5, a significant difference ($p < 0.05$) between groups can be observed only in domains no. 2 and 5, i.e. layout and assignments, only in Subject No. 1. However, due to the risk of type I error inflation, it is always necessary to adjust the alpha significance level when performing multiple statistical tests. For this purpose, an adjusted p-value (Holm's correction) was calculated. With 8 statistical tests, the above-mentioned p-values no longer reached statistical significance, adjusted $p = 0.32$, and $p = 0.17$, respectively.

The implemented Bayesian analysis in (Table 6) 10 out of 15 tested scales clearly indicates that the difference between EG and CG is not due to the differing e-

Table 5 Test of mean differences between EG and CG posttest

S	1				2				3			
D	*t*	*df*	*p*	Δ	*t*	*df*	*p*	Δ	*t*	*df*	*p*	Δ
1	−1.82	154.70	0.07	0.29	1.16	41.71	0.25	0.35	−0.95	87.98	0.35	0.20
2	2.02	152.20	*0.05*	0.32	0.07	40.43	0.94	0.02	0.05	91.00	0.96	0.01
3	−1.08	155.00	0.28	0.17	0.17	40.86	0.87	0.05	−0.73	90.75	0.47	0.15
4	−0.83	148.30	0.41	0.13	0.52	38.72	0.61	0.16	−0.79	90.66	0.43	0.16
5	−2.32	145.70	*0.02*	0.37	–	–	–	–	–	–	–	–
6	−0.24	154.30	0.81	0.04	–	–	–	–	–	–	–	–
7	−1.47	154.90	0.15	0.23	–	–	–	–	–	–	–	–
$\sum$	−0.99	154.90	0.32	0.16	0.59	40.19	0.56	0.18	−0.75	89.40	0.46	0.16

Table 6 Bayesian t-test for independent samples posttest

S	1			2			3		
D	BF_{01}	BF_{10}	error %	BF_{01}	BF_{10}	Error %	BF_{01}	BF_{10}	Error %
1	1.305	0,766	1.772e − 8	1.968	0,508	2.171e − 4	3.095	0,323	3.678e − 4
2	0.89	1123	1.078e − 8	3.379	0,296	2.030e − 4	4.589	0,218	3.314e − 4
3	3.408	0,293	5.150e − 8	3.345	0,299	2.029e − 4	3.634	0,275	3.637e − 4
4	4.225	0,236	6.330e − 8	3.029	0,330	2.032e − 4	3.495	0,286	3.661e − 4
5	0.531	1883	5.087e − 9	–	–	–	–	–	–
6	5.653	0,176	8.241e − 8	–	–	–	–	–	–
7	2.177	0,459	3.228e − 8	–	–	–	–	–	–
$\sum$	3.703	0,270	5.589e − 8	2.94	0,340	2.040e − 4	3.587	0,279	3.646e − 4

learning delivery method of the course content distribution. For the other five scales, the data do not have sufficient evidential value. None of the effects represented by the Cohen's d constant across the assessment domains listed in Table 5 are strong enough (Cohen's $d > 1$), nor are they at the alpha level of statistical significance. Thus, we cannot express the belief that these effects actually exist and that they can be expected in the population. This is confirmed by the Bayes factors above, which do not have sufficient evidential value in all domains of assessment in favour of the alternative hypothesis Ha. Thus, the data presented above provide empirical support for the conclusion that there is no difference in students' course evaluations using the Moodle and Google platforms across all subjects, and a rejection of hypothesis H2 can be established.

Overall, it can be concluded that the data from the conducted experiment very consistently and clearly indicate a null effect between the choice of e-learning distribution, i.e. between the LMS Moodle and Google Apps platforms, on demonstrated knowledge and subjective evaluation across qualitatively diverse courses. Given the results of the analyses, this observation has (by the standards of the social and behavioural sciences) robust empirical validity.

6 Discussion

There are many ways, forms, and methods in which technology can be integrated into education. In the research presented here, technology is presented in the form of blended learning and thus a combination of traditional forms of teaching and e-learning. Blended learning, also known as technology-mediated instruction, web-enhanced instruction, or mixed-mode instruction, is an approach to education that combines online educational materials and opportunities for interaction online with physical place-based classroom methods. Technology, its specifics such as accessibility, appearance, structure, and communication tools, play a key role in this form of teaching. The research worked with two environments, and by specific e-learning distribution, we meant distribution via Google Apps, which had never been used before in e-learning support for face-to-face teaching and was the first time students encountered this form of distribution.

The data from the statistical analysis provided us with a robust empirical basis for the null effect of the choice of e-learning distribution in relation to students' knowledge acquisition and students' evaluation of the e-learning course. It is thus irrelevant to look for the reasons for the emergence of such small effects, and instead, the question of why a significant statistical difference in knowledge and evaluations did not occur comes to the fore.

In our hypotheses, we predicted that students using Google Apps would achieve better results in the final knowledge tests and would evaluate the course more positively in the specified areas, given the current form of the Web, its services, programmes and Web 2.0 applications, in which static content has been replaced by a space for collaborative content creation, sharing, and comments in the context of the LMS Moodle, which works on the Web 1.0 principle. In our research, we therefore offered an alternative to Google Apps, which are characterized by their simplicity, constant online accessibility, support for sharing, community creation, comments section, personalization of services, high integration, large storage space, and ease of accessibility.

In our opinion, there are several relevant reasons for the absence of a significant difference in knowledge and assessment between the electronic environments examined. Both hypotheses were tested at the FE of UP in Presov in the study programme Pre-school and Elementary Pedagogy and Teaching in Primary Education. When formulating the research problem, based on the current knowledge about Web 2.0 or Generation Z students, we assumed that the students at FE of UP in Presov are average users of Google Apps and, after training, will have no problem accessing learning content, creating activities, and communicating through Google Apps.

Today, however, we can conclude that one possible reason for the absence of differences could have been partial unfamiliarity with Google technologies in the research group. In all the research subjects, even after the initial training, but also during the consultations throughout the semester, problems arose with the Google Drive application, which formed the basis of our 'alternative LMS'. Issues arose in comprehending the mutual integrity of all the Google Apps. Students often actively use only the asynchronous communication tool, Gmail, and are unaware of the existence of the other apps that they can access through their Google account. After multiple explanations, not all students understood the technology of sharing via email without editing rights. After sharing all the study materials, they attempted to access them using their group email or were unable to access the study materials at all. Thus, based on the above findings, we suggest that Google Apps is as challenging a technology for students as the LMS Moodle, which resulted in a minimal effect size across all qualitatively different subjects.

In this context, we would like to point out, about Generation Z, that their computer skills are probably overrated. This assumption is also confirmed by some of the research studies monitoring the level of computer skills of the aforementioned generation. Gallardo et al. [29] argue that there is no certainty that when the digital generation has computer skills, they can also apply them in relation to education. Conducting research among three groups of students aged 14–18 confirmed that even though the digital generation is growing up in a world of digital technologies, their ability to critically assess information and use it in education is limited [30]. Similarly, the results of the computer skills testing of students conducted in 2015 on a sample of 532 students showed a relatively low success rate in matters related to collaborative tools and social networks (54%). Considering that, according to research on mobile technology use [9], out of a sample of 473 respondents, as many as 381 had a Google account, and similarly, all faculty students have group accounts on Gmail, with all five of the assignments in this area focusing on Google and Google Drive, this fact was surprising. We assume that the above facts may have been reflected in the research we conducted.

In this context, the role of the teacher appears to be crucial, as the students' existing computer skills, often at the level of amateur, self-studied skills, should be developed by the teacher into a form that can be used in tertiary education, whether in the LMS Moodle or Google Apps. It cannot be automatically assumed that the student will master and use the new tools.

7 Conclusion

Our aim was to determine the impact of a specific electronic learning environment on the knowledge of students at the Faculty of Education of the University of Presov in Presov and on the attitudinal evaluation of the e-course by students at the Faculty of Education of the PU in Presov. The conducted research clearly showed that the tested environments do not influence the specific knowledge and attitudes of students. However, LMS Moodle is already fully integrated at the Faculty of Education PU in Presov, used by students and teachers. Moreover, due to the complexity of the tools it offers for the management of education, it seems to be more suitable than Google Apps. It would be advisable to focus in the future on the connection of the best of both systems. One of the solutions could be continuous, multi-level training in the use of managed learning systems, not only for teachers, which has already been implemented by the Faculty of Education of the University of Presov through a series of trainings in LMS Moodle, but also for students, who are the end users of this product.

It would be useful to develop multimedia guidelines and tutorials focusing on basic user tasks related to the use of the electronic environment. Following the results of the research, which equalized LMS Moodle and selected Google Apps according to students' evaluation and attitudes, we also recommend integrating Google Apps into selected courses or combining LMS Moodle with selected Web 2.0 tools. These applications are simple solutions to the issues that are associated with teaching through LMS Moodle. As mentioned above, in recent years, it has been possible to enrich LMS Moodle with various plugins or repositories and various new kinds of Web 2.0 tools. Thus, it is possible to integrate cloud applications into Moodle to expand the virtual space (Google Drive, One Drive), communication tools to support modern synchronous communication (Viber, WhatsApp), blogs (Google Blogger), portfolios, and in this way to get closer to the generation of contemporary students.

References

1. Howe, N., Strauss, W.: Generations: The History of America's Future, 1584 to 2069. Quill, New York (1991)
2. Tapscott, D.: Growing Up Digital: The Rise of the Net Generation. McGraw-Hill Book, New York (1998)
3. Prensky, M.: Digital Natives, Digital Immigrants Part 1. Horiz. 9, 1–6 (2001). https://doi.org/10.1108/10748120110424816
4. Zhao, E., Liu, L.: China's Generation Y: Understanding the workforce. In: 2008 4th {IEEE} International Conference on Management of Innovation and Technology. Institute of Electrical and Electronics Engineers ({IEEE}) (2008)
5. Rosen, D., Nelson, C.: Web 2.0: A New Generation of Learners and Education. Comput. Sch. 25, 211–225 (2008). https://doi.org/10.1080/07380560802370997
6. Bullen, M., Morgan, T.: Digital learners not digital natives. La Cuestión Univ. 7, 60–68 (2011)
7. Oyedemi, T.: Beyond access: Youth and digital skills. Communicatio. 40, 137–154 (2014). https://doi.org/10.1080/02500167.2014.907191
8. Helsper, E.J., Schneider, L.S., van Deursen, A.J.A.M., van Laar, E.: The youth Digital Skills Indicator, https://doi.org/10.5281/zenodo.4476540,(2021)
9. Adamkovičová, M., Burgerová, J., Kochová, H., Piskura, V.: Pilot Testing of It Skills of Preschool and Elementary Education Students by the It Fitness Test. Int. J. Inf. Commun. Technol. Educ. 6, 40–48 (2017). https://doi.org/10.1515/ijicte-2017-0004
10. Peacock, J.G., Grande, J.P.: An online app platform enhances collaborative medical student group learning and classroom management. Med. Teach. 38, 174–180 (2016). https://doi.org/10.3109/0142159X.2015.1020290
11. Karabayeva, K.: Didactic Potential Of Cloud Technology In Creating A Unified Educational "Space{"} In A Sla Classroom. In: Titrek, O and Mikelsone, I and Pavitola, L and Gultekin, GS (ed.) 2ND INTERNATIONAL CONFERENCE ON LIFELONG LEARNING AND LEADERSHIP FOR ALL (ICLEL 2016). pp. 516–519. ICLEL CONFERENCES, SAKARYA UNIV FAC EDUC, SAKARYA, 54300, TURKEY (2016)
12. Sureephong, P., Singjai, A.: SaaS for Education: A Case Study of Google Apps in Software Engineering Class. Presented at the (2016)

13. Holmes, M.R., Tracy, E.M., Painter, L.L., Oestreich, T., Park, H.: Moving from Flipcharts to the Flipped Classroom: Using Technology Driven Teaching Methods to Promote Active Learning in Foundation and Advanced Masters Social Work Courses. Clin. Soc. Work J. 43, 215–224 (2015). https://doi.org/10.1007/s10615-015-0521-x
14. Ilie, M.M., Almasan, B.H.: CASE STUDY ON CONSTRUCTIVIST LEARNING USING GOOGLE APPS AT UNIVERSITY OF BUCHAREST ODL DEPARTMENT CREDIS. In: Roceanu, I and Ciolan, L and Holotescu, C and Radu, C and Stanescu, I and Logofatu, B (ed.) RETHINKING EDUCATION BY LEVERAGING THE ELEARNING PILLAR OF THE DIGITAL AGENDA FOR EUROPE!, VOL. II. pp. 512–517. CAROL I NATL DEFENCE UNIV PUBLISHING HOUSE, PANDURI ST, 68–72, BUCHAREST, 00000, ROMANIA (2015)
15. Ilie, M.M.: PSYCHO-PEDAGOGICAL ASPECTS OF CONSTRUCTIVIST LEARNING THRU GOOGLE APPS AT ODL DEPARTMENT CREDIS OF UNIVERSITY OF BUCHAREST. In: Roceanu, I and Ciolan, L and Vlada, M and Chicioreanu, T and Ionita, A and Beligan, D (ed.) ELEARNING VISION 2020!, VOL II. pp. 109–114. CAROL I NATL DEFENCE UNIV PUBLISHING HOUSE, PANDURI ST, 68-72, BUCHAREST, 00000, ROMANIA (2016)
16. Chircu, S., Neacsu, I.: USING E-LEARNING TOOLS IN THE GROWTH OF STUDENTS' MOTIVATION REGARDING FORMAL EDUCATION. In: Roceanu, I and Ciolan, L and Holotescu, C and Radu, C and Stanescu, I and Logofatu, B (ed.) RETHINKING EDUCATION BY LEVERAGING THE ELEARNING PILLAR OF THE DIGITAL AGENDA FOR EUROPE!, VOL. II. pp. 121–128. CAROL I NATL DEFENCE UNIV PUBLISHING HOUSE, PANDURI ST, 68–72, BUCHAREST, 00000, ROMANIA (2015)
17. Min, K.S., Yamin, F.M., Ishak, W.H.W.: Design, Purpose of Usage and the Impact of LMS on Student Learning: A Preliminary Findings. In: Baharom, F and Mahmuddin, M and Yusof, Y and Hashim, NL and Hassan, S and Yusop, NI and Ishak, WHW and Saip, MA (ed.) PROCEEDINGS OF KNOWLEDGE MANAGEMENT INTERNATIONAL CONFERENCE (KMICE) 2012. pp. 673–676. UNIV UTARI MALAYSIA-UUM, COLL ARTS & SCI, INFOR TECHNOL BLDG, SINTOK, KEDAH 06010, MALAYSIA (2012)
18. Maldonado Berea, G.A., Vega Gea, E.: ATTITUDE OF UNIVERSITY STUDENTS TOWARDS THE MOODLE PLATFORM. PIXEL-BIT- Rev. MEDIOS Y Educ. 105–117 (2015)
19. Adamkovičová, M., Burgerová, J.: Vybrané aspekty komunikačnej dimenzie e-learningu. Prešovská univerzita v Prešove, Pedagogická fakulta, Prešov (2014)
20. Mohd Kasim, N.N., Khalid, F.: Choosing the Right Learning Management System (LMS) for the Higher Education Institution Context: A Systematic Review. Int. J. Emerg. Technol. Learn. 11, 55 (2016). https://doi.org/10.3991/ijet.v11i06.5644
21. Herrera, J.L. de, Herrero-Tejedor, T., Pérez-Martín, E., Conejo-Martín, M.Á., Martín-Romero, J.L., Velasco-Gómez, J.: Tutorial 2.0 on Technical Drawing 3D and Visualization. Procedia Comput. Sci. 25, 108–112 (2013). https://doi.org/10.1016/j.procs.2013.11.013
22. Rodríguez Ruibal, A.: Utilización de Google Hangouts en e-learning. Hist. y Comun. Soc. 19, (2014). https://doi.org/10.5209/rev_HICS.2014.v19.46567
23. Daraghmi, E.Y., Hsiao, C.H., Yuan, S.M.: A New Cloud Storage Support and Facebook Enabled Moodle Module. In: 2014 7th International Conference on Ubi-Media Computing and Workshops. pp. 78–83. IEEE (2014)
24. Ornat, C., Kallas, Z.: MOODLE, WEB 2.0 & "IFRAME{"}. In: Chova, LG and Martinez, AL and Torres, IC (ed.) INTED2012: INTERNATIONAL TECHNOLOGY, EDUCATION AND DEVELOPMENT CONFERENCE. pp. 1200–1203. IATED-INT ASSOC TECHNOLOGY EDUCATION A& DEVELOPMENT, LAURI VOLPI 6, VALENICA, BURJASSOT 46100, SPAIN (2012)
25. Castro Fernandez, F.: RUBRICS AND PEER ASSESSMENT IN A VIRTUAL ENVIRONMENT: A CASE STUDY. In: Chova, LG and Martinez, AL and Torres, IC (ed.) EDULEARN14: 6TH INTERNATIONAL CONFERENCE ON EDUCATION AND NEW LEARNING TECHNOLOGIES. pp. 5035–5041. IATED-INT ASSOC TECHNOLOGY

EDUCATION A& DEVELOPMENT, LAURI VOLPI 6, VALENICA, BURJASSOT 46100, SPAIN (2014)
26. Bettoni, M., Bernhard, W., Eggs, C., Schiller, G.: Participative Faculty Development With an Online Course in eCollaboration. In: Balcaen, P (ed.) PROCEEDINGS OF THE 6TH INTERNATIONAL CONFERENCE ON E-LEARNING. pp. 493–496. ACAD CONFERENCES LTD, CURTIS FARM, KIDMORE END, NR READING, RG4 9AY, ENGLAND (2011)
27. Barnard-Ashton, P.: USING WIKI'S IN PROBLEM BASED LEARNING: E-LEARNER BEHAVIOUR AND ATTITUDE. In: Chova, LG and Belenguer, DM and Torres, IC (ed.) EDULEARN10: INTERNATIONAL CONFERENCE ON EDUCATION AND NEW LEARNING TECHNOLOGIES. pp. 581–586. IATED-INT ASSOC TECHNOLOGY EDUCATION A& DEVELOPMENT, LAURI VOLPI 6, VALENICA, BURJASSOT 46100, SPAIN (2010)
28. Wright, C.R.: Learner Questionnaire for Online or eLearning Courses, https://www.scribd.com/doc/36958035/Learner-Questionnaire-for-Online-or-eLearning-Courses-C-R-Wright
29. Gallardo-Echenique, E.E., Marqués-Molías, L., Bullen, M., Strijbos, J.-W.: Let's talk about digital learners in the digital era. Int. Rev. Res. Open Distrib. Learn. 16, (2015)
30. Aberšek, M.: The Digital Natives Generation and their Prerequisites for Learning Online. In: DIVAI 2016 11th International Scientific Conference on Distance Learning in Applied Informatics. pp. 55–66. Wolters Kluwer, Praha (2016)

RespectNET Initiative: Respectful Communication Through Media Education Network

Anca Draghici, Maria Elena Boatca, Valerij Dermol, and Caius Luminosu

1 Introduction

1.1 Brief Overview of the Research Context

Respectful communication in the Digital Era faces a unique set of hurdles due to the spread and access to technology and the acceptance of using it. The new generations (Millennials, Gen Z, and most of Gen Alpha) are dependent (or even addicted) to the Internet, social media, and Artificial Intelligence (AI) software [1–6]. Thus, many changes are already announced and present in education, science and technology, and organizational practices, as these changes affect people's work styles and living. All these phenomena have been associated with the acceleration of digital transformation at all organizational levels [1]. The scientific literature, policy makers, and nongovernmental reports present the actual challenges but also their main impact on the communication process, as presented in the following:

- Loss of nonverbal cues: Text-based communication lacks facial expressions, body language, and voice tone, which can lead to errors in communication and misunderstandings. This can negatively impact the interlocutors' behaviors due to sarcasm or humorous situations, which can be misinterpreted [7, 8].
- Distance and anonymity: The physical and emotional distance created by "screens" can make people less inhibited. This can lead to saying things online that one wouldn't say in person, fostering a harsher tone [9–11].

A. Draghici (✉) · M. E. Boatca · C. Luminosu
Politehnica University of Timisoara, Timisoara, Romania
e-mail: anca.draghici@upt.ro; maria.boatca@upt.ro; caius.luminosu@upt.ro

V. Dermol
International School for Social and Business Studies, Celje, Slovenia

M. Turčáni (ed.), *15th International Scientific Conference on Distance Learning in Applied Informatics*, EAI/Springer Innovations in Communication and Computing,
https://doi.org/10.1007/978-3-031-81261-3_5

- Echo chambers and filter bubbles: Algorithms on social media platforms can create personalized feeds that reinforce existing beliefs. This can limit exposure to diverse viewpoints and foster hostility toward those with differing opinions [10, 12].
- The speed of online communication: The ease and speed of reaction and online response can generate impulsive situations without considering the potential impact of words [13, 14].
- Misinformation and disinformation: The spread of false information online can create confusion and distrust, making respectful dialogue difficult [15, 16].

In this context of profound changes in internal and external communication processes related to HEIs, the literature underlines the necessity of regulating communication behavior with an adequate code of conduct or internal decisions, thus encouraging ethical behavior [17, 18]. Moreover, during and after the Covid-19 pandemic, universities have strengthened their role as active communicators of events, results, and successful projects carried out for local, regional, national, and international communities, thus becoming true "Voices of the city" (consolidating "the third mission") [8, 10, 13]. The main ways to cultivate respectful communication online and using different types of media in the HEIs context are recognized by specialists and discussed at the individual level [8, 13, 15, 16, 19–22], most often and less at the organizational level.

In conclusion, creating a respectful online environment (not stressful and safe) for all is a requirement (practically a tendency in each network or online community) of the new Digital and AI Era. With this scope in mind, the RespectNET project was developed, financed by the Erasmus+ program as a cooperation partnership at the higher education level.

In this context, the present research initiative addresses the topics of "Democracy and inclusive democratic participation," "Media literacy and tackling disinformation," and the "Prevention of bullying" (according to the Erasmus+ Programme Guide [23]. Preliminary studies on the challenges of the Digital and AI Era at the international consortium level [24] have underlined that the digital capabilities of the higher education sector are currently insufficient to react adequately to phenomena like attacks on scientific thinking, bullying over digital media, and the distortion of public discourse through conspiracy narratives, fake news, and hate speech (e.g., violence in media, violence online, on social media, etc.) [1–6].

A jump in the level of digital media literacy must change this. Next to the already good technical digital capacities of universities must be added the appropriate social and communication skills (so-called soft skills) to adequately use digital media like social media, video, multimedia, and news portals and others [25]. These are currently used by a civil society that is characterized by a growing segregation and distrust, hate speech, and bitterness, which reflects negatively on universities, as on the one hand, members of universities, staff, and students alike, are part of this societal phenomena [15–18]. On the other hand, universities should enlighten society about the standards of good communication and real, respectful, and ethical debate, and they can only do this if they represent a role model for such standards in society, generating new approaches and positively influencing academic and social behavior with understanding and acceptance of social, ethnic, linguistic, and cultural diversity. Thus, the need for new approaches is emerging to improve competitiveness and improve the dialogue between the university and civil society actors, especially at the regional and local level with the recognition of academic excellence and the role of the university as an active actor in society [7, 8, 14].

Therefore, our initiative's objectives are developing a matrix of such competencies, developing a training program, and developing material for a transmedia campaign aimed at fostering the horizontal priorities [23] of: "Common values, civic engagement and participation" and "Addressing digital transformation through development of digital readiness, resilience and capacity." Universities and the actors of civil society need to reflect on the common European values of enlightenment, equal rights, respect for the individual, diversity, and safety [26].

The Covid-19 pandemic period has provided many cases when scientists (belonging to universities) were attacked by civil society members or organizations. Participation in civil society and the free discussion in universities is discouraged by negative behavior and attitudes, which are boosted by the outreach and algorithms of the most common digital media, up to phenomena where scientists are physically threatened and attacked, and civil society activists are discouraged by bitterness, personal attacks, and bullying on social media [27, 28].

Strengthening media literacy among young people will aim to help them form their own opinions, make informed decisions, and fully exercise their freedom of expression, being able to participate actively in the civic society and contributing to the creation of an informed, multicultural, and engaged community. Universities must be "Laboratories of Democracy," where the rules and standards of respectful dialogue must be identified, developed, and communicated. As such, universities are critical for setting the standard for public dialogue and must serve civil society by teaching such standards. As the public discourse today is dominated by the communication standards that are set by the new digital media, in terms of speed, format, and expected content, while universities mostly stick to quite traditional forms of scientific communication and publication, Universities must enhance their digital readiness, also with respect to science communication and the general methods of communication and university–civil society interaction.

Therefore, it is the objective of the RespectNET initiative (an educational research approach) to identify the needed competencies for communication and digital media use by university members, to develop them via a targeted modular training program, and to improve their impact on universities and the civil society digital transmedia dialogue via a transmedia campaign. Consequently, the feedback collected will demonstrate the impact of the RespectNET initiative in the universities to empower and encourage a respectful and ethical dialogue with civil society stakeholders.

1.2 RespectNET Project Presentation

Therefore, the present research context is given by the activities developed by the international consortium implementing the RespectNET project (2021–1-IT02-KA220-HED-000027578, https://respectnet.eu/). Four partners (Universita Telematica Pegaso, Italy; International School for Business and Social Studies, Slovenia; Politehnica University of Timisoara, Romania, and Verein für sozialwissenschaftliche Beratung und Forschung e.V., Germany) have agreed to collaborate on the project implementation, share their knowledge and experiences, and create innovative and open educational resources (OERs) for training and the transmedia campaigns supporting common objectives [24]. RespectNET aims to develop a training program with practices related to a common code of conduct for responsible media use and media and communication literacy training multipliers as a resource for university personnel, as well as civil society activists as role models for the younger generation in universities and society [24]. The target groups are training designers and responsible people in universities (e.g., teaching and research staff, marketing, and ethics responsible persons), civil society, policymakers, and other stakeholders.

The project objectives are [24]: (1) Developing the media literacy competencies of university staff and students to foster a culture of respectful scientific dialogue within the university and as members of civil society; (2) Increasing the exchange of experiences and interchange with the wider regional civic society to build a knowledge partnership between universities and civil society for a respectful societal dialogue; (3) Developing a code of respectful dialogue in university and (social) media for all university members; (4) Developing multimedia modules on media competencies and respectful communication, available on a created transmedia learning platform on media and communication competencies for university members.

These concrete objectives are meant to contribute in a wider context to support: (1) the contribution of science to mitigating societal tensions, as expressed in phenomena like conspiracy narratives, hate speech, and cyberbullying; (2) consideration of European values in the further internal development of universities and in their collaboration with civil society. Furthermore, one of the European values that the RespectNET project carefully addresses is the relation to multidimensional diversity in all its activities. The final beneficiaries of the platform of transmedia learning on media and communication competencies are university staff, students, and other members of civil society [24].

2 The Research Methodology

It was expected by the RespectNET consortium members that the research results would provide a realistic response to the challenges of digitized communication, which has disrupted traditional communication styles, especially in terms of the

Table 1 The research methodology for developing the RespectNET initiative

Research stage (work packages and activities)	Methods and tools
1. Matrix of media competencies and respectful communication for university staff (Analysis of reports on COVID impact and influence of social media on in-university communication in each country; development of focus groups in each country; identification of five good practices/country; survey among the university staff members in each country; finalize matrix of media competencies and respectful communication for university staff in skill development recommendation report and infographic format)	Literature review; best practice analysis; focus group; survey based on a designed questionnaire; data process with excel; SPSS software solution for the statistical data processing; Canva design tool
2. Multimedia modules on media competencies and respectful communication (Review competencies matrix and survey results to deduce training priorities according to the local needs of target groups; synthesize and translate draft version of 10 multimedia modules; get feedback from partner organizations and improve draft versions; elaboration of final version of modules, translation, and introduction of content to online portal/e-learning platform; international training of civil-socially active students in advanced semesters from partner organizations with multimedia modules (3 days F2F and 8 h online))	Critical analysis of the preliminary existing content; webinars for critical analysis and feedback; PowerPoint and word docs for content development; graphical design; multimedia tools for short videos; Canva; Moodle; YouTube; web page
3. Online international training for university staff and transmedia learning platform content (Elaborate framework didactical concept based on transmedia storytelling; technical preparation of content templates; developing the transmedia campaign materials on relevant topics in infographics, stories, snippet info, and adapting it to be used in a web format; registering users, building user community, facilitating use of platform; online international training for university members (virtual blended learning, 8 h online, followed by a feedback survey with analysis))	Webinars and interactive trainings via ZOOM; PowerPoint and word docs for content development (short articles, graphics, posters, infographics); graphical design; feedback survey with data analysis using excel; multimedia tools for short videos; Canva; Moodle; YouTube; web page; speaker; Spotify, etc.

university's relationship with civil society and with local and regional communities (stakeholders). Thus, the complex research methodology consists of qualitative and quantitative methods, mixed or followed by creative developments, web design, graphical design, etc. The research approach is associated with project implementation activities, as described in Table 1.

In the following section, there will be detailed the results achieved in each methodological stage and will be provided arguments for the transferability of these results into higher education institutions' practice.

3 Research Results

3.1 RespectNET Matrix of Competencies

The approach of defining the matrix of competencies consists of a series of consultations, discussions, survey, and analysis to refine the initial list of competencies (see the research stages description in Table 1) to a schema that reflects the reality in many higher education institutions and that can be quickly transferred and developed via the training program. Figure 1 presents the result of the applied approach with relevant details on the competencies considered to be most needed to be developed via a training program.

The initial six competencies' clusters were identified: respectful communication in science; public speaking, engagement, and motivation; individual and organizational communication in university; combatting anti-respectful communication; respectful networking; respectful communication for sustainable development awareness. These categories were (re)grouped after the survey results into four categories, as follows: (1) critical media competencies, (2) science communication, (3) general communication, and (4) democracy competence. This "grouping action" proved to be particularly useful for the structured characterization of training needs and, subsequently, for the design of the RespectNET training program. The definition of the competency matrix is particularly useful in the judicious design of training programs, and the impact of the research has provided a relevant and realistic diagnosis of the current way of university communication.

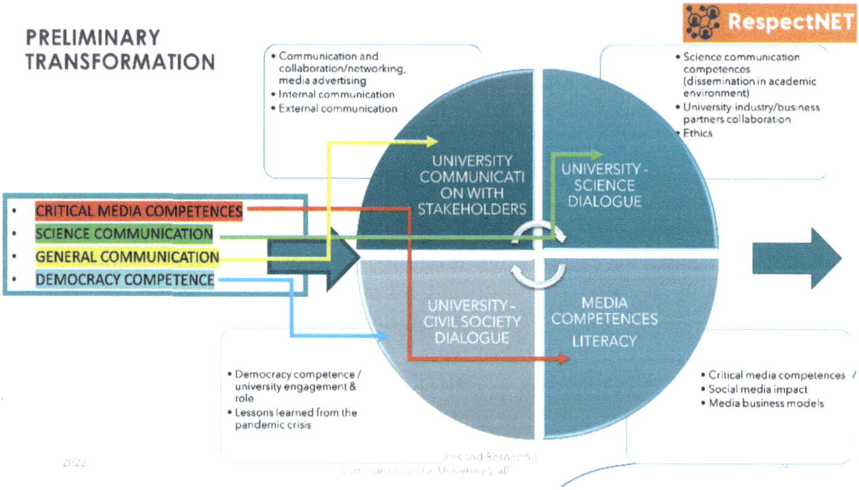

Fig. 1 RespectNET matrix of competencies

The research results of this stage were collected and structured in the "Respect-NET Matrix of Competencies and Policy Recommendation Report," available online at: https://respectnet.eu/matrix-of-competencies/.

3.2 *The E-Learning Platform*

According to the matrix of competencies content, for each competence, there have been associated several descriptors based on the existing information in the literature and the European framework of Competences for Democratic Culture [29]. The refinement and adaptation of the general matrix of competencies to the considered target groups' specifics have been filtered by the partners' competencies and expertise. The result is shown in Fig. 2.

The e-learning platform was developed under the Moodle application, and its graphic aspect was particularly carefully designed to ensure easy use and easy identification of the educational resources (OERs). The largest time (and financial) budget was allocated to the development of the digital content of the training program, which consists of 13 dedicated lessons grouped into four modules, as follows: (1) media competencies (with 4 lessons); (2) communication with universities' stakeholders (with 3 lessons); (3) university and civil society dialogue (with 3 lessons); and (4) university and science society dialogue (with 3 lessons).

The results of this research stage can be visualized online at: https://elearningproject.eu/all-courses/?course_catid=99. In addition, the creative developments of multimedia materials are on the YouTube channel's specific playlist: https://www.youtube.com/playlist?list=PL7Ij3-xTzrJJzL-cV7-AH-y5_GSVPje7p.

The created OERs (under Creative Commons Licenses) support not only the students' competencies development but also the teaching staff (trainers) by pro-

Fig. 2 RespectNET skill card (the modules and courses/lessons of the training program)

viding a mix of materials (explicative text with PowerPoint presentations, videos, additional references, etc.) ready to be used. Thus, RespectNET OERs are quickly transferable into education curricula or university's courses (syllabus).

3.3 RespectNET Transmedia Campaigns

The most valuable result of the RespectNET project implementation is the creation of four transmedia campaigns on the most important and actual topics (established during a focus group with partners' representatives and important stakeholders from their community): Artificial Intelligence (AI) in University Communication (developed by the Slovenian partner); Multicultural Respectful Communication (developed by the Romanian partner); Representing Environmental Sustainability in the University System (developed by the Italian partner); University Third Mission Activities to Support Refugees (developed by the German partner).

The research approach consisted of selecting the best bibliographic sources (based on best practices analysis from the scientific literature and the gray literature, and by considering the European recommendations and competencies frameworks). Thus, a lot of types of materials were created (to be easily understood by different subgroups in the target groups) such as: articles (1 page), guidelines, posters, infographics, videos (posted on YouTube, TikTok, Twitter, Instagram, etc.), Snippet-sized and other kinds of multimedia messages, social media messages (FB, Twitter, Instagram), and podcasts (posted on Spotify, Specker). The distribution of the materials, associated with the transmedia campaign development, has been done on various channels to be sure that as many people as possible from the target groups will be touched (web page, YouTube, FB, TikTok, Twitter, Instagram, Spotify, Specker, etc.). The aim of the transmedia campaigns was to create a dominance of respectful attitudes as a standard of good practice and behavior at universities and in civil society.

- The creative approach of the transmedia campaign materials is explained on the project webpage: https://respectnet.eu/
- The results of this research stage can be completely visualized on the e-learning platform (designed as a separate training course with a mix of propaganda materials): https://elearningproject.eu/all-courses/?course_catid=99
- The multimedia materials of the campaigns are posted on the YouTube channel (a specific playlist was created): https://www.youtube.com/playlist?list=PL7Ij3-xTzrJJzL-cV7-AH-y5_GSVPje7p
- Podcasts on specific topics included in the transmedia campaigns' materials are on Speaker channel: https://www.spreaker.com/search/episode?query=RespectNET and on the Spotify channel: RespectNET Official Podcast | Podcast on Spotify

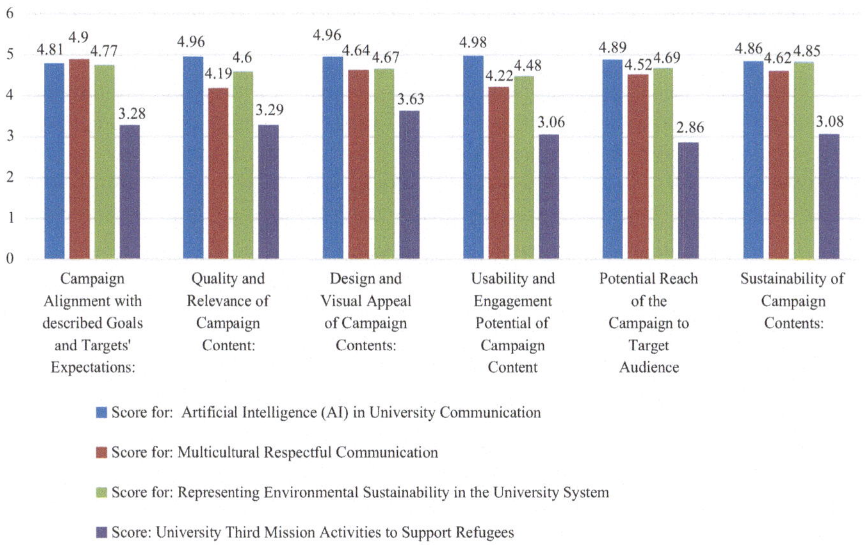

Fig. 3 Feedback collected for the transmedia campaigns

The impact of the transmedia campaigns has been analyzed based on a feedback survey with an online questionnaire (14 questions) aiming to assess the content, format, and impact on "Shaping Narratives for Social Impact." The feedback was collected mainly from target groups in the project partners' countries, and 664 valid questionnaires were processed (228 from Italy, 217 from Romanian, 121 from Slovenia, and 98 from German), showing a pronounced interest in the campaign topics. Furthermore, 34% of the respondents were teaching and research staff, 54% were students, 7% were representatives from civil society, and 5% belonged to university leadership. The research results on the campaigns' evaluation are presented in Fig. 3 (there are presented the first set of questions invites participants to assess the campaign against stated criteria on a Likert scale (5—Very much, . . . , 1—not at all)).

All the campaigns enjoyed positive reviews from the respondents (in the evaluation of the six criteria), except for the "university third mission" campaign, where 129 respondents opined that it was not relevant for their organization. Also, all respondents claimed that they would recommend studying the materials to others, motivating its usefulness in the current context of the Digital Era and the widespread use of AI applications, but also considering global warming and migration phenomena, together with concerns about the existence of several war zones.

4 Conclusion

The Digital Era has introduced issues like the loss of nonverbal cues, anonymity, echo chambers, and misinformation, which impact respectful communication in universities and with their stakeholders or civil society actors. The presented initiative, approach, and results achieved have a high degree of transferability in the practice of European higher education institutions and beyond. The created OERs related to the RespectNET training program can be used in the case of teaching and research staff and students from different specializations to encourage and support ethical and responsible behavior in academic, scientific communication and in civil society. The initiative, approach, and presented results have a high degree of transferability in the practice of European higher education institutions and beyond. Relevant aspects supporting this statement are presented in the following.

Defining the skills matrix is particularly useful for knowing the skills gap and orienting programs and training courses in the direction of elucidating unknown or less known aspects in the field of communication, ethics, and for creating a respectful dialogue between the university and civil society stakeholders. Thus, the competency matrix is the result of a relevant and realistic diagnosis of the current way of university communication style.

The created OERs related to the RespectNET training program can be used in the case of teaching and research staff and students from different specializations to encourage and support ethical and responsible behavior in academic, scientific communication and in civil society. Thus, we recommend the transfer of RespectNET OERs in the case of communication courses, research ethics, ethics and academic integrity, culture and civilization, sociology, etc.

The four transmedia campaigns designed, developed, and implemented in the context of the RespectNET project partners have proven extremely attractive and useful in modeling ethical and responsible behavior in terms of: (1) use of AI; (2) multicultural communication in the context of the internationalization of universities; (3) pro-sustainability communication; and (4) growing the intensity of defining activities for the "third mission" of universities.

The RespectNET initiative will help universities reflect on negative communication styles and identify which skills need improvement. Moreover, the initiative will help define the skills to understand and respond to the anti-scientific, conspiratorial narrative and will encourage reasoned, evidence-based communication through essential skills of effective science communication. The RespectNET initiative has all the means to ensure flexibility in application, but also sustainability. The training materials (available under Creative Commons) and those related to the transmedia campaigns were well perceived by the students' communications from the various higher education institutions, which entitles us to affirm that the impact of the project and the initiative were very strong.

In addition, the teaching and research staff of universities have declared that the RespectNET initiative's impact has increased the attractiveness of education and training programs, in line with individual needs and expectations, and it has positively affected the dynamic, committed, and professional environment inside the department/faculty/laboratory, which was ready to integrate good practices and new methods, including digital capabilities, into daily activities. Thus, RespectNET has been appreciated for providing guidelines to universities for improving communication and dialogue with civil society actors.

The limits of this approach are associated with the translation and adaptation (where necessary) of all the materials related to the transmedia campaign into other languages, to ensure a better understanding of the transmitted messages. Thus, depending on cultural and social specifics, some topics discussed in the training lessons and promoted in the transmedia campaigns may not relate to the local, regional, or national reality, and there could be some missing aspects, too. Thus, the presented educational approach could only be applied successfully in the realities of Italy, Slovenia, Germany, and Romania in improving the university–civil society dialog (refinements and feedback were primarily collected from project partners' countries, which may not fully represent a broader audience).

Acknowledgments The content of this article is related to the project RespectNET project—"Respectful Communication through Media Education Network" implementation activities (Erasmus+, KA2 Strategic Partnership project, 2021-1-IT02-KA220-HED-000027578, https://respectnet.eu/), which was co-financed by the European Commission. Thus, the content of the article reflects only the views of the authors, and the Commission cannot be held responsible for any use that may be made of the information contained therein.

References

1. Troitiño, D. R.: The European Union facing the 21st century: The digital revolution. TalTech Journal of European Studies, 12(1), 60–78 (2022).
2. Pawar, A. B., Gawali, P., Gite, M., Jawale, M. A., William, P.: Challenges for hate speech recognition system: approach based on solution. In 2022 International conference on sustainable computing and data communication systems (ICSCDS) (pp. 699–704). IEEE (2022, April).
3. Rahmatullah, A. S., Mulyasa, E., Syahrani, S., Pongpalilu, F., Putri, R. E.: Digital era 4.0: The contribution to education and student psychology. Linguistics and Culture Review, 6(S3), 89–107 (2022).
4. Buchholz, B. A., DeHart, J., Moorman, G.: Digital citizenship during a global pandemic: Moving beyond digital literacy. Journal of Adolescent & Adult Literacy, 64(1), 11–17 (2020).
5. Yu, S., Abbas, J., Draghici, A., Negulescu, O. H., Ain, N. U.: Social media application as a new paradigm for business communication: the role of COVID-19 knowledge, social distancing, and preventive attitudes. Frontiers in Psychology, 13, 903082 (2022).
6. Zhou, Y., Draghici, A., Abbas, J., Mubeen, R., Boatca, M. E., Salam, M. A.: Social media efficacy in crisis management: effectiveness of non-pharmaceutical interventions to manage COVID-19 challenges. Frontiers in psychiatry, 12, 626134 (2022).
7. Mukherjee, S.: Advancing Communication Skills in the Digital Era: A Comprehensive Review of Modern Techniques, Challenges, and Innovations for Effective Interpersonal and Media-Based Interactions. International Journal for Multidimensional Research Perspectives, 2(3), 28–43 (2024).
8. Venter, E.: Challenges for meaningful interpersonal communication in a Digital Era. HTS: Theological Studies, 75(1), 1–6 (2019).

9. Panjeti-Madan, V. N., Ranganathan, P.: Impact of screen time on children's development: cognitive, language, physical, and social and emotional domains. Multimodal Technologies and Interaction, 7(5), 52 (2023).
10. Reynolds Community College—Library, Fake News: Separating Fact from Fiction, https://libguides.reynolds.edu/fakenews/bias, last accessed 2024/05/14.
11. Medium, Online Disinhibition and the Role of Anonymity in Social Media, https://waivlengthdev.medium.com/online-disinhibition-and-the-role-of-anonymity-in-social-media-809e8309da12, last accessed 2024/05/14.
12. Reuters Institute for the Study of Journalism, Echo chambers, filter bubbles, and polarization: a literature review, https://reutersinstitute.politics.ox.ac.uk/echo-chambers-filter-bubbles-and-polarisation-literature-review, last accessed 2024/05/14.
13. Ahmad, H., Rydén, I.: Time is Tik(Tok)ing: An Insight on Contemporary Media Consumer Behavior-An exploratory study on user-generated content's potential impact on contemporary media consumption (2023).
14. Lohmann, R. C., Smith, A. C.: 15-Minute Focus: Digital Citizenship: Supporting Youth Navigating Technology in a Rapidly Changing World. National Center for Youth Issues (2023).
15. Hefner, D., Vorderer, P.: Digital stress: Permanent connectedness and multitasking. In The Routledge handbook of media use and well-being (pp. 237–249). Routledge (2016).
16. Nguyen, M. H.: Managing social media use in an "always-on" society: Exploring digital wellbeing strategies that people use to disconnect. Mass Communication and Society, 24(6), 795–817 (2021).
17. Gottardello, D., Karabag, S. F.: Ideal and actual roles of university professors in academic integrity management: a comparative study. Studies in Higher Education, 47(3), 526–544 (2022).
18. Mijwil, M. M., Hiran, K. K., Doshi, R., Dadhich, M., Al-Mistarehi, A. H., Bala, I.: ChatGPT and the future of academic integrity in the artificial intelligence era: a new frontier. Al-Salam Journal for Engineering and Technology, 2(2), 116–127 (2023).
19. Nadeem, M. U., Kulich, S. J., Zabrodskaja, A., Bokhari, I. H.: The impact of empathy, sensation seeking, anxiety, uncertainty, and mindfulness on the intercultural communication in China during the COVID-19. Frontiers in Public Health, 11, 1223215 (2023).
20. Masrani, T. W., Jamieson, J., Yamashita, N., He, H. A.: Slowing it down: Towards facilitating interpersonal mindfulness in online polarizing conversations over social media. Proceedings of the ACM on Human-Computer Interaction, 7(CSCW1), 1–27 (2023).
21. Akinrinola, O., Okoye, C. C., Ofodile, O. C., Ugochukwu, C. E.: Navigating and reviewing ethical dilemmas in AI development: Strategies for transparency, fairness, and accountability. GSC Advanced Research and Reviews, 18(3), 050–058 (2024).
22. Potegal, M., Nordman, J. C.: Non-angry aggressive arousal and angriffsberietschaft: A narrative review of the phenomenology and physiology of proactive/offensive aggression motivation and escalation in people and other animals. Neuroscience & Biobehavioral Reviews, 147, 105110 (2023).
23. EU programme for education, training, youth and sport, Programme Guide, https://erasmus-plus.ec.europa.eu/erasmus-programme-guide, last accessed 2024/05/14.
24. RespectNET - Respectful Communication through Media Education Network, https://respectnet.eu/, last accessed 2024/05/14.
25. Volkov, A., Rishko, Y., Kostyukhin, Y., Sidorova, E., Boboshko, D., Savinova, D., Ershova, V.: Using digital tools to teach soft skill-oriented subjects to university students during the COVID-19 pandemic. Education Sciences, 12(5), 335 (2022).
26. Veugelers, W, de Groot, I., Stolk, V.: Research for CULT Committee—Teaching Common Values in Europe, European Parliament, Policy Department for Structural and Cohesion Policies, Brussels (2017), https://www.europarl.europa.eu/RegData/etudes/STUD/2017/585918/IPOL_STU(2017)585918_EN.pdf, last accessed 2024/04/08.

27. Robescu, D., Reiner, S., Trunk, A.: Toward a matrix of competences for respectful communication in the university-civil society context (2023). https://doi. org/10.21125/edulearn, last accessed 2024/04/08.
28. Draghici, A., Repanovici, A., Ng, P. K.: Digital challenges to empower universities' implication in the community. Human Systems Management, 42(2) (2023).
29. Council of Europe, Model of Competences, https://www.coe.int/en/web/reference-framework-of-competences-for-democratic-culture/model, last accessed 2024/05/10.

On Suitability of Renko Charts for Algorithmic Trading

Martin Drozda, Maros Cavojsky, and Patrik Sebes

1 Introduction

Profitable trading of stocks, commodities, indexes, derivatives, and cryptocurrencies is one of the last enigmatic challenges of artificial intelligence (AI). AI has recently become dominant in the fields previously believed to be a human supremacy domain, such as Go (game) [16], computer games [10], or vehicle driving [7]. AI approaches such as convolutional neural networks (CNNs), long short-term memory neural networks (LSTMs), transformers, reinforcement learning (RL), etc., have gained the capability to excel where humans are prone to fail; see [19] for a review on deep learning approaches in stock market prediction.

There are two basic approaches of how to analyze assets for their ability to generate trading profits:

- Fundamental analysis, where traders evaluate the intrinsic value by examining related economic, financial, and other qualitative and quantitative factors
- Technical analysis, where traders attempt to evaluate and predict the future price movements based on historical price data and trading volumes

Both approaches can provide an input for discretionary or algorithmic trading. In discretionary trading, decisions are made based on the trader's judgment and intuition, rather than relying solely on predefined rules or algorithms. On the other hand, algorithmic trading involves computer programs and algorithms to automatically execute trades based on a set of predefined rules and parameters.

M. Drozda (✉) · M. Cavojsky · P. Sebes
Faculty of Electrical Engineering and Information Technology, Slovak University of Technology, Bratislava, Slovakia
e-mail: martin.drozda@stuba.sk; maros.cavojsky@stuba.sk; patrik.sebes@stuba.sk

M. Turčáni (ed.), *15th International Scientific Conference on Distance Learning in Applied Informatics*, EAI/Springer Innovations in Communication and Computing,
https://doi.org/10.1007/978-3-031-81261-3_6

In technical analysis, traders can apply various indicators that reflect how the price or volume evolves or evolved in the past. The look-back time interval can range from seconds to months depending on the applied trading strategy. For example, Gerard Appel suggested to use the 12-day and 26-day EMA (exponential moving average) for his MACD (moving average convergence divergence) indicator [3]. In this case, the expectation would be to hold several days or weeks on to a trade. MACD can however be successfully applied to much shorter or longer time frames.

Let us assume that we evaluate a stock chart with 20 technical indicators such as EMA, RSI (relative strength indicator), MACD, OBV (on-balance volume), or CMF (Chaikin money flow); see [3, 5, 11, 18] and the references therein for various technical indicators and trading techniques. The task for a trader is to find out which indicators are relevant in a given market state. In this model case the trader has to choose a subset of these 20 indicators and decide whether market conditions are welcoming for any trading strategy.

The trader however might not only need to consider these 20 indicators, but also their various temporal instances, for example, their 19 past values, which leads to $20 \times 20 = 400$ pieces of information. As we know, a set of 400 items has $2^{400} - 1 \approx 2.58225 \times 10^{120}$ distinct non-empty subsets. This leads to understanding that a successful trader may not be able to choose which subset of indicators is relevant in the given market state. How is then possible that there are elite traders [14] that seem to be able to trade the markets successfully?

A plausible answer to the above question is that many traders (i) focus on a small number (1–5) of indicators and become experts in their interpretation and (ii) wait patiently until these indicators align with the market conditions. It is precisely their ability to wait for a favorable market state that increases their success rate. For example, a trader might wait until the 21-, 34-, 55-, and 89-day EMAs are nondecreasing and ordered such that the 21-day EMA is above the 34-day EMA, and so on. Additionally, another indicator reflecting current asset volatility might be necessary to trigger a trade, where low volatility could suggest a continuation trade in the direction of the EMAs.

Since any trade involves some level of risk, a stop loss is often used to limit the maximum loss from a trade to a predetermined value (e.g., -2%). Stop loss is a prime example of risk management. It reflects the reality that a trade based on a "perfect" indicators' setup still has high chances of failure. Another example of risk management is portfolio diversification, which suggests to divide funds among several assets, so that a complete failure of an asset does not lead to complete funds liquidation.

The goal of this chapter is to provide insights into whether Renko charts can be applied in algorithmic trading and, if so, what advantages and disadvantages they introduce. We show that Renko charts have a property that allows for training a classifier with Renko charts; however, any future predictions can be done on a candlestick basis, i.e., after each price close, which is key for continuous algorithmic trading.

We apply convolutional neural networks (CNNs) connected with Renko charts in order to predict the next asset price move. We construct a matrix (image) that encodes various indicators and their historical values. In addition, we create row permutations of these indicators across several layers, where each matrix row represents the values of a single indicator. By permuting the matrix rows, we aim to reduce the bias that may arise from having a single arrangement of neighboring indicators.

This document is organized as follows. In Sect. 2 we discuss the related work. In Sect. 3 we introduce our methodology based on Renko charts. In Sect. 4 we present the applied data set and the experimental parameters. In Sect. 5 we provide insights on how Renko charts can be used in algorithmic trading, and finally, in Sect. 6 we conclude.

2 Related Work

Stock market prediction (with direct applicability to algorithmic trading) has been approached by a variety of AI techniques including CNNs, LSTMs, transformers, autoencoders, GANs (generative adversarial networks), etc.; see [19] for a detailed survey.

Sezer and Ozbayoglu [15] introduced an approach based on CNN that encodes values in 15×15 pixel images, where each rows correspond to 15 indicators and columns correspond to 15 values of these indicators. In the image they grouped the applied indicators by whether they belong to trend indicators or oscillators. Before the training process, the created images are labeled by observing how the price for an asset behaves in the future. They report results for several assets, for example, for SPY they report an annualized gain of 10.77%, which is better than buy-and-hold approach that would gain 4.63%. The authors use daily historical data.

Salkar et al. [13] investigated whether Renko charts can be used in algorithmic trading. They combine Renko charts with either MACD or RSI indicator. The rules for trading are based on the expert knowledge of the authors, i.e., they do not apply a machine learning approach. They report gains of 5−10% over a time period of 3 months, what might be a too short period for any reliable testing.

Surprisingly, Renko charts are rarely applied in algorithmic trading. The above-mentioned survey on deep learning approaches [19] does not include any approach that would take advantage of Renko charts.

3 Methodology

3.1 Renko Charts

In order to filter out the noise of market data, in our experiments we use Renko charts. A Renko (small brick in Japanese) gets created once the asset price moves by a certain amount in any direction (up or down). The price move that implies a new Renko is computed by a volatility measure, and the default option is ATR (average true range) [18]. There are two standard ways to compute a Renko: (i) Price has to close outside the existing Renko so that a new Renko gets created and (ii) price needs to be outside the existing Renko in order to create a new Renko. Herein we apply the former approach. Whether the price closes outside the existing Renko depends on the applied aggregation time period, for example, for an OHLC data item and a 30-min aggregation time period, the aggregation may start at 12:00 and end at 12:30. OHLC stands for Open, High, Low, and Close price, respectively.

An example of how a Renko is computed is demonstrated in Fig. 1. The figure assumes that the asset price is represented as candlesticks in OHLC format. Notice that candlestick colors black/red indicate whether the price increased or decreased, respectively, compared to the open price. Black and red Renkos indicate that the Renko is created upward or downward, respectively.

A single Renko contains at least one candlestick. Often, it can contain a large number of candlesticks, for example, the first Renko in Fig. 1 contains nine candlesticks. In this example, for simplicity, we set the Renko height to 6USD; however, the height can change based on the current ATR value. When a Renko gets created and concluded, we are unable to tell how many candlesticks are contained within. Renkos in a Renko chart are shown to have the same width. As Renko is a key concept in our study, we also provide its formal definition.

Definition 1 (Renko) Let $P = \{p_j, \ldots, p_l, p_{l+1}\}$ be a time series of close prices, where $\{p_j, \ldots, p_l\}$ are already processed within one or more Renkos and p_{l+1} is the price close yet to be processed. Let R^i be the last created Renko characterized by the minimum price value p^i_{min}, the maximum price value p^i_{max}, and the height $\Delta P^i = p^i_{\text{max}} - p^i_{\text{min}}$. If $p_{l+1} > p^i_{\text{max}}$ or $p_{l+1} < p^i_{\text{min}}$, then at least a single new Renko R^{i+1} is created above or below R^i, respectively.

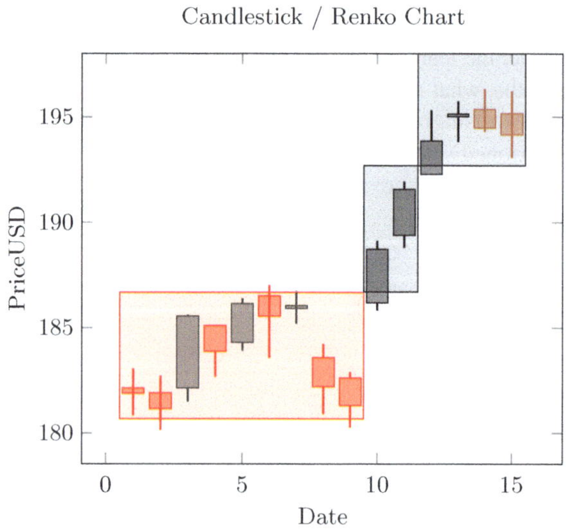

Fig. 1 Candlestick chart with 15 candlesticks converted into a Renko chart with three Renkos. The data is for Apple (ticker: AAPL) from Jan. 4–25, 2024

The $p_{\min}^{i+k}$ and $p_{\max}^{i+k}$ for the k-th new Renko R^{i+k} can be expressed as follows:

$$p_{\min}^{i+k} = p_{\max}^{i+k-1}, \quad p_{\max}^{i+k} = p_{\max}^{i+k-1} + \Delta P^{i+k} \quad \text{if} \quad p_{l+1} > p_{\max}^{i+k-1},$$
$$p_{\max}^{i+k} = p_{\min}^{i+k-1}, \quad p_{\min}^{i+k} = p_{\min}^{i+k-1} - \Delta P^{i+k} \quad \text{if} \quad p_{l+1} < p_{\min}^{i+k-1},$$

where $k \in \mathbb{Z}^+$ is the number of Renkos that get created until it holds that $p_{l+1} < p_{\max}^{i+k}$ or $p_{l+1} > p_{\min}^{i+k}$.

Notice that ΔP^i can be distinct for each R^i. Given the above definition, if the close price is above $p_{\max}^i + \Delta P^{i+1}$, then at least two Renkos will be created. Renkos are created until it is possible to create one, and this implies that in the case of a large price move, several Renkos can be created after price close. The first Renko R^0 is created by setting $p_{\min}^0 = p_0$ and $p_{\max}^0 = p_0 + \Delta P^0$ assuming that ΔP^0 is known at that time point.

3.2 *Renkos and Classifier Training*

Renko charts have an interesting property that they are either increasing or decreasing. If the price volatility remains low, then candlesticks are added to an existing Renko. This is a property with far-reaching consequences. When the task is to predict whether a new Renko will be created upward or downward, we train a classifier with the implicit knowledge that excludes the possibility of price stagnation.

When classifier training is done in a supervised way, it is necessary to have labels for training/validation data. For example, when training a classifier to recognize dogs and cats, it is necessary to label training/validation images with these two labels.

Renko charts allow for a specific way to label training/validation data. Let us assume that we are looking at the Renko chart shown in Fig. 2. It shows a specific way of computing the reward associated with this chart, where an upward Renko

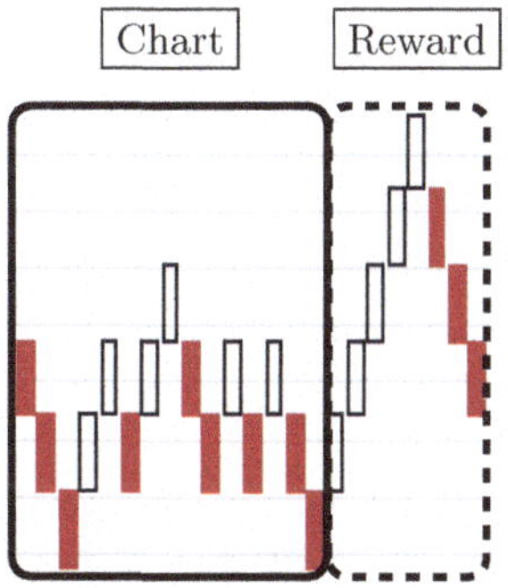

Fig. 2 Reward for a chart is computed from future Renko bricks

could be valued $+1$ and a downward Renko could be valued -1. In this case, the computed reward would be 2.0.

In general, future rewards Γ can be discounted and defined in the following way:

$$\Gamma = \sum_{t=0}^{\rho} \gamma^t r_t,$$

where $0.0 < \gamma \leq 1.0$ is the discount rate, $r_t \in \{-1, +1\}$ is the reward derived from the i-th Renko orientation, and ρ is the time horizon for future reward computation ($\rho = 7$ in Fig. 2). This approach to computing discounted future rewards is motivated by reinforcement learning, where the discounted return needs to be computed in order to decide which action should be selected next [17].

3.3 Input Representation

We use convolutional neural networks (CNNs) as the classifier for predicting the reward associated with a Renko chart. The input is a matrix $n \times m$, where n is the number of indicators and m is the number values for each indicator.

As we use Renko charts, the parameter m represents the number of Renkos that we consider; in other words, m is the chart width in Renkos. For example, the width of the chart shown in Fig. 2 is 15.

All applied indicators get normalized. For example, RSI, which has values ranging from 0.0 to 1.0, where 0.0 indicates oversold market conditions, 1.0 indicates overbought market conditions, and 0.5 indicates neutral market conditions, gets normalized to [0.5, −0.5]. In most cases, normalization is straightforward as they have well-established maximum, minimum, and neutral values except for average directional index (ADX) that ranges from 0.0 to 100.0 and its neutral value is 25.0. OBV for the k-th step is normalized as $\frac{OBV_k - OBV_{k-1}}{SMA(50)}$. This is necessary due to very different volume levels for various assets.

Each Renko can contain several candlesticks; therefore it is necessary to decide which indicator value will be attached to a Renko. For example, if we had two candlesticks contained in a Renko, then we have two values for the RSI indicator. Our strategy is to apply the value that represents a greater extreme. For RSI indicator with 0.5 neutral value, having two values, for example 0.7 and 0.4, the former would represent a greater extreme. In a more formal way, we choose the indicator value as follows:

$$\begin{aligned} q &= \arg\max_j \{|v_j|\}, \\ v_{\max} &= v_q, \end{aligned} \tag{1}$$

where v_j is the indicator value that is already normalized to 0.0 neutral value for the j-th candlestick within a Renko and $v_{\max}$ is the indicator value with the largest absolute value over candlesticks within this given Renko. Note that $v_{\max}$ can have both positive and negative values.

3.4 *CNN Architecture*

The CNN architecture is shown in Fig. 3. It applies a normalization layer, a rearrangement (implemented as a lambda function) layer that permutes the positions of the indicators, three convolutional layers, a flatten layer, two dense layers with a 0.25 dropout, and an output layer. The rearrangement is fixed and shown in Table 1. The purpose of this rearrangement is to bring various indicators to neighboring positions, i.e., to remove the systemic bias caused by a single indicator ordering. The filter size for the first convolutional layer is (2×2), and (3×3) for the remaining layers. The stride is set to 1.

As activation functions we use ReLU [2] for the convolutional layers, *tanh* for the dense layers, and linear activation for the output layer. The output of this architecture is a prediction of the reward for the given Renko chart that is represented as a matrix. Hereafter we refer to this prediction value as α.

We use Adam optimizer [9] and the Huber loss function [8].

3.5 *Trading Strategy*

When trading, it is necessary to decide, when an asset will get bought, sold, or a neutral position will be assumed. For interpretation reasons, we use a simple strategy that derives a market action from the predicted value for reward $\alpha \in \mathbb{R}$ that is an output of our CNN architecture:

$$\text{action} = \begin{cases} \text{Buy} & \text{if } \ell_{\max} \geq \alpha > \ell, \\ \text{Sell} & \text{if } -\ell_{\max} \leq \alpha < -\ell, \\ \text{Neutral} & \text{otherwise,} \end{cases} \tag{2}$$

where ℓ and $\ell_{\max}$ are experimental parameters. The reason for using $\ell_{\max}$ is that more extreme predictions of $|\alpha|$ indicate that there will be a pronounced move up in either direction, however, such moves are more obvious to notice by other market participants, and therefore, the market might already be reflecting this in its behavior.

The buy action results in a *long* position, while the sell action results in a *short* position. If a long position is followed by a sell action, we assume that the given asset will be sold and immediately shorted. Conversely, if a short position is followed by a buy action, we assume that the given asset will be bought (to cover the short) and immediately bought again to switch to a long position.

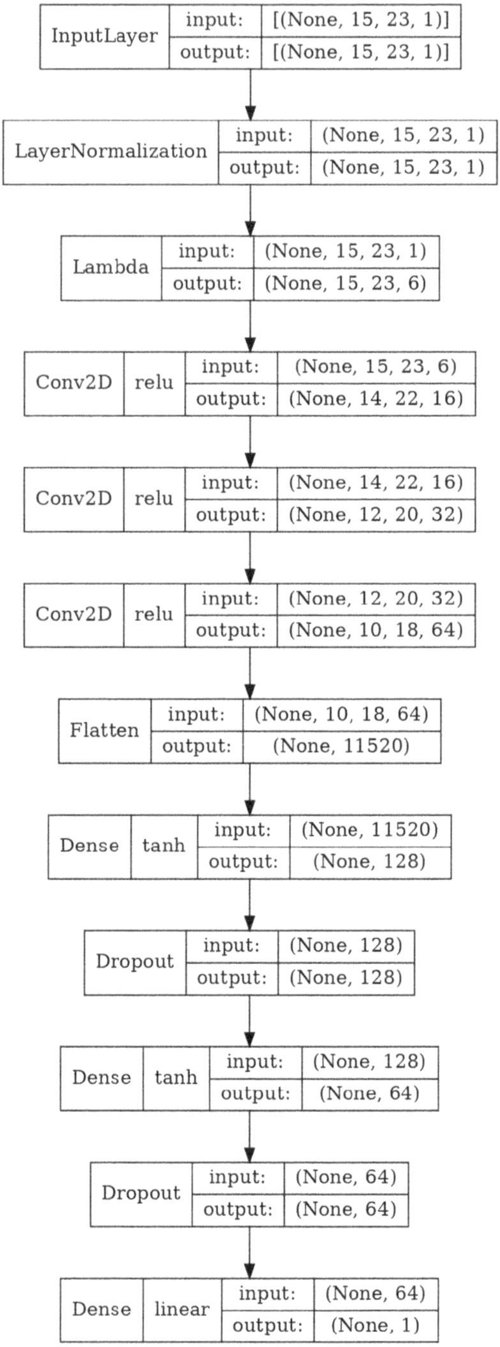

Fig. 3 Architecture of the CNN model

Table 1 Index matrix with rearrangement of indicators

Original	Transformed
0–22	12–22, 0–11
	18–22, 12–17, 6–11, 0–5
	21–22, 18–20, 15–17, 12–14, 9–11, 6–8, 3–5, 0–2
	6–11, 0–5, 18–22, 12–17
	9–11, 6–8, 3–5, 0–2, 21–22, 18–20, 15–17, 12–14

Obviously, there are many alternatives to this basic trading strategy. For example, one could employ only Buy and not Sell, reflecting the fact that an asset may be difficult to short.

4 Experimental Setup

4.1 Data Set

We use Kibot [12] as the source of historical market data, which provides data for various time aggregation periods. In our experiments, we utilize 30-min data for the 50 stocks and ETFs (Exchange-Traded Funds) with the highest trading volumes, as determined by Kibot. Although Kibot's market data dates back to 1998, some stocks and ETFs were introduced later, so their respective market data may start at a later date.

We trained with the historical data for the following stocks and ETFs (the tickers are in alphabetical order): AAPL, AMZN, BAC, BIDU, BRK, CAT, COP, CSCO, C, CVX, DIA, EEM, EFA, EWZ, FAS, FCX, F, FXI, GDX, GE, GLD, GOOGL, GS, HAL, IBM, INTC, IWM, JNJ, JPM, MRK, MSFT, MS, NFLX, ORCL, PFE, PG, QCOM, QQQ, SLB, SLV, SPY, SSO, TLT, TNA, T, VWO, WFC, XLE, XLF, and XOM. We only list their tickers, as their full names are easy to look up. For training, we use any historical data before 2021, and for testing we use historical data for 2021 and 2022.

For training purposes, we concatenate these 50 stocks and ETFs into a single training data set.

4.2 Indicators

In our experiments we apply a number of various indicators: relative strength index (RSI) [18], moving average convergence divergence (MACD), simple moving average (SMA) with several lengths, exponential moving average (EMA) with several lengths, TTM squeeze [5], pivot pocket [11], rate of change (ROC), Bollinger bands [4], Keltner channels, true strength index (TSI), money flow index (MFI), volume weighted average price (VWAP), average true range (ATR) [18],

Table 2 The list of applied indicators and their parameters

Indicator	Parameters
RSI	Window $= 20$
MACD	(12,26,9)
SMA	{20, 50, 100, 200}
EMA	{20, 50, 100, 200}
TTM squeeze	Window $= 20$, ATR window $= 20$, multiplier $= 1.5$
Pivot pocket	Length $= 10$, size $= 1.3$
ROC	Window $= 20$
Bollinger bands	Window $= 20$, std. dev. $= 2.0$
Keltner channels	Window $= 20$, ATR window $= 10$, multiplier $= 2.0$
TSI	Slow window $= 25$, fast window $= 13$
MFI	Window $= 14$
VWAP	Window $= 14$
ATR	Window $= 14$
ADX	Window $= 14$
OBV	No parameters
CMF	Window $= 14$
AROON	Window $= 25$

average directional index (ADX), on-balance volume (OBV), Chaikin money flow (CMF), AROON; see [6] for the definitions of indicators that we apply. Each indicator can be controlled by various input parameters; the applied parameters are shown in Table 2.

4.3 Parameters

We apply the following parameters: $\gamma = 0.9$, $\rho = 7$, $\ell = 0.1$, $\ell_{\max} = 3.1$, and the stop loss is set to -3%. For the stop loss we apply a rather high value as an asset can gap up or down overnight. This usually happens when some unexpected market news emerges.

The number of epochs when training is 80 and the batch size is 4,096.

4.4 Implementation Details

Our code is implemented in Python 3.10 and Keras 2.14.0 [1]. Keras is a high-level neural networks API that can be used with Python.

We use a PC with Ubuntu OS 22.04.4 LTS with Intel i9–13900KF CPU, 64GB DDR5 RAM, 4TB M.2 SSD, and NVIDIA GeForce RTX 4090 GPU, which has 24GB GDDR6X VRAM and 16,384 stream processors.

5 Discussion

A Renko chart can be viewed as a time series $\tau = R_i$, where each Renko R_i has an orientation $\{+1, -1\}$ relative to R_{i-1} and contains at least a single candlestick; see Definition 1. Reward prediction α is done for each concluded R_i, with the caveat that in the real world, we do not know how many candlesticks are going to be included in a Renko, i.e., we do not know when to run a prediction. Therefore, there is a practical difference between Renko-based prediction and candlestick-based prediction.

Nevertheless, prediction based on Renkos offers a range of possibilities. A key question to ask is, why to use candlestick data for training if our aim is to predict whether the next price step will be significantly upward or downward, often estimated in terms of ATR? Once a classifier is trained, it can be applied each time a candlestick is added to a Renko. This creates a win-win situation: On one hand, we train for either an increase or decrease in price, and on the other hand, we retain the option to run a prediction after each price close.

Figure 4 shows a sample result for trading FXI, an ETF that tracks large-capitalization Chinese companies. It indicates that a trading profit of 55.2% could be achieved over a period of two years, assuming that we know whether the given candlestick is the last one added to a Renko. The performance of FXI over the same period is −35.4%. Note that we assume trading fees are 0%, as many brokers

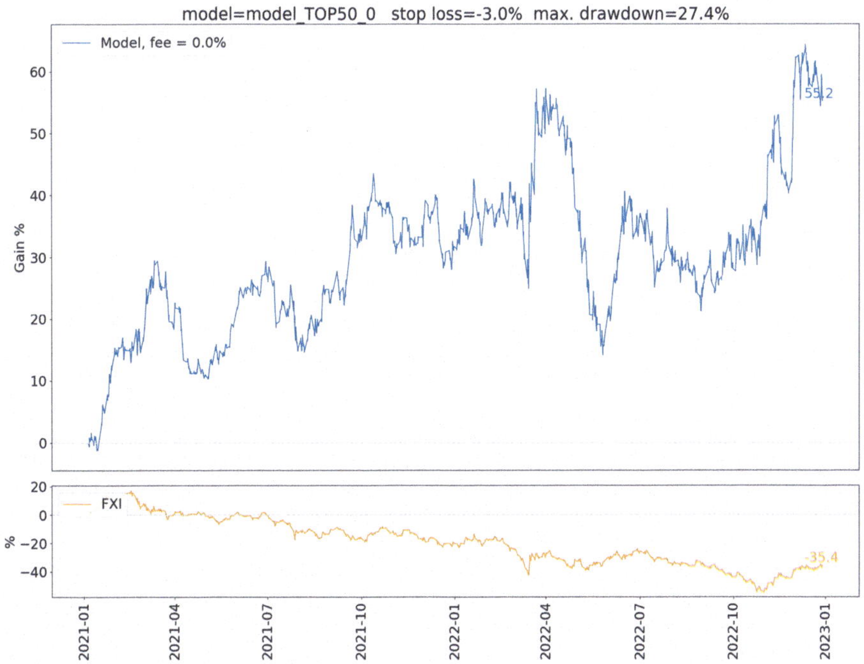

Fig. 4 The result of Renko trading for FXI

nowadays offer commission-free trading. Note also that the drawdown is 27.4%, i.e., it might be challenging for a human to remain optimistic about the future trading performance.

6 Conclusion

Our ambition was to present Renko charts as a viable starting point for algorithmic trading. We describe in detail how Renko charts can be converted into input for CNNs. We analyze what it means to train with Renko charts (including associated indicators) and conclude that Renko charts have the key property that any later prediction process can be done on a candlestick basis. This is necessary if a prediction needs to be made in a continuous way after each price close.

We apply a simple trading strategy where we either buy, sell, or do nothing (remain neutral) after each price close. While continuous trading might be the goal of algorithmic trading, this approach is rare in discretionary trading. In the future, we expect to evaluate different trading strategies that align more closely with those used by discretionary traders.

References

1. Keras, https://www.keras.io/
2. Agarap, A.F.: Deep learning using rectified linear units (ReLU). CoRR **abs/1803.08375** (2018), http://arxiv.org/abs/1803.08375
3. Appel, G.: Technical analysis: power tools for active investors. FT Press (2005)
4. Bollinger, J.: Using Bollinger bands. Stocks & Commodities **10**(2), 47–51 (1992)
5. Carter, J.F.: Mastering the trade: Proven techniques for profiting from intraday and swing trading setups. McGraw-Hill (2012)
6. Colby, R.W.: The Encyclopedia of Technical Market Indicators, Second Edition. McGraw-Hill Education (2002)
7. Grigorescu, S., Trasnea, B., Cocias, T., Macesanu, G.: A survey of deep learning techniques for autonomous driving. Journal of Field Robotics **37**(3), 362–386 (2020)
8. Huber, P.J.: Robust estimation of a location parameter. Annals of Mathematical Statistics **35**(1), 73–101 (1964)
9. Kingma, D.P., Ba, J.: Adam: A method for stochastic optimization. arXiv preprint arXiv:1412.6980 (2014)
10. Mnih, V., Kavukcuoglu, K., Silver, D., Graves, A., Antonoglou, I., Wierstra, D., Riedmiller, M.: Playing Atari with deep reinforcement learning. arXiv preprint arXiv:1312.5602 (2013)
11. Morales, G., Kacher, C.: Trade like an O'Neil disciple: How we made over 18,000% in the stock market. John Wiley & Sons (2010)
12. Oricsoft: Kibot historical data, https://www.kibot.com/
13. Salkar, T., Shinde, A., Tamhankar, N., Bhagat, N.: Algorithmic trading using technical indicators. In: 2021 International Conference on Communication Information and Computing Technology (ICCICT). pp. 1–6 (2021). doi: https://doi.org/10.1109/ICCICT50803.2021.9510135

14. Schwager, J.D.: Market wizards, updated: Interviews with top traders. John Wiley & Sons (2012)
15. Sezer, O.B., Ozbayoglu, A.M.: Algorithmic financial trading with deep convolutional neural networks: Time series to image conversion approach. Applied Soft Computing **70**, 525–538 (2018). doi: https://doi.org/10.1016/j.asoc.2018.04.024
16. Silver, D., Schrittwieser, J., Simonyan, K., Antonoglou, I., Huang, A., Guez, A., Hubert, T., Baker, L., Lai, M., Bolton, A., et al.: Mastering the game of go without human knowledge. Nature **550**(7676), 354–359 (2017)
17. Sutton, R.S., Barto, A.G.: Reinforcement learning: An introduction. MIT Press (2018)
18. Wilder, J.W.: New concepts in technical trading systems. Greensboro, NC (1978)
19. Zou, J., Zhao, Q., Jiao, Y., Cao, H., Liu, Y., Yan, Q., Abbasnejad, E., Liu, L., Shi, J.Q.: Stock market prediction via deep learning techniques: A survey. arXiv preprint arXiv:2212.12717 (2023)

Low-Energy Attendance System Based on IoT Technology

Jan Francisti and Zoltán Balogh

1 Introduction

The importance of attendance in an organizational context is undeniable. Current challenges related to monitoring attendance in businesses, schools, and institutions demand efficient solutions. Technological advancements in attendance management systems, such as mobile applications, web platforms, and RFID (Radio Frequency Identification) systems, aim to enhance accuracy and efficiency in attendance tracking. RFID technology, which utilizes radio waves for identification, minimizes errors and enables real-time data reporting, thereby supporting faster and more reliable attendance monitoring.

Traditional paper documentation consumes a significant amount of space and restricts quick access to information, which can be problematic for organizations with multiple departments or locations. Automated attendance systems utilizing RFID technology address this issue by ensuring efficient data processing. However, considering potential cybersecurity threats and the implementation costs associated with such systems, it is crucial to ensure adequate data protection and evaluate their environmental impact. A survey and analysis of various approaches to IoT-based attendance systems will provide a comprehensive overview of their benefits, drawbacks, and practical applications in the modern organizational environment.

J. Francisti (✉)
Constantine the Philosopher University in Nitra, Nitra, Slovakia

Z. Balogh
Constantine the Philosopher University in Nitra, Nitra, Slovakia

Kandó Kálmán Faculty of Electrical Engineering, Óbuda University, Budapest, Hungary
e-mail: zbalogh@ukf.sk

M. Turčáni (ed.), *15th International Scientific Conference on Distance Learning in Applied Informatics*, EAI/Springer Innovations in Communication and Computing,
https://doi.org/10.1007/978-3-031-81261-3_7

2 Related Works

A low-energy attendance system based on IoT technology can significantly enhance the efficiency and accuracy of attendance management while reducing operational costs and energy consumption. Various studies have explored different approaches to implementing such systems. For instance, the use of ESP32 cameras for facial recognition, combined with RFID controllers, provides a robust solution that ensures attendance can still be recorded even in cases of sensor failure or facial injuries [1]. Another approach involves using the ESP8266 Wi-Fi module and R307 Fingerprint Sensor, which has demonstrated high efficiency and 99.9% accuracy, completing attendance in an average of 7.86 s [2]. Additionally, integrating temperature measurement and hand sanitization modules can further enhance the system's functionality, especially in the context of health and safety during the COVID-19 pandemic [3, 4]. The use of OpenCV for face detection and recognition on a Raspberry Pi, as demonstrated in high school settings, shows the potential for automating attendance while providing real-time data to improve student performance [5]. Moreover, combining RFID modules with cloud computing technologies like AWS can offer real-time attendance tracking and reporting, accessible via web or mobile applications, making the system more reliable and cost-effective [6]. Finally, integrating both biometric and facial detection methods into a single system connected to a cloud server ensures data can be accessed anytime, providing a comprehensive solution for educational institutions [7]. These diverse approaches highlight the potential of IoT-based attendance systems to offer low-energy, efficient, and accurate solutions for various organizational needs.

Implementing a low-energy attendance system offers numerous benefits, including enhanced efficiency, cost savings, and improved accuracy. Such systems, which often utilize technologies like Bluetooth Low Energy (BLE) beacons, facial recognition, and RFID, streamline the attendance process by reducing the time and effort required for manual recording [8, 9]. These systems are designed to be user-friendly and portable, making them suitable for both educational and commercial purposes [10]. By leveraging BLE technology, these systems can ensure that attendance is marked only when students or employees are physically present, thereby minimizing the risk of proxy attendance [11]. Additionally, the integration of cloud servers and mobile applications allows for real-time data synchronization and remote access, further enhancing management efficiency [12]. The use of low-energy components, such as those in BLE beacons and optimized power supply modules, significantly reduces energy consumption, making these systems more sustainable and cost-effective in the long run [13]. Moreover, the incorporation of multiple authentication methods, including fingerprint sensors and facial recognition, adds layers of security and ensures the integrity of attendance records. The ability to operate on both PC and mobile platforms provides flexibility and convenience for users, allowing for seamless integration into existing workflows. Overall, low-energy attendance systems not only improve the accuracy and reliability of attendance tracking but also contribute to energy conservation and operational efficiency, making them a valuable asset for modern institutions and enterprises.

The paper presents the design and implementation of an IoT-based attendance system, leveraging modern technology to enhance accuracy, reliability, ease of use, security, and cost-effectiveness. The ESP32 microcontroller plays a crucial role in various systems due to its versatile features and capabilities. It is utilized in real-time monitoring systems for tracking energy and water consumption, pH value detection, and data transmission to databases through Wi-Fi [14]. The ESP32 microcontroller offers Wi-Fi and Bluetooth functionality, independent timers, ADCs, DACs, and more, making it suitable for a wide range of projects in different formats like development boards and integrated watches [15]. Moreover, the ESP32 is popular in embedded applications such as robotic joints and smart devices, with a rich software ecosystem including Espressif's SDK and third-party projects supporting various programming languages like MicroPython [16]. Studies have shown that the ESP32 microcontroller demonstrates good data processing capabilities with minimal errors, making it a reliable choice for control systems [17]. The system employs the ESP32 microcontroller as the central unit, the RC522 RFID reader for person identification, and an OLED display for showing attendance information. This approach addresses the limitations of traditional attendance systems, which are often prone to errors and inefficiencies [1, 6]. The system's architecture includes both hardware and software components, with the hardware comprising the RFID module connected to the ESP32 and the software built using Python Django, hosted on the AWS cloud for real-time tracking and reporting [18]. The use of RFID technology simplifies the attendance process, making it user-friendly and reducing manual entry errors. Additionally, the system incorporates biometric technology, such as fingerprint sensors, to enhance security and prevent impersonation, ensuring a high level of accuracy and efficiency [2, 5]. The integration of facial recognition technology using an ESP32 camera further ensures that attendance can be recorded even in cases of facial injuries, providing a robust backup method [19]. The system's design also includes the use of IoT and cloud computing to automate attendance logging, reducing the workload on staff and allowing for real-time data access from anywhere [5]. The implementation of this system in a real-world scenario demonstrated its effectiveness, with high accuracy and quick processing times, making it a reliable solution for various organizations [20, 21]. Overall, the proposed IoT-based attendance system successfully meets the essential requirements, streamlining the attendance tracking process and offering a comprehensive solution for modern attendance management.

3 Materials and Methods

The section "Materials and Methods" focuses on the design and methodology of an attendance system utilizing Internet of Things (IoT) technology. The introduction defines the requirements for the proposed attendance system and describes its hardware and software components.

The hardware section includes the ESP32 microcontroller as the central unit, an RFID RC522 reader for person identification, and an OLED display for showing attendance-related information. The software part of the system involves a cloud platform for data storage and analysis.

The subsequent section provides a detailed description of the development and implementation methodology, including testing. The IoT-based attendance system meets the following key requirements:

- Accuracy and reliability: The system precisely records all arrivals and departures.
- User-friendliness: The system is intuitive and easy to use for both employees and administrators.
- Security: The system is protected against unauthorized access, and sensitive user data is secured through encryption and authentication.
- Low operating and maintenance costs: The system is cost-effective and requires minimal maintenance.

The detailed methodology, presented step by step, allows the reader to gain a clear understanding of the development and implementation process. Within the methodology, the following subchapters are included:

1. System architecture proposal, containing a software development flowchart for the operation of an attendance system device and a wiring diagram.
2. Microcontroller programming, which includes setting up the environment and implementing the program for the ESP32 microcontroller.
3. Creation of a web application, which involves user registration, adding a new user, displaying attendance, and exporting data from the attendance system.

3.1 System Architecture

The implementation of the proposed attendance monitoring system is based on criteria and metrics such as equipment availability, cost, and programming simplicity. Additionally, an important factor is low power consumption. The software and hardware requirements, along with the implementation steps, are depicted in Fig. 1. The architecture of the proposed system draws from Internet of Things (IoT) technologies.

To create the presented system, the following software and hardware components are required:

- RFID card with unique identification number: This card contains an RFID tag with a unique identification number.
- RFID reader connected to ESP32 development board: The RFID reader is connected to the ESP32 development board using pins.

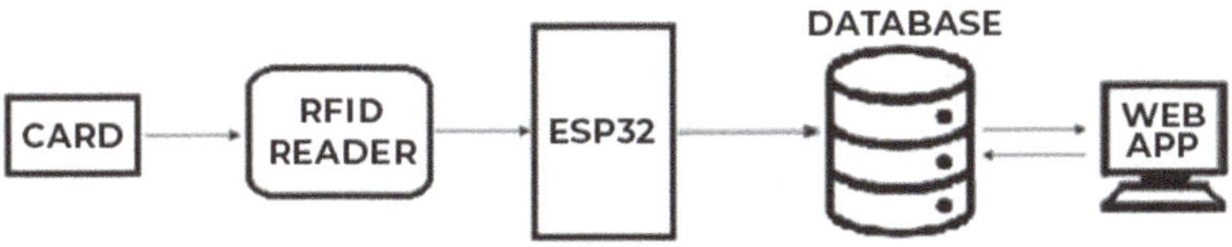

Fig. 1 System architecture diagram

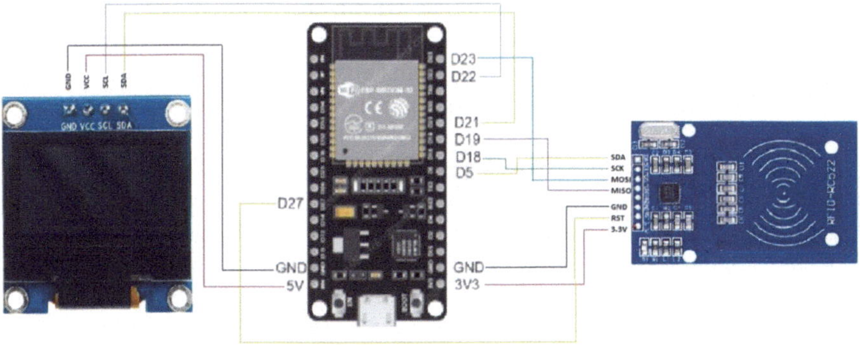

Fig. 2 Hardware component connection diagram

- Micro-USB cable for powering the ESP32 board via battery: Use a Micro-USB cable to power the ESP32 board through a battery.

The connection for interaction with the server is established using an encrypted wireless connection. Data is stored and processed using MySQL, PHP, and an XAMPP server, which allows for viewing in a web application created using front-end development. Administrators and authorized personnel can use the web application to display records of user arrivals and departures. This application also allows for filtering records according to selected parameters and exporting them to Excel format.

3.2 *Microcontroller Programming*

In the first step, it was necessary to download and install the Arduino IDE program, which serves as the main tool for software development for the ESP32. Subsequently, a library was added to support the NodeMCU ESP32 Wi-Fi + Bluetooth development board in the Arduino IDE, enabling the programming of the microcontroller (Fig. 2).

After successfully setting up the environment, it was necessary to configure the board. In the Arduino IDE, the ESP32 Dev Module was selected as the target board, and the correct parameters, such as the programmer and upload speed, were set according to the board's specifications. This ensured the successful uploading of the program to the ESP32. Special functions and libraries for the ESP32 were utilized, enabling communication with the RFID reader and OLED display. Functionality for reading RFID cards and displaying data on the OLED display was implemented.

At the beginning of the program, it was necessary to define the individual libraries. Imports allow the use of various libraries and functions needed for working with Wi-Fi, HTTP communication, time, the RFID reader, the OLED display, and other peripherals. The library MFRC522.h was used for communication with the chip card reader, and the libraries Adafruit_GFX.h and Adafruit_SSD1306.h were used for communication with the OLED display. These libraries needed to be additionally installed and were obtained from GitHub.

In the first part of the program, specifically in the setup() section, various components and functions are initialized and configured. The initial step involves setting up serial communication using Serial.begin(9600), which enables communication between the microcontroller and the computer via the serial port. Next, the OLED display and RFID reader are initialized. If the OLED display initialization fails, the program stops and displays an error message. During the OLED display initialization, the text "Ready" is shown on the display. Subsequently, the RFID reader is initialized, and the device connects to a Wi-Fi network. Finally, a time offset is set, and time synchronization occurs using NTP servers.

In the main program loop(), the primary functionality of the program is executed. First, it checks the Wi-Fi connection. If not connected, the connectToWiFi() function is called, attempting to reconnect to Wi-Fi. Next, it verifies whether a certain time interval (intervalLock) has elapsed, resetting the timer if necessary. Additionally, it checks whether 5 s have passed since the last RFID card read. If so, the old card identification number is reset. If no new RFID card is present, the program waits.

When a new card is detected, its identification number is read and compared with the previous number. If they differ, the SendCardID() function is triggered, sending the identification number to a server via an HTTP request. The SendCardID() function handles the transmission of the card's identification number, creating a URL address with the card and device identifiers, and then sending the request to the server.

The connectToWiFi() function is responsible for the Wi-Fi network connection. Initially, a message about the Wi-Fi connection is displayed. The device then attempts to connect to the specified SSID and password. Upon successful connection, the OLED display shows the message "Connected."

3.3 Web Application Creation

The user interface of the attendance system represents a web application that efficiently provides visualization and data management in a simple and intuitive interface. This interface consists of tables and other elements that present information from a MySQL database. The design and development of the user interface were implemented using programming languages such as HTML, CSS, JavaScript,

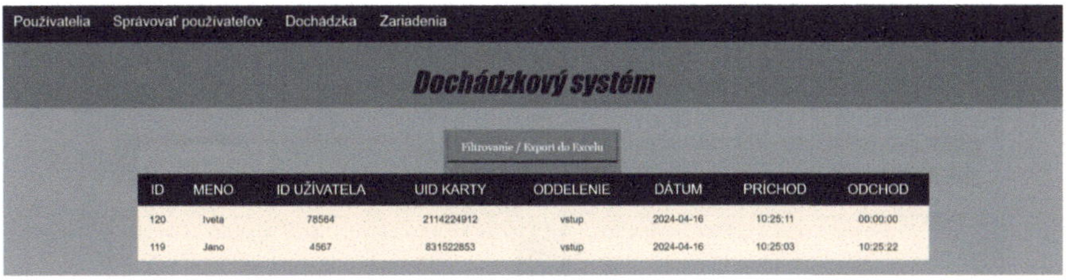

Fig. 3 User interface

and the scripting language PHP for data manipulation and management within the database. MySQL queries are used to load and process results in the user interface (Fig. 3).

The system is being developed and tested on a local XAMPP server for development purposes. The user interface (frontend) is created using HTML, CSS, and JavaScript, while server-side tasks (backend) are handled with PHP and MySQL. During the development of the user interface, it was crucial to consider appropriate technologies and choose those that best meet the project's requirements. In this case, HTML, CSS, JavaScript, PHP, and MySQL were evaluated as suitable for creating a functional, intuitive, and responsive web environment that would be usable on various devices. Ensuring a clear and intuitive design was important to provide easy usability for different users.

The display of data from the attendance system is protected against unauthorized access and is accessible only to registered users who log into the system.

User Registration

User registration is a pivotal element in the attendance system. The integration of a chip card reader module with the ESP32 microcontroller allows for secure and reliable identification of users based on the unique identification number of their chip card. Before scanning the identification card, it is essential to assign a unique identification number to each user, enabling unambiguous identification in the database. After successful registration, the user's details are stored in the central database using PHP, ensuring uniform and efficient data management.

During the registration process, it is also possible to enter and update additional user information, such as contact details, employment information, and other personal data. The system interface allows administrators to add new users, update existing records, and, if necessary, remove user accounts. Ensuring reliable registration and data management is key to the flawless operation of the attendance system. The functions of adding, updating, and removing users provide administrators with the necessary flexibility in managing user accounts, thereby ensuring effective attendance tracking and efficient personnel management within the attendance system.

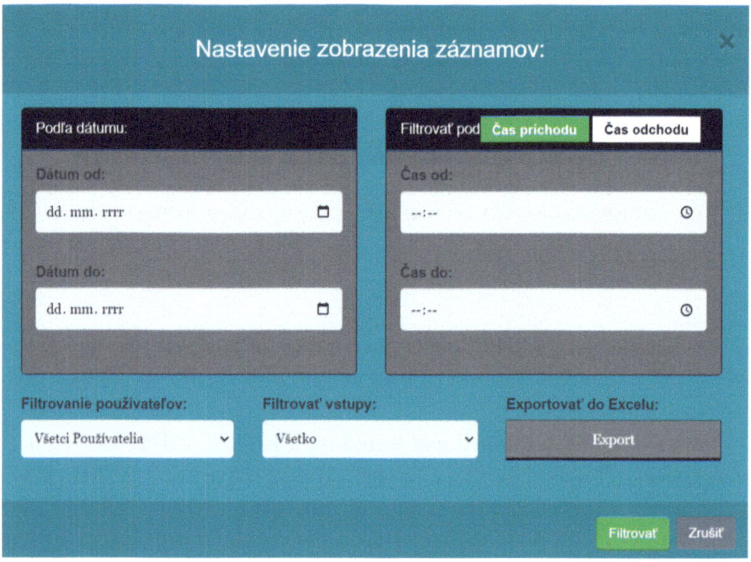

Fig. 4 Filtering the desired data

Export of Data from the Attendance System

Monitoring user records and data export are key aspects of efficient management and analysis of attendance in the system. These features allow administrators and managers to obtain crucial information about employee presence and working hours, which is essential for effective workforce planning and management.

Thoroughly recording and storing all user activities in the central system repository provides the opportunity for detailed tracking of arrival and departure times for each employee. This process is further supported by the ability to export data to an Excel file, enabling additional options for analysis and information processing (Fig. 4).

Exporting data to an Excel file provides the flexibility to further customize and format information according to specific requirements and organizational needs. This approach allows for better adaptation of analytical processes to the specific conditions and demands of the business.

4 Results

The implementation of an attendance system based on IoT technology began by connecting individual components to create a functional and integrated system. The first step involved preparing the hardware components that form the core of the system, including the ESP32 microcontroller, the RC522 reader, and the OLED display.

After obtaining these components, the physical integration process followed. To facilitate cable connections to the individual components, the pins to which the cables were attached were soldered. This step was crucial because proper connections ensure reliable communication between the components, and the mechanical stability of the connections is essential for the long-term functionality of the system.

Once the pins were successfully connected to the components, the next phase involved wiring the individual components according to the schematic mentioned in the "Materials and Methods" section. To enhance the reliability of cable connections, the joints were secured with electrical tape to prevent them from coming apart during handling. Precision and thoroughness were necessary during this step to avoid errors in the wiring, which could lead to system malfunction or damage to the components.

Following the successful assembly of the components, the next step was the initialization and configuration of the software, which included programming the ESP32 microcontroller and ensuring communication between the RC522 reader and the OLED display. This process required a thorough familiarity with the C/C++ programming language and the Arduino IDE environment used in software development.

Finally, the system underwent testing and debugging to ensure its correct functionality. This included testing the reading and writing of data using the RC522 reader and verifying the correct display of information on the OLED display. The goal of this process was to ensure the reliable operation of the system.

Since the designed attendance system was created with specific components, it was necessary to design and create a box to house the hardware components. A box was created to serve as a protective case for the integrated system. With the help of Inventor software, a 3D model of the box was created that meets the needs and requirements of the solution.

4.1 Description of the Created System

The attendance system is based on a process initiated by applying a Mifare-type chip card to a reader. Upon card application, a unique identification number (UID) and identifier (ID) are extracted, which are stored in memory block 1 of sector 4. These data are subsequently verified against their assignment to a user in the database. This mechanism ensures that only authorized users have access, thereby recording their arrivals and departures. It is important to emphasize that this mechanism serves as a protective element against unauthorized duplication of chip cards, increasing the overall security of the system.

The verification of the ID in block 1 of sector 4 acts as a security barrier against card cloning using the UID. This ensures that each card must be associated with a specific ID in the database for authentication and recording of employee entries and exits. This procedure guarantees that, in the event of card loss or theft, there will be no unauthorized intrusion into the system or its misuse.

Connecting the microcontroller to the network via a secure wireless connection enables the implementation of the IoT concept. This feature allows the system to communicate with the central server, where attendance data are stored and processed. This enables the integration of data into the IoT ecosystem and ensures their availability from anywhere and at any time.

If the card ID is associated with a user in the database, an entry or exit is recorded. These records are then displayed in a web application, where they can be further filtered and exported to Excel according to criteria such as date and time. This allows for efficient management of attendance and analysis of work trends.

The web application also provides the capability for user and chip card management. The administrator has the ability to add new users, assign them chip cards, and modify or remove user data as needed. This feature allows for simple administration of the system and its adaptation to the current requirements of the organization.

The comprehensive attendance system ensures accurate recording of employee attendance and provides efficient tools for managing and monitoring work activity. The system allows for effective tracking of attendance, ensuring transparency and accuracy in the recording of working hours. Thanks to wireless connectivity and a central server, continuous access to data, its processing, and analysis from any location and at any time are ensured. The ability to filter and export data from the web application enables the performance of analyses and the generation of reports according to management needs, which aids in better work planning, trend monitoring, and identification of areas for improvement.

Overall, the result is an integrated system that meets the requirements for tracking attendance, provides efficient tools for managing and monitoring work activity, and ensures the security of building entry. The system contributes to increased efficiency and transparency in the work environment and simplifies administrative processes associated with recording employee attendance (Fig. 5).

4.2 *Testing of the Attendance System*

During the implementation process of the proposed attendance system, significant attention was devoted to testing and evaluation to ensure functionality and identify potential weaknesses. Testing occurred in several stages and encompassed various aspects of the system, including hardware and software functionality, security, and user-friendliness. Combinations of standardized testing methods and specific testing procedures designed for the system were employed.

During testing, several issues were identified and subsequently addressed. One of the problems involved excessive rapid data writing from the chip card, where the reader occasionally registered the same card multiple times, resulting in unintended simultaneous arrival and departure records. To prevent this issue, a time delay was

Fig. 5 Attendance system model

implemented, allowing users to remove the card from the reader before duplicate records could be created.

Another challenge arose when the microcontroller program became stuck due to unsuccessful data retrieval from the card. This state led to improper system functioning because the microcontroller did not provide adequate feedback and was unprepared for subsequent card readings. As a solution, a return function was implemented in the program, ensuring that the system remains functional even in the event of an error.

Testing and evaluation were essential to ensure the smooth operation of the attendance system. Identifying and resolving issues during the testing phase resulted in a robust and reliable system that meets user requirements and expectations.

5 Discussion

In the current work environment, employee attendance management is an integral part of effective process management and performance optimization. Within attendance systems, two primary approaches are distinguished: systems based on Internet of Things (IoT) technology and systems utilizing biometric data. Each of these approaches has specific advantages and disadvantages that need to be considered when selecting the optimal solution for a particular organization.

Comparing the attendance system based on chip cards to biometric verification systems, the IoT-based system that utilizes RFID cards offers several advantages:

- Ease of use: Employees simply need to place their chip card on the reader, and their arrivals and departures are automatically recorded in the system. This

process is quick and efficient, minimizing the possibility of errors in attendance recording.
- Access rights management: Chip cards allow for easy management of access rights and prevention of unauthorized entry into the building. Administrators can quickly and effectively add, update, or revoke employee access rights.
- Reader placement flexibility: The system allows employees to access it from multiple entry points, which is particularly advantageous in large workspaces.

While chip card-based systems offer several advantages, it is essential to consider their limitations compared to biometric systems:

- Potential loss or misuse of chip cards: In the event of card loss, there is a risk of unauthorized manipulation of attendance records. Lost cards could fall into the wrong hands, potentially allowing someone to impersonate an employee and record false entries.
- Data forgery possibility: Compared to biometric systems that utilize unique identifiers such as fingerprints or facial recognition, chip cards are susceptible to data forgery. If someone copies or concurrently uses a chip card, it could lead to falsified attendance data.

In conclusion, the suitability of a specific type of attendance system depends on the individual needs and preferences of the organization. While a biometric system may offer a higher level of security and accuracy, an IoT-based system using chip cards may be preferred for its simplicity in implementation and management. It is crucial to choose a solution that best aligns with specific requirements and the company's environment, ensuring proper implementation and administration.

6 Conclusion

The presented article delves into the topic of attendance systems and explores the possibilities of improvement through Internet of Things (IoT) technology. Based on predefined objectives, the article analyzes IoT technology and its application in the field of attendance systems. Existing solutions are compared with a proposed custom solution. Additionally, a comprehensive attendance system leveraging IoT components is designed and implemented.

The outcome of this research is a functional and efficient attendance system. It enables the recording of employees' arrivals and departures using RFID cards and a central unit in the form of a reader. The software component of the system processes and stores the recorded data in a database, presenting it in a user-friendly manner through a web application.

Individual subgoals were successfully achieved during this study. The article highlights the benefits of implementing IoT devices across various sectors while also identifying security risks associated with IoT usage in attendance systems. Furthermore, a hardware solution for the attendance system was proposed and realized, meeting the specified objectives. The software component was effectively developed, facilitating efficient processing and display of attendance data.

It is also important to emphasize that the created attendance system is available and ready for deployment in a real work environment. Its implementation can contribute to increased efficiency and accuracy of attendance records, which positively impacts the management of work processes and employee performance optimization.

The research part of the article successfully achieved the set goals and contributed to the development and utilization of IoT technology in the field of attendance systems. The identified problems and created solutions provide a foundation for further improvements and innovations in this area. The developed attendance system has the potential to enhance attendance management processes and ensure better organization and efficiency of work activities.

Acknowledgments This work was supported by the Scientific Grant Agency of the Ministry of Education of the Slovak Republic (ME SR) and the Slovak Academy of Sciences (SAS) under contract No. VEGA 1/0385/23, as well as by the University Grant Agency under contract No. VII/7/2024.

References

1. Swathi MN, Pallavi G, Jyothika KS, et al (2023) An IoT Based Attendance System using PI Cam and RFID Controller. International Journal of Advanced Research in Science, Communication and Technology (IJARSCT) International Open-Access, Double-Blind, Peer-Reviewed, Refereed, Multidisciplinary Online Journal 3:. https://doi.org/10.48175/IJARSCT-11648
2. Gujarkar P, Lonkar S, Jain T, et al (2023) IoT based Smart Attendance System. 2023 International Conference on Emerging Smart Computing and Informatics, ESCI 2023. https://doi.org/10.1109/ESCI56872.2023.10099839
3. Kaveri K, Jervila R, Alhaseena KM, Naskath J (2023) RFID Attendance System-Enabled Automated Hand Sanitizer Dispenser using IoT. Proceedings of the 2nd International Conference on Applied Artificial Intelligence and Computing, ICAAIC 2023 1394–1399. https://doi.org/10.1109/ICAAIC56838.2023.10140315
4. Meenakshi A, Leelarani K, Shopika S, Rajasekaran M (2022) An Intelligent Sensor based Automatic Attendance Management System Using IoT. International Journal on Recent and Innovation Trends in Computing and Communication 10:29–35. https://doi.org/10.17762/IJRITCC.V10I2S.5909
5. Netinant P, Akkharasup-Anan N, Rakhiran M (2023) Class Attendance System using Unimodal Face Recognition System based on Internet of Educational Things. Proceedings of the 6th IEEE Eurasian Conference on Educational Innovation 2023: Educational Innovations and Emerging Technologies, ECEI 2023 67–70. https://doi.org/10.1109/ECEI57668.2023.10105374
6. Samaddar R, Ghosh A, Sarkar SD, et al (2023) IoT & Cloud-based Smart Attendance Management System using RFID. International Research Journal on Advanced Science Hub 5:111–118. https://doi.org/10.47392/IRJASH.2023.020
7. Sri Madhu BM, Kanagotagi K, Devansh (2018) IoT based Automatic Attendance Management System. International Conference on Current Trends in Computer, Electrical, Electronics and Communication, CTCEEC 2017 83–86. https://doi.org/10.1109/CTCEEC.2017.8455099

8. Vinod VM, Murugesan G, Mekala V, et al (2021) A Low-Cost Portable Smart Card Based Attendance System. IOP Conf Ser Mater Sci Eng 1012:012046. https://doi.org/10.1088/1757-899X/1012/1/012046
9. Narayanan DrVA (2022) Smart Attendance Marking System Using Bluetooth Low Energy and Facial Recognition Technology. Int J Res Appl Sci Eng Technol 10:493–500. https://doi.org/10.22214/IJRASET.2022.40305
10. Menezes AG, Sa JMDDC, Llapa E, Estombelo-Montesco CA (2020) Automatic Attendance Management System based on Deep One-Shot Learning. International Conference on Systems, Signals, and Image Processing 2020-July:137–142. https://doi.org/10.1109/IWSSIP48289.2020.9145230
11. Apoorv R, Mathur P (2017) Smart attendance management using Bluetooth Low Energy and Android. IEEE Region 10 Annual International Conference, Proceedings/TENCON 1048–1052. https://doi.org/10.1109/TENCON.2016.7848166
12. Noguchi S, Niibori M, Zhou E, Kamada M (2015) Student attendance management system with bluetooth low energy beacon and android devices. Proceedings - 2015 18th International Conference on Network-Based Information Systems, NBiS 2015 710–713. https://doi.org/10.1109/NBIS.2015.109
13. Jung J, Lim J, Lee S, et al (2014) A low-energy video event data recorder using dual image/video codec. 11th IEEE International Conference on Advanced Video and Signal-Based Surveillance, AVSS 2014 277–282. https://doi.org/10.1109/AVSS.2014.6918681
14. Lu Z, Wang G, Wei X, Li J (2024) A Real-time Faint Space Debris Detector with Learning-based LCM. IEEE Trans Instrum Meas. https://doi.org/10.1109/TIM.2024.3390203
15. Cameron N (2023) ESP32 Microcontroller. 1–54. https://doi.org/10.1007/978-1-4842-9376-8_1
16. Howard M, Bruce Irvin R (2023) ESP32: QEMU Emulation within a Docker Container. Lecture Notes in Networks and Systems 815 LNNS:63–80. https://doi.org/10.1007/978-3-031-47457-6_5
17. Wahyu Pratama E, Kiswantono A, Ahmad Yani No J, et al (2022) Electrical Analysis Using ESP-32 Module In Realtime. JEECS (Journal of Electrical Engineering and Computer Sciences) 7:1273–1284. https://doi.org/10.54732/JEECS.V7I2.21
18. Ramani G, Pradeepkumar G, Palanisamy PN, et al (2023) Smart Attendance Monitoring System Using IoT. 2023 9th International Conference on Advanced Computing and Communication Systems, ICACCS 2023 1099–1104. https://doi.org/10.1109/ICACCS57279.2023.10112850
19. Jeffrey Vaz FA, Anusuyadevi G, Indhu Priya AM, et al (2023) Smart Attendance System using RFID and Raspberry Pi. Proceedings of the 2023 2nd International Conference on Electronics and Renewable Systems, ICEARS 2023 1450–1455. https://doi.org/10.1109/ICEARS56392.2023.10085186
20. Zengin M, Albayrak Z Designing a New Data Encryption Algorithm Using a Genetic Code Method. Acta Polytechnica Hungarica 19:2022–235
21. Kovari A, Rajcsanyi-Molnar M Mathability and Creative Problem Solving in the MaTech Math Competition. Acta Polytechnica Hungarica 17:2020–147

Using an AI-Based Language Model to Generate Synthetic Statistical Data

Mikuláš Gangur and Olga Martinčíková Sojková

1 Introduction

The use of artificial intelligence (AI) in a wide range of fields, along with the application of machine learning (ML) algorithms, has not escaped education. As in other fields, the availability of large datasets is a critical element in many applications of educational practice. However, data availability often faces challenges such as privacy concerns, security concerns, lack of information, and the comparability of the difficulty of data manipulation, which is particularly important in the knowledge and skill verification phase of learning.

One solution to this problem is the controlled generation of synthetic datasets. Artificially generated synthetic datasets are a solution to these problems and can offer similarly complex datasets while maintaining the desired statistical properties. In this way, datasets for analysis in areas such as finance, corporate data, and healthcare, where data is realistically unavailable due to its sensitivity and security, can also be created for practical learning needs. These data, algorithmically created without the need for real- world collection, have thus become a response to the escalating need for large-scale, adaptive data that adheres to privacy and availability constraints.

In the research described in this chapter, two language models, ChatGPT and Bing in MS copilot, were investigated. Finally, the research focused on working with MS copilot. This tool was used to generate synthetic tabular data in a record × variable format for multivariate data analysis tasks. Considering the different tasks of multivariate analysis and the methods used in the fields of statistical data analysis, data mining, and machine learning, an appropriate data structure was also required

M. Gangur (✉) · O. M. Sojková
Faculty of Economics, University of West Bohemia, Pilsen, Czech Republic
e-mail: gangur@fek.zcu.cz; sojkova@fek.zcu.cz

M. Turčáni (ed.), *15th International Scientific Conference on Distance Learning in Applied Informatics*, EAI/Springer Innovations in Communication and Computing,
https://doi.org/10.1007/978-3-031-81261-3_8

concerning the data types and assumptions of the application of each method. The chapter shows the problems in generating synthetic data, especially concerning the design of the prompts, but also shows the advantages of the described approach, which are evident from the outputs of the copilot activity that implements the generating process.

2 Related Work

Many publications focus on different methods of generating synthetic data. In this section, we focus on the generation of synthetic statistical data for educational purposes, particularly on a small number of applications of synthetic data generation using AI-based language models.

The generated synthetic data for teaching purposes is offered on some websites created using embedded generators. The data is widely used by teachers to teach statistics. For example, two websites generate data in this way. One tried-and-trusted source was created by industrial/organizational (I/O) psychologist Richard Landers [1]. The other site is offered by social psychologist Andrew Luttrell [2]. Both provide interactive, user-friendly generation of statistical datasets for further use in statistics education. Another example of the use of prepared, noninteractive synthetic data in teaching can be found in Mobahi [3], for example.

Hassani and Silva [4] show the potential of ChatGPT to automate various aspects of data science processes such as data cleaning, preprocessing, model training, and result interpretation. They highlight the potential of ChatGPT to provide new insights and improve decision-making processes by analyzing unstructured data. The paper discusses the advantages of the ChatGPT architecture, such as its ability to be tuned for linguistic tasks and its ability to generate synthetic data. The authors also acknowledge the limitations and concerns regarding plagiarism when using ChatGPT and highlight the need for ethical use.

The paper by Lingo [5] explores the construction and use of synthetic datasets, particularly in the field of telematics, and shows the potential of artificial intelligence models such as ChatGPT in improving data availability. It also shows the iterative process of generating a synthetic telematics dataset using ChatGPT, highlighting the importance of quality parameters such as diversity, relevance, and coherence. The research highlights the importance of synthetic datasets in solving problems related to data preservation, data scarcity, and control over variables, opening up new research opportunities in complex industries such as telematics.

Solomon et al. [6] are particularly concerned with augmenting datasets for Natural Language Processing (NLP) purposes using ChatGPT under low-resource scenarios. They show an important dependency of the quality of the augmented data on the efficiency of the manually created prompts.

In their work, Kim et al. [7] are developing a Python-based educational dataset library that addresses educators' needs with diverse and accessible datasets. Utilizing random forest modeling, the library generates high-quality synthetic datasets validated by an expert panel. Evaluations show that the library's datasets accurately mirror the original data, confirming their effectiveness.

As Stenger et al. [8] state in their paper, while image synthesis is well-studied, time series synthesis is crucial for practical applications, such as in industry. This is demonstrated by Sasiaowapak et al. [9], who present a methodology to generate synthetic datasets for industrial products such as bolts and screws using a segment model and a stable diffusion technique to create accurate representations.

Hansen et al. [10] examine the use of data-centric AI techniques to guide synthetic data generation, highlighting the pitfalls of ignoring data profiles. They propose a framework to evaluate the integration of these profiles and empirically assess 5 models on 11 datasets, offering insights into their performance. Consequently, they provide recommendations for improving synthetic data generation, focusing on classification performance, model selection, and feature selection.

Despite the research conducted on the use of AI language models such as ChatGPT for generating synthetic data, there is a knowledge and experience gap in research on the possibilities of generating synthetic data for learning purposes using AI language models. Thus, the current status requires further research on the potential benefits and challenges of using AI-based language models for generating synthetic data for educational purposes.

3 Methodology

An ideal synthetic dataset requires careful replication of the inherent patterns present in the original data. However, a dataset generated using an AI language model may not meet such rigorous standards. Concerning generating data from randomized sources, there is an inevitable trade-off in the depth and breadth of information compared to a complex synthetic dataset. Importantly, the primary goal is not to mirror reality flawlessly, but rather to improve features that are suitable for teaching and research, thereby supporting the educational process in relevant domains.

The synthetic datasets generated as examples described in this chapter are not as exhaustive concerning reality but provide controlled data. These data allow researchers and students to conduct experiments, explore, and derive insights while ensuring privacy and confidentiality [5].

The aim of the proposed synthetic data generation process is:

1. To design the structure of tabular data according to the specified domain (cars, employees, schools, books, etc.) and in the required structure concerning data types (numeric, categorical).
2. Generation of random values:
   (a) Numeric variables in the specified range or according to the specified distribution concerning the values of other numeric and categorical variables.

(b) Categorical variables by real permutations, which may be possibly tied to permu tations of other categorical variables.

3. For both previous steps, the output is required in the form of Matlab program code (the programming environment of the automatic parameterized problem generator). In the code, not only are data structures created that are populated with permutations of categorical variables, but also program sequences for generating random values of nu merical variables according to specified requirements.

When experimenting with the language model, one of the difficult things became the implementation of the requirements for binding the allowable values of one variable with respect to the values of other variables. By adding requirements for additional bindings between variable values, the prompt became more complex, and the model tended to simplify the answer. Thus, the eventual requirement for completely real data in all variables could not be achieved, which, according to Lingo [5], is not the primary goal of generating synthetic training data with the AI model.

The original idea was the direct generation of data by the AI model with storage in a file. The file is part of the output of an automatic parameterized task generator with a focus on statistical analysis [11]. The location of the data file then depends on the format of the output and subsequent use. In the case of saving assignments to the LMS e-course assignment bank, the file is stored in the LMS repository [12], and in the case of generating assignments as output to PDF files, the file is stored in the cloud [13].

When working with the language AI module, one of the features that emerged was the variety (variability) of the AI model's responses to the same prompts at different times. The user has to guide the model to the same or similar results by "preparatory" questions (prompts) that help to define the role in which the model should perform. This shortcoming was particularly evident in the design of the data structure of the required spreadsheet data. The resulting data structure sometimes varied so significantly from one assignment to another at different times that one of the requirements for synthetic data, namely equal difficulty of the generated tasks, was not met.

Given this shortcoming and the uncertain response of the AI model, the approach of outputting (responding to a given prompt) from the AI model as program code that randomly generates the required data and stores it in a temporary file for later processing was ultimately chosen (see Sect. 3 in the list of objectives).

4 Results

The work with MS Copilot, built on top of the Bing language model, is shown in the following example of generating data about cars. Prompts for the language model are marked in a box, and outputs are in the form of Matlab code in a different font.

The initial idea was to directly generate the source data and save it in an output CSV file. This is implemented by the following prompt.

Create attributes for table cars with data type and generate 100 examples and save them to a CSV file

Considering the ambiguous behavior of the model (see below), an output in the form of Matlab code was added, which is then used as part of the automatic generator of parameterized tasks.

Create attributes for table cars with data type and create Matlab code for generation of 100 examples and save it to CSV file

The output of the command is the following Matlab code.

```
% Generate random car data n = 100; % Number of cars

% Create random data for each attribute CarID = (1:n)';
Manufacturer = randManufacturer(n); % Replace with actual
                                     manufacturer names
Model = randModel(n); % Replace with actual model names
                        Year = randi([2000, 2022], n, 1);
Trim = randTrim(n); % Replace with actual trim levels
BodyType = randBodyType(n); % Replace with actual body types
EngineType = randEngineType(n); % Replace with actual engine types
                      Horsepower = randi([100, 500], n, 1);
FuelEfficiency = randFuelEfficiency(n); % Replace with actual fuel
                                                   effi- ciency values
Price = randPrice(n); % Replace with actual price ranges

% Create a table
carsTable = table(CarID, Manufacturer, Model, Year, Trim,
BodyType, En- gineType, Horsepower, FuelEfficiency, Price);
% Save to CSV file writetable(carsTable, 'cars_data.csv');
```

In this case, the addition of the required categorical pronoun values, especially the brand and model names, is left to the user and their manual addition. The generation of random but real values of categorical variables was one of the main reasons for using the language module to generate the datasets. Therefore, the prompt used was refined. It is important to note that the following query, and especially the result, was implemented in the language model mode Balanced.

Create data structure for cars with factor and cardinal attributes with a minimum of 10 values for makes and a minimum of 3 models for every makes including random examples in Matlab code

```
% Generate random car data
nMakes = 10; % Number of car makes nModelsPerMake = 3;
        % Number of models per make

% Create random data for makes and models
makes = {'Toyota', 'Honda', 'Ford', 'Chevrolet', 'Nissan',
   'Volkswagen', 'BMW', 'Hyundai', 'Mercedes', 'Subaru'};
models = {'Camry', 'Civic', 'Mustang', 'Malibu', 'Altima', 'Jetta',
   '3 Series', 'Elantra', 'C-Class', 'Outback'};

% Initialize empty arrays for other attributes year = zeros(nMakes
     * nModelsPerMake, 1); trim = cell(nMakes * nModelsPerMake, 1);
horsepower = zeros(nMakes * nModelsPerMake, 1); fuelEfficiency =
     zeros(nMakes * nModelsPerMake, 1); price = zeros(nMakes
        * nModelsPerMake, 1);

% Generate random data for each car for i = 1:nMakes
  for j = 1:nModelsPerMake
     idx = (i - 1) * nModelsPerMake + j; year(idx) = randi([2000,
     2022]); trim{idx} = ['Trim', num2str(j)]; horsepower(idx) =
        randi([100, 500]); fuelEfficiency(idx) = randi([20, 40]);
          price(idx) = randi([15000, 60000]);
  end
end

% Create a table
carsTable = table(makes, models, year, trim, horsepower,
fuelEfficiency, price);
```

This command was used to generate the required brands and models, but as the code shows, the links between the brand and the models are not addressed, even though this requirement is specified in the prompt. The Toyota brand can then be assigned a Mustang model that belongs to Ford. One solution is to switch the language model to Exact and specify the same command.

```
% Define the car makes
makes = {'Toyota', 'Honda', 'Ford', 'Chevrolet', 'Mercedes', 'BMW',
    'Au- di', 'Hyundai', 'Nissan', 'Volkswagen'};

% Define the models for each make models = {
  {'Camry', 'Corolla', 'Highlander'}, % Toyota
  {'Civic', 'Accord', 'CR-V'}, % Honda
  {'F-150', 'Escape', 'Mustang'}, % Ford
  {'Silverado', 'Equinox', 'Malibu'}, % Chevrolet
  {'C-Class', 'E-Class', 'S-Class'}, % Mercedes
  {'3 Series', '5 Series', '7 Series'}, % BMW
  {'A4', 'A6', 'Q5'}, % Audi
  {'Elantra', 'Sonata', 'Tucson'}, % Hyundai
  {'Altima', 'Rogue', 'Sentra'}, % Nissan
  {'Golf', 'Passat', 'Tiguan'} % Volkswagen
};       % Create a struct array for cars
cars = struct('Make', {}, 'Model', {});

% Populate the struct array with car makes and models
```

```
    for i = 1:length(makes)
  for j = 1:length(models{i})
    car = struct('Make', makes{i}, 'Model', models{i}{j});
      cars = [cars; car];
  end
end      % Display a random example
example = cars(randi(length(cars)));
```

This command solved the problem of assigning the corresponding models to the correct brand; however, the design of other required attributes of the data file structure disappeared from the code completely. A subsequent modification of the prompt solves this problem:

> Create data structure for cars with 3 factor and 5 cardinal attributes with 10 values for makes and 3 models for every makes.
> Generate 100 examples in Matlab code.

```
% Define the makes and models
makes = {'Toyota', 'Honda', 'Ford', 'Chevrolet', 'Nissan', 'Hyundai',
'Volkswagen', 'Subaru', 'Mercedes-Benz', 'BMW'};
models = {{'Camry', 'Corolla', 'Prius'}, {'Civic', 'Accord', 'CRV'},
{'F-150', 'Mustang', 'Explorer'}, {'Silverado', 'Impala', 'Malibu'},
{'Altima', 'Maxima', 'Rogue'}, {'Elantra', 'Sonata', 'Tucson'},
{'Golf', 'Passat', 'Tiguan'}, {'Forester', 'Outback', 'Impreza'},
{'C-Class', 'E-Class', 'S-Class'}, {'3 Series', '5 Series',
 '7 Series'}};

% Define the colors
colors = {'Red', 'Blue', 'Black', 'White', 'Silver'};

% Initialize the cars structure
cars = struct('Make', {}, 'Model', {}, 'Color', {}, 'Year', {},
   'Price', {}, 'Mileage', {}, 'Horsepower', {}, 'Weight', {});

% Populate the cars structure with random examples for i = 1:100
  % Select a random make and model make_idx = randi([1,
                                   numel(makes)]);
  model_idx = randi([1, numel(models{make_idx})]);

% Select a random color
  color_idx = randi([1, numel(colors)]);

% Generate random cardinal attributes year = randi([2000, 2024]);
  price = randi([5000, 50000]); mileage = randi([0, 200000]);
    horsepower = randi([100, 500]); weight = randi([1000, 5000]);

% Add the car to the structure
cars(i) = struct('Make', makes{make_idx}, 'Model', mod-
  els{make_idx}{model_idx}, 'Color', colors{color_idx}, 'Year', year,
  'Price', price, 'Mileage', mileage, 'Horsepower', horsepower,
```

```
            'Weight', weight);
    end
```

One of the problems of working with an AI language model is shown in the following example, where the previous prompt was given to the same language model, in the same mode, after a new login, causing the model to start from the beginning (zero-shot prompt). The result shows an imaginary "crashing" of the system.

```
% Define the car structure

car = struct('Make', {}, 'Model', {}, 'Factor1', {}, 'Factor2', {},
'Factor3', {}, 'Attribute1', {}, 'Attribute2', {}, 'Attribute3', {},
'Attribute4', {}, 'Attribute5', {});

% Define possible makes and models
makes = {'Make1', 'Make2', 'Make3', 'Make4', 'Make5', 'Make6',
'Make7', 'Make8', 'Make9', 'Make10'};
models = {'Model1', 'Model2', 'Model3'};

% Generate 100 examples
....
```

At that point, it is necessary to start learning the model again, and by progressing from simpler to more complex commands, reach the previous level of "intelligence" in the subject (chain-of-thought prompt). That is why we try to impose all our requirements on the model with the following prompt.

> Create data structure for cars with 3 factors with word levels and 5 cardinal attributes.
> Named every factor a cardinal attribute with the real name.
> Generate 10 real makes and 3 real correspond models with unique names for every makes.
> Generate 3 real motorizations for every model.
> Generate the Price for every model.
> Generate 100 examples in Matlab code.

```
% Define the car structure
car = struct('Make', {}, 'Model', {}, 'Motorization', {}, 'FuelType',
{}, 'Transmission', {}, 'DriveTrain', {}, 'Price', {}, 'Mileage', {},
'Color', {}, 'EngineSize', {}, 'Warranty', {});

% Define possible makes and corresponding models
makes = {'Toyota', 'Ford', 'Chevrolet', 'Honda', 'Mercedes',
'BMW', 'Audi', 'Hyundai', 'Nissan', 'Volkswagen'};
models = {{'Camry', 'Corolla', 'Highlander'}, {'F150', 'Mustang',
'Explorer'}, {'Impala', 'Malibu', 'Tahoe'}, {'Civic',
 'Accord', 'CR-V'},
{'C-Class', 'E-Class', 'S-Class'}, {'3 Series',
 '5 Series', '7 Series'},
```

```
{'A4', 'A6', 'Q5'}, {'Elantra', 'Sonata', 'Tucson'}, {'Altima',
'Maxima', 'Rogue'}, {'Golf', 'Passat', 'Tiguan'}};

% Define possible motorizations for each model
motorizations = {{'1.5L', '2.0L', '2.5L'}, {'3.5L', '5.0L', '6.2L'},
{'1.4L', '1.8L', '2.0L'}, {'1.5L', '2.0L', '2.4L'}, {'2.0L', '3.0L',
'4.0L'}, {'2.0L', '3.0L', '4.4L'}, {'2.0L', '3.0L', '4.2L'}, {'1.6L',
'2.0L', '2.4L'}, {'2.5L', '3.5L', '4.0L'}, {'1.4L', '2.0L', '3.0L'}};

% Define possible levels for factors
fuelTypes = {'Gasoline', 'Diesel', 'Electric'}; transmissions =
{'Manual', 'Automatic', 'CVT'}; driveTrains = {'FWD', 'RWD', 'AWD'};

% Generate 100 examples for i = 1:100
  % Randomly select a make and corresponding model makeIndex =
  randi([1, length(makes)]);


  make = makes{makeIndex};
  model = models{makeIndex}{randi([1, length(models{makeIndex})])};
  motorization = motorizations{makeIndex}{randi([1, length(motoriza-
tions{makeIndex})])};

% Randomly select factors
FuelType = fuelTypes{randi([1, length(fuelTypes)])}; Transmission =
transmissions{randi([1, length(transmissions)])}; DriveTrain =
driveTrains{randi([1, length(driveTrains)])};

% Generate random attributes
Price = randi([10000, 50000]); % Price range from 10,000 to 50,000
Mileage = randi([1, 10]);
Color = randi([1, 10]); EngineSize = randi([1, 10]); Warranty =
randi([1, 10]);

% Add to car structure
car(i) = struct('Make', make, 'Model', model, 'Motorization',
motoriza- tion, 'FuelType', FuelType, 'Transmission', Transmission,
'DriveTrain', DriveTrain, 'Price', Price, 'Mileage', Mileage,
'Color', Color, 'En- gineSize', EngineSize, 'Warranty', Warranty);
end
```

As you can see from the code, the system fulfilled most of the requirements; however, some of them were forgotten, such as generating name values for the categorical variable color. Similarly, in most cases, the model failed to assign the current price to each model, which was also pointed out at the beginning of the output of its work. Only in one case, which could not be repeated even with identical input settings, did the model generate the corresponding real prices for each model.

5 Conclusion

This chapter shows how to work with AI-based language models that were used in the development of the Matlab code. This code is then applied as part of an automatic.

parameterized task generator, which is used to generate statistical processing and data analysis tasks.

From the above experiments, which in selected cases were performed on both the Bing and ChatGPT Java models simultaneously and compared, it became clear that to achieve at least similar output, the model must first be learned, and only then can one expect outputs approaching the desired ones. Also, the mode in which language models process the commands given is important. It turns out that even in the case of a "learned" model, the system's responses degrade to simple outputs when a certain complexity is reached.

These findings led to a change in the original intention to use a language model to generate datasets online. Instead, we focused on creating Matlab code that then generates synthetic data as part of an automatic parameterized task generator and stores it in a data file that is part of the task. This Matlab code should eventually be compiled from the output of multiple prompts.

The main contribution of the language model is in the area of generating real values for the proposed categorical variables. Consequently, the constructed prompts primarily focus on generating the values of categorical variables. This study does not address the generation of numerical variable values according to specified characteristics and properties or following a required distribution. The generation of numerical variable values based on these specified properties will be the subject of future research.

The chapter shows a very small part of the use of the possibilities of AI language models. The use of language models opens up a lot of room for research into their capabilities and research into ways of communicating to achieve the best-desired outcomes.

References

1. Landers, R.: Dataset Generator for Learning Introductory Statistics, https://rlanders.net/dataset-generator, last accessed 2024/03/20.
2. Luttrell, A.: Data Generator for Teaching Statistics, http://andyluttrell.com/datagen.html, last accessed 2024/03/15.
3. Mobahi, H., Min, H., Wojtusiak, J.: Synthetic Data for Teaching Data Integration in Informatics Graduate Program. In: AMIA 2019, American Medical Informatics Association Annual Symposium (2019).
4. Hassani H., Silva S. E.: The Role of ChatGPT in Data Science: How AI-Assisted Conversational Interfaces Are Revolutionizing the Field. Big Data and Cognitive Computing 7(2), 62 (2023).

5. Lingo, R.: Exploring the Potential of AI-Generated Synthetic Datasets: A Case Study on Telematics Data with ChatGPT. arXiv preprint. (2023).
6. Solomon Ubani and Suleyman Olcay Polat and Rodney Nielsen. ZeroShotDataAug: Generating and Augmenting Training Data with ChatGPT. arXiv preprint. (2023).
7. Kim, Seul-ki, and Yong-ju Jeon. Development of a Python Library to Generate Synthetic Datasets for Artificial Intelligence Education. International Journal on Advanced Science, Engineering and Information Technology 14(3), 936–45 (2024).
8. Stenger, M., Leppich, R., Foster, I., Kounev, S., & Bauer, A. Evaluation is key: a survey on evaluation measures for synthetic time series. Journal of Big Data 11(1), (2024).
9. Sasiaowapak, T., Boonsang, S., Chuwongin, S., Tongloy, T., & Lalitrojwong, P. Generative AI for Industrial Applications: Synthetic Dataset. In: 15th International Conference on Information Technology and Electrical Engineering (ICITEE), pp. 1–6. IEEE, Chiang Mai, Thailand (2023).
10. Hansen, L., Seedat, N., Schaar, M., & Petrović, A. Reimagining Synthetic Tabular Data Generation through Data-Centric AI: A Comprehensive Benchmark. Advances in Neural Information Processing Systems (36), 33781–33823 (2023).
11. Gangur, M.: Automated generation of statistical tasks. In: DIVAI 2018: 12th International Scientific Conference on Distance Learning in Applied Informatics, pp. 47–58 (2018).
12. Gangur, M., Sova Martinovský, V.: On possibility of automatic generation of data files and their use in tasks of descriptive statistics. In: CSEDU 2019: proceedings of the 11th International Conference on Computer Supported Education. 2, pp. 446–453 (2019).
13. Gangur, M.: The use of cloud for generated statistics data file storage in statistics tasks. In: DIVAI 2022: 14th International Scientific Conference on Distance Learning in Applied Informatics, pp. 373–383 (2022).

Pillar 3: Did COVID-19 and the War in Ukraine Affect Interest in Mandatory Disclosure?

Lívia Kelebercová and Michal Munk

1 Introduction

The financial crisis of 2008–2009 had fatal consequences for the banking sector and revealed many weaknesses in the Basel II regulatory framework, which was supposed to serve to ensure the stability of banking institutions in times of crisis. Extensive economic turbulence has caused a decrease in confidence in financial systems, a lack of liquidity, and a deepening of the debt crisis within the whole of Europe. In response to these events, governments, regulators, as well as dedicated teams of researchers came together to reform regulatory standards. In this way, the new Basel III regulatory framework was created. Like its predecessor, Basel III was built on three pillars. As part of Pillar 1, better forms of capital, higher capital ratios, and capital reserves were introduced, and the minimum capital ratio was adjusted. In the context of Pillar 2, risk management requirements were increased, and greater demands were placed on the responsibility of supervisory authorities to monitor and assess the risks of banks and their capital plans. Under Pillar 3, the scope and detail of published information increased, including better quality and quantitative data on risks, capital reserves, and liquidity indicators. Standardized formats for disclosure were introduced to improve comparability between banks, and a requirement for regular updates of published information was also introduced. The idea behind the implementation of these changes was to strengthen market discipline and enable market participants to better understand the risks and capital positions of banks in order to prevent future financial distress due to the fact that mandatory disclosure is an effective tool to promote the adoption of desired policies [1, 2].

L. Kelebercová (✉) · M. Munk
Department of Informatics, Faculty of Natural Sciences and Informatics, Constantine the Philosopher, University in Nitra, Nitra, Slovakia
e-mail: lkelebercova@ukf.sk

M. Turčáni (ed.), *15th International Scientific Conference on Distance Learning in Applied Informatics*, EAI/Springer Innovations in Communication and Computing,
https://doi.org/10.1007/978-3-031-81261-3_9

The aim of our study is to analyze how crisis periods affected interest in information published under Pillar 3. We analyze changes in interest in Pillar 3 information by quarters in 2016–2022 and interpret the results with respect to quarters, as well as historical periods. This analysis is based on knowledge obtained from the log files of a specific bank. The results of our study provide insight into the effectiveness of reforms aimed at increasing transparency and can serve as a basis for further improvements in regulatory frameworks in the future.

The Related Works section describes existing studies that have addressed the issue of the effectiveness of mandatory disclosure. In the Materials and Methods section, we describe the methodology of verification of interest, and in Sect. 4, we present the achieved results, which are interpreted in more detail in the Discussion section.

2 Related Works

Pillar 3 disclosures are a necessary component of banking regulation, aimed at enhancing market discipline by requiring banks to disclose detailed information about their activities [3]. Pillar 3 disclosures should attract the interest of stakeholders by providing key information on capital, risk, as well as risk assessment processes that affect institutions' capital adequacy [4]. Understanding how stakeholders' interests in these disclosures change during and after market turbulence is essential for evaluating the effectiveness of these regulations. The topic of modeling user behavior on web parts containing Pillar 3 information is addressed in several studies.

In 2013, the authors [5] presented a project whose purpose was to verify the fulfillment of the objectives of Basel II, Pillar 3—market discipline during the financial crisis. The authors of the project provided a methodological procedure for the preparation of data from web log files, including heuristic methods for completing paths, which we were also inspired by. A similar study [6] focused on assessing the efficacy of disclosing financial and risk-related data under the third pillar for commercial bank users. The authors employed behavioral modeling techniques to analyze visitor engagement with web sections containing third pillar information. Their findings revealed consistently low interest in this type of information, particularly in financial and risk-related data, as highlighted by the authors. In 2017, there was a study conducting the analysis of the interest in Pillar 3 web sections by quarter, revealing that stakeholder engagement with disclosures declined after the turbulent period in 2009. Interest peaked in the first quarter and alongside annual reports, with relatively lower engagement observed for standalone information from the third pillar. Studies from 2021 [7–9] also agree with the conclusions of this study.

Related studies that analyzed the period of the financial crisis (2008–2009) and after the financial crisis in terms of modeling user behavior on bank websites agreed on the fact that stakeholder interest in Pillar 3 disclosures was highest in the first quarter of each year and fell during a period of market turmoil. Stakeholders preferred comprehensive data that includes annual reports and group information, rather than relying solely on Pillar 3 disclosures. This fact can be influenced by the statement of the authors [10, 11], who claim that annual reports also include social and management information of a structured and semi-structured nature, which play a fundamental role in financial transparency and are an effective means of communication between the company and its stakeholders. Our research question was to determine whether recent crisis periods, such as the COVID-19 pandemic or the war in Ukraine, have reduced interest in mandatory disclosure compared to periods of stability (Pillar 3 review period) as well as decreased interest in mandatory disclosure in crisis periods (period global financial crisis) in 2009–2010 according to related studies.

3 Materials and Methods

This section is dedicated to the description of the data file and the data preparation methods used.

3.1 The Dataset

The source of our data was the log files of the selected bank for the years 2016–2022. The log files were available in MIIS format and consisted of the following fields: *date*, *time*, *s-ip*, *cs-method*, *cs-uri-stem*, *c-ip*, *cs(User-Agent)*, *cs(Referrer)*, *sc-status*, *sc-substatus*, *sc-bytes*, *cs-bytes*, and *time-taken*. Figure 1 shows the header of both types of web logs. We provide IP addresses in hashed form.

3.2 The Data Preprocessing

The log files were preprocessed in accordance with the studies using the Python programming language. To preprocess the log files, a custom algorithm was created

```
#Software: Microsoft Internet Information Services 8.5
#Version: 1.0
#Date: 2017-01-01 00:00:01
#Fields: date time s-ip cs-method cs-uri-stem c-ip cs(User-Agent) cs(Referer) sc-status sc-substatus sc-bytes cs-bytes time-taken
2017-01-01 00:00:00 97f612e2759d092462baee3bdad2b5c2 GET /apphealth/mon.aspx 3fc8fc856bc1c151835eb430515f8530 - - 200 0 405 74 0
2017-01-01 00:00:03 97f612e2759d092462baee3bdad2b5c2 GET /apphealth/mon.aspx e4d51dc5cfc0cbcc442fa8552d31a535 - - 200 0 405 74 2
2017-01-01 00:00:05 eef0abbc16b29c721ae79e64ef97f70f GET /apphealth/mon.aspx 3fc8fc856bc1c151835eb430515f8530 - - 200 0 405 74 2
2017-01-01 00:00:08 97f612e2759d092462baee3bdad2b5c2 GET /apphealth/mon.aspx e4d51dc5cfc0cbcc442fa8552d31a535 - - 200 0 405 74 0
2017-01-01 00:00:11 97f612e2759d092462baee3bdad2b5c2 GET /apphealth/mon.aspx 3fc8fc856bc1c151835eb430515f8530 - - 200 0 405 74 0
2017-01-01 00:00:13 97f612e2759d092462baee3bdad2b5c2 GET /apphealth/mon.aspx e4d51dc5cfc0cbcc442fa8552d31a535 - - 200 0 405 74 0
```

Fig. 1 Log file example

to process the files quarterly for each year. The algorithm consisted of a series of the following steps: loading log files for the relevant year and quarter into the DataFrame structure, removing unnecessary fields (columns in the DataFrame) from the log files, removing rows with robot accesses, removing rows with accesses to unnecessary files, removing rows with unsatisfactory status codes, removing lines with inappropriate methods, calculating Unix time, identifying users, identifying sessions, and completing paths.

We processed the log files separately for each year at the quarter level, thus obtaining 4 data matrices for each year. The following description of the preprocessing steps was therefore applied separately for each of the matrices.

To reduce the size of the matrices, we removed columns (fields of log files that will not be part of preprocessing and are not the subject of our analysis). Specifically, these were the fields *cs-uri-query*, *sc-substatus*, *sc-bytes*, *cs-bytes*, and *time-taken*.

The next step implemented was the removal of robot access. The *cs(User-Agent)* field was monitored for each record in the respective quarter. In case the field contained any of the values *bot, robot, spider,* and *crawl*, we identified a robot access and deleted the record. Eliminating bot access by monitoring the *cs(User-Agent)* field may not be sufficient. To effectively remove robots, it is necessary to define a set of IP addresses accessing the robots.txt file. Records that contained IP addresses accessing this file were removed.

The next step of data preprocessing was the removal of accesses to unnecessary files. More precisely, records were removed that contained, in the *cs-uri-stem* or *cs(Referer)* field, the path to files with extensions such as: *.bmp*, *.jpg*, *.jpeg*, *.png*, *.gif*, *.css*, *.flv*, *.ico,* *.swf*, *.rss*, *.xml*, *.cur*, *.js*, *.json*, *.svg*, *.woff*, *.tff*, *.eot*, *.aspx*, *.action*, and *.txt*, *.jsp*.

Subsequently, we filtered the records based on the *sc-status* field. In the context of the analysis leading to the fulfillment of the objectives of the work, the records whose return code had the value 200 were sufficient for us. The other records were removed.

The next step of preprocessing the log files was filtering based on the methods field. Records whose methods field value was not *get* were removed from the file. Subsequently, we calculated the Unix time from the date and time fields. Unix time is the number of seconds that have elapsed since 00:00:00 Coordinated Universal Time (UTC), Thursday, January 1, 1970 (excluding leap seconds). It provides a standardized way of expressing and calculating time in different systems because it is not affected by time zones or daylight saving time. The next step was to identify the users. We identified users by creating a categorical variable *user-id* created from the fields *cs(UserAgent)* and *cs-ip*. For each user, we tracked the time difference between two consecutive lines. We kept this difference in a column we called *length*.

We then identified the sessions. We identified sessions using the Reference Length heuristic [12]. To test the possibility of using Reference Length as a method to identify sessions, we needed to look at the length variable. The distribution of the length variable is exponential (Fig. 2), and as a result, we can use the Reference Length method to identify sessions.

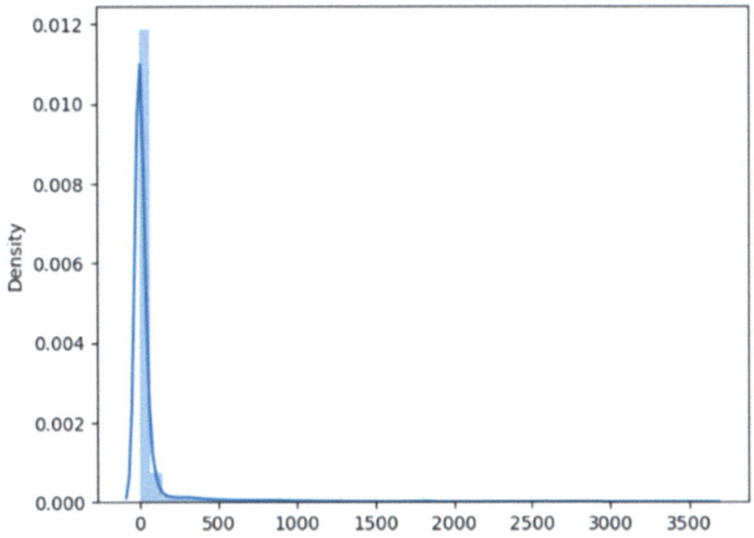

Fig. 2 The distribution of time spent by users on the page represented by the length variable

Table 1 Number of records of log files 2016–2022

Year	Before preprocessing	After preprocessing
2016	707,493,083	23,765,214
2017	952,136,791	18,288,655
2018	1,204,958,521	31,874,983
2019	707,493,083	31,662,016
2020	888,358,789	16,592,952
2021	605,218,153	11,950,811
2022	587,482,940	17,090,826

Completing paths (missing referrer records due to pressing the back button) requires a sitemap. Since we have been analyzing log files from 2016, it should be noted that the sitemap in the sitemap.xml file may have changed over the years. For this reason, we created a site map by crawling the URL addresses in the log file. We supplemented the paths according to the most suitable heuristics.

Finally, we combined the matrices for the quarters in individual years. Table 1 describes the number of records before and after preprocessing.

3.3 Web Parts Categories

We divided web pages into larger categories and subcategories following the pattern of the existing study [8]. For each record in the preprocessed dataset, we recorded the *category* and *subcategory* variables. The *category* variable represents the broadest possible category of the visited web part and can take the values *Pillar 3 disclosure requirements*, *Pillar 3 related*, or *Other*. The *subcategory* variable represents a narrower category. If it is a category *Other*, the subcategory can take the values *Other*, *Information service*, *Products and services for customers*, *About us*, *Press center*, and *Social responsibility*. In the case of the *Pillar 3 disclosure requirements* category, we recognize the subcategories *Information about bank* and *Financial statement*. In the case of the *Pillar 3 related category*, we recognize the subcategories *Annual reports*, *Information for shareholders*, *Financial reports*, *Covered bonds*, *Financial reports*, *Root financial indicators*, and *Information for investors except shareholders*.

4 Results

To analyze the interest in Pillar 3 information, we used association analysis, specifically the Apriori algorithm [13]. The aim of the association analysis was to examine the patterns in the visitation of the subcategories belonging to the Pillar 3 disclosure requirements and Pillar 3-related categories with respect to the sessions. We extracted tuples of rules, with the support set to 1%.

4.1 Differences Between Quarters

The first research task was to look at the differences in applicability across individual quarters. To test differences between quartiles, we used Cochran's Q-test [13–16] (Table 2), which was applied to the binarized values of the association rule support metric. The Null Hypothesis (*H0*) that we wanted to verify can be formulated as follows: *There are no statistically significant differences between occurrences of association rules in quarters in the periods from 2016 to 2022.* Given that the title of our post attempts to answer the question of whether the COVID-19 pandemic and the war in Ukraine affected users' interest in mandatory disclosure, it was first necessary to determine whether there were statistically significant differences in the support metric representing traffic converted to sessions between the individual quarters of the observed years.

Table 2 Cochrane Q-test for testing differences between quarters in 2016–2022

Quarter	*Q*-value	*df*	*p*
Q1	267.7985	6	0.0000
Q2	214.8356	6	0.0000
Q3	135.1380	6	0.0000
Q4	293.5407	6	0.0000

Based on the results of the Cochrane Q-test (Table 2), where the p-value is lower than the chosen level of significance (0.05), we reject the null hypothesis *H0*. This means that there are statistically significant differences between the occurrences of association rules in the quarters during the examined period from 2016 to 2022, while this period represents the years after the financial crisis (2016), the revision of pillar 3 (2017–2019), the pandemic COVID-19 (2020–2021), and the war in Ukraine (2022).

4.2 Differences Between Categories

In this section, we looked at the comparison of average support and lift values between the Pillar 3 Disclosure Requirements and Pillar 3 Related categories. Our aim was to identify whether there were statistically significant differences between the mean values of support and lift in individual quarters between the two categories. Due to the violation of the assumption of normal data distribution in some cases, using the Shapiro-Wilk test ($p < 0.05$), we chose the Mann-Whitney U test. We used this test to test the hypothesis of equal mean support and lift values between the two categories for each quarter. The null hypothesis, which we would like to verify within these Mann-Whitney U tests, can be formulated as follows: Between the groups of the "disclosure" category and the "related" category, there is no statistically significant difference in the distribution of the values of the support and lift metrics in individual quarters. We reject the null hypothesis at the 0.05 level for all cases (Tables 3 and 4). It is therefore not necessary to interpret the following results about the averages of the support and lift metrics for each of the categories separately.

Table 3 Mann-Whitney U test results for the disclosure category and related category in each quarter for the support metric

Quarter	U-statistic	p-value
Q1	21	0.701165
Q2	21	0.699134
Q3	29	0.605287
Q4	22	0.588745

Table 4 Mann-Whitney U test results for the disclosure category and related category in each quarter for the lift metric

Quarter	U-statistic	p-value
Q1	9	0.329004
Q2	8	0.246753
Q3	13	0.294872
Q4	10	0.428571

5 Discussion

In the context of the bank's websites, which contain mandatory published information under Basel II Pillar 3, we can see interesting trends in the values of support and lift in individual years and quarters (Table 5). An increased support and lift average in the first quarter may indicate a start to the year with a greater emphasis on transparency and disclosure related to Basel II Pillar 3. A drop in the support and lift values in the second quarter may be a result of a peak in activity or a reaction to previous increases. A further decline in the third quarter may indicate a decline in interest in these subcategories or a decline in activity during this period. The return to higher values of support and lift in the fourth quarter may be associated with increased efforts before the end of the year to ensure sufficient transparency and make the necessary information available. However, the values remain overall at a low level, indicating a cautious approach by users and interest only to a limited extent.

The year 2016 was followed by a period of revision of the third pillar, which was characterized by relatively stable values of support and uplift. This may reflect established procedures and reactions to regulations in the area of Basel II Pillar 3 after the period of financial crisis. In 2017, lift values reached significant highs, which may be related to increased interest in updates related to the revision of the third pillar. These periods may have been marked by increased confidence in the banking sector and the pursuit of transparency.

The COVID-19 pandemic in 2020 and 2021 may have caused increased uncertainty. The decrease in the association values between the relevant rules in these years indicates a possible response to this uncertainty and changes in the behavior of website visitors. In the period at the beginning of the pandemic (first and second quarters of 2020), the decline in support and lift may be associated with increased uncertainty and reduced demand for information during this period. The subsequent increase in the third and fourth quarters could reflect increased demand for information and transparency in response to the pandemic and related uncertainty. In 2021, the high values of support and lift continued to hold for a period of three quarters, which may signal the continued demand for information

Table 5 Average values of support and lift for quarters

	Q1 Support	Q2 Support	Q3 Support	Q4 Support	Q1 Lift	Q2 Lift	Q3 Lift	Q4 Lift
2016	0.00068	0.00055	0.00013	0.00072	18.216	67.29217	8.91361	50.94251
2017	0.0008	0.00068	0.00079	0.00063	41.44663	24.75072	51.29109	98.37327
2018	0.00037	0.00033	0.00045	0.00045	13.09806	10.66204	10.22112	8.30294
2019	0.00036	0.00032	0.00025	0.0001	10.67	12.29745	13.7276	15.55692
2020	0.00002	0.00001	0.00003	0.00001	12.17114	6.76027	142.07462	9.96283
2021	0.00002	0.00000	0.00004	0.00000	1.80657	13.72136	10.32314	1.00000
2022	0.00000	0.00000	0.00000	0.00000	1.00000	1.00000	1.65135	4.12123

Fig. 3 Average support values in individual quarters

and the stabilization of the position of financial institutions during the pandemic. The decline in support and lift values during the pandemic period compared to previous periods may have been influenced by the COVID-19 pandemic itself, as users may have focused on other areas and become less interested in Basel II Pillar 3 details due to the current health and economic situation.

In 2022, we observed stabilization of support and lift values at a relatively low level. This decline may be linked to uncertainty arising from the geopolitical situation, as well as general uncertainty and concern among users. Users may have focused on other aspects in the context of current events.

Overall, the drop in support during the periods of the COVID-19 pandemic as well as the energy crisis and the war in Ukraine (Fig. 3) can be justified by related studies that showed low interest in Pillar 3 information in times of crisis. On the other hand, during the period of stability (2016–2019), the values of support and lift were higher.

5.1 User Interest in Pillar 3 Information Regarding the Quarter

Given that the highest interest was shown in the first and fourth quarters, we will take a closer look at the behavior of users in relation to Pillar 3 information in these quarters.

In the first quarter (Fig. 4), the same trend can be seen for each subcategory in individual years, except for the subcategory Information for investors (except shareholders). The highest support was in the year of stability (2017), i.e., during the period of the revision of Pillar 3. The highest drop was in the period at the beginning of the COVID-19 pandemic (2020). The highest support was achieved by the Covered bonds subcategory, followed by Root financial indicators, Annual reports,

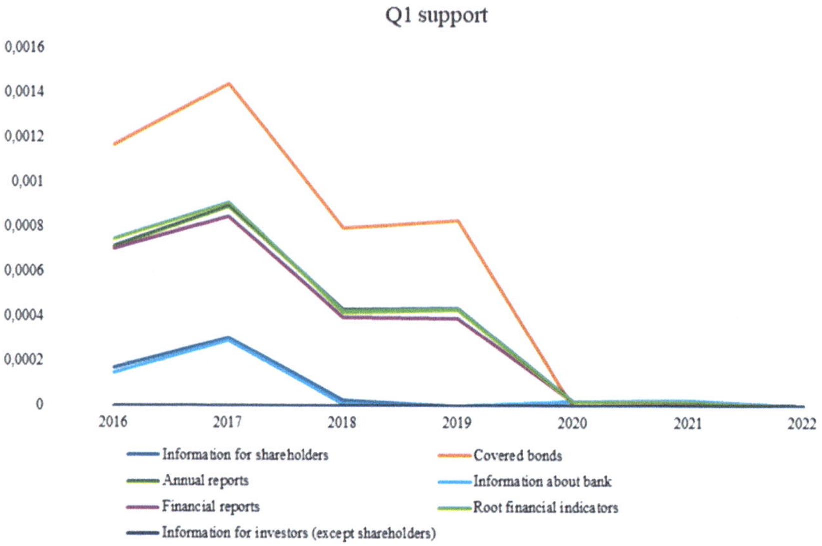

Fig. 4 Support for individual subcategories in the first quarter

Financial reports, Information for shareholders, and Information for investors (except shareholders).

From Fig. 5, it can be seen that Annual reports and Financial reports most often appeared together within the session. Conversely, the thinnest lines lead to Information about bank, which clearly indicates that it is not a frequent subcategory within the sessions.

In the case of the fourth quarter (Fig. 6), the subcategory Information for investors (except shareholders) again had zero support in all years. The remaining subcategories had the highest support in 2016 and the lowest in 2020. The trend in both the first and fourth quarters clearly shows that the interest in Pillar 3 information has a downward tendency during periods of crisis.

The most frequented set was again Financial reports and Annual reports. An interesting finding is the fact that in the fourth quarter, Information about bank, which is our key mandatory disclosure under Pillar 3, occurs more frequently within the session. Users visit them most often together with the Annual report, Information for shareholders, and Financial statements (Fig. 7).

6 Conclusion

Our study examined the impact of crisis periods on stakeholders' interest in Pillar 3 disclosures from 2016 to 2022, utilizing web log data from a commercial bank. The results highlight significant fluctuations in stakeholder engagement with these

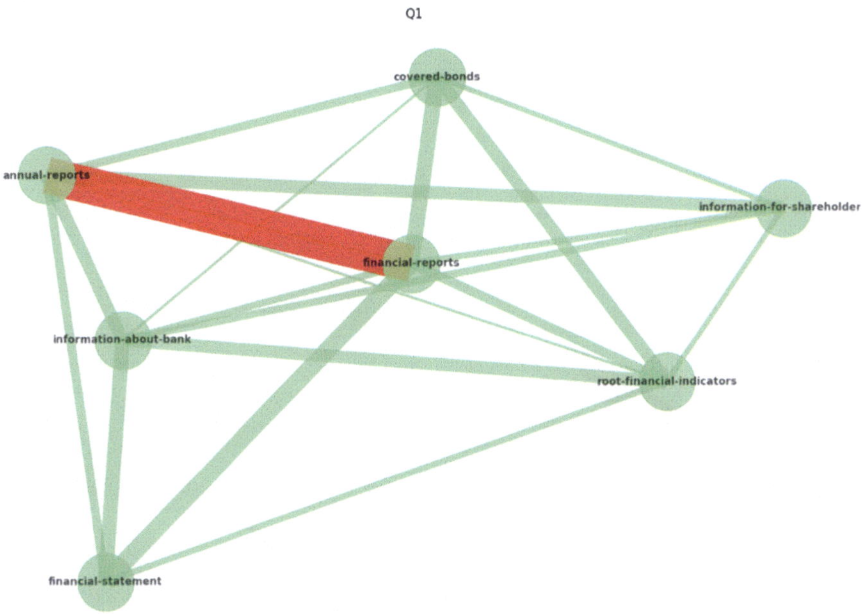

Fig. 5 Frequented parts in the first quarter based on the lift

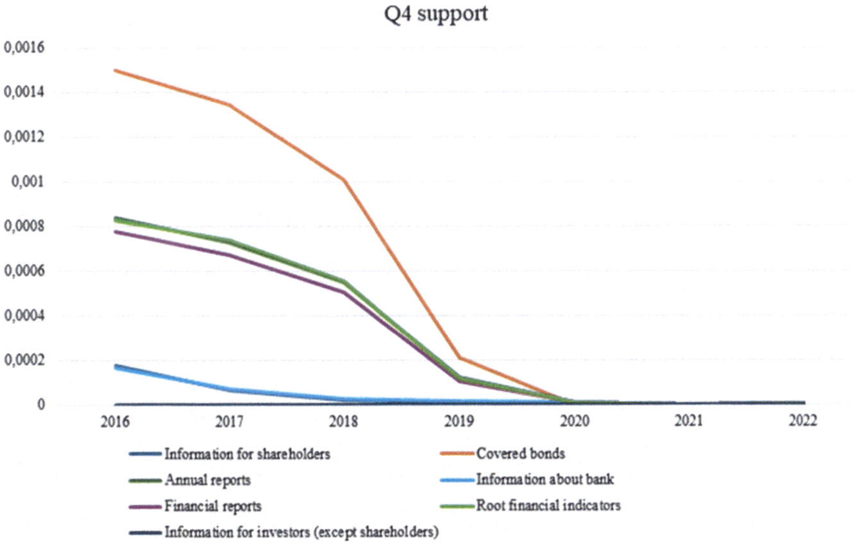

Fig. 6 Support for individual subcategories in the fourth quarter

disclosures, driven largely by external economic conditions and market turbulence. Key findings reveal that interest in Pillar 3 information peaks in the first quarter of each year.

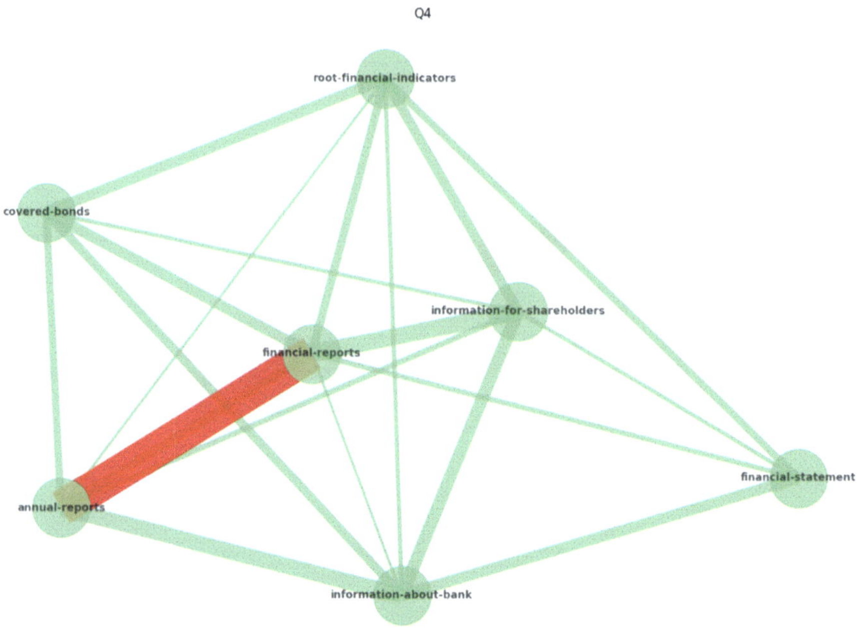

Fig. 7 Frequented parts in the fourth quarter based on the lift

However, this interest sharply declines during periods of financial instability, such as the COVID-19 pandemic and geopolitical crises like the war in Ukraine. These observations suggest that while stakeholders value transparency and comprehensive reporting, their engagement with specific regulatory disclosures diminishes during times of uncertainty. To maintain stakeholder engagement and enhance market discipline, regulators might consider revising disclosure policies to ensure that information remains relevant and accessible during turbulent times. This could involve integrating more timely and contextually pertinent data that addresses current market conditions and stakeholder concerns.

Overall, our findings provide valuable insights into the effectiveness of Basel III reforms, particularly regarding Pillar 3's role in promoting market discipline. By understanding patterns of stakeholder engagement, regulators and financial institutions can better tailor their disclosure practices to sustain transparency and trust, even in challenging economic climates. Future research should continue to explore the dynamic relationship between market conditions and regulatory disclosures to further optimize the regulatory framework for financial stability.

Acknowledgments This work was supported by the Scientific Grant Agency of the Ministry of Education of the Slovak Republic (ME SR) and the Slovak Academy of Sciences (SAS) under contract No. VEGA-1/0734/24.

References

1. Boland, C.M., Hogan, C.E., Johnson, M.F.: Motivating Compliance: Firm Response to Mandatory Disclosure Policies. Accounting Horizons. 32, 103–119 (2018). https://doi.org/10.2308/acch-52037.
2. Alvarez, F., Barlevy, G.: Mandatory disclosure and financial contagion. Journal of Economic Theory. 194, 105237 (2021). https://doi.org/10.1016/j.jet.2021.105237.
3. Calver, P., Owladi, J.: Pillar 3 Disclosures: Looking Back and Looking Forward. In: Bank of England Quarterly Bulletin (2017).
4. Balthazar, L.: Pillar 3: Market Discipline. In: From Basel 1 to Basel 3. pp. 95–100. Palgrave Macmillan UK, London (2006). https://doi.org/10.1057/9780230501171_8.
5. Munk, M., Pilková, A., Drlík, M., Kapusta, J., Švec, P.: Verification of the fulfilment of the purposes of Basel II, Pillar 3 through application of the web log mining methods. Acta Univ. Agric. Silvic. Mendelianae Brun. 60, 217–222 (2013). https://doi.org/10.11118/actaun201260020217.
6. Pilkova, A., Munk, M., Svec, P., Medo, M.: Assessment of the Pillar 3 Financial and Risk Information Disclosures Usefulness to the Commercial Banks Users. In: Huang, D.-S. and Han, K. (eds.) Advanced Intelligent Computing Theories and Applications. pp. 429–440. Springer International Publishing, Cham (2015). https://doi.org/10.1007/978-3-319-22053-6_46.
7. Munk, M., Pilkova, A., Benko, L., Blazekova, P., Svec, P.: Web usage analysis of Pillar 3 disclosed information by deposit customers in turbulent times. Expert Systems with Applications. 185, 115503 (2021). https://doi.org/10.1016/j.eswa.2021.115503.
8. Pilková, A., Munk, M., Benko, Ľ., Blažeková, P., Kapusta, J.: Pillar 3: Does banking regulation support stakeholders' interest in banks financial and risk profile? PLoS ONE. 16, e0258449 (2021). https://doi.org/10.1371/journal.pone.0258449.
9. Blažeková, P., Benko, Ľ., Pilková, A., Munk, M.: Is Pillar 3 a Good Tool for Stakeholders in CEE Commercial Banks? In: Kryvinska, N. and Poniszewska-Marańda, A. (eds.) Developments in Information & Knowledge Management for Business Applications. pp. 421–440. Springer International Publishing, Cham (2021). https://doi.org/10.1007/978-3-030-76632-0_15.
10. Dias, A., Rodrigues, L.L., Craig, R.: Global financial crisis and corporate social responsibility disclosure. SRJ. 12, 654–671 (2016). https://doi.org/10.1108/SRJ-01-2016-0004.
11. Oncioiu, I., Popescu, D.-M., Aviana, A.E., Şerban, A., Rotaru, F., Petrescu, M., Marin-Pantelescu, A.: The Role of Environmental, Social, and Governance Disclosure in Financial Transparency. Sustainability. 12, 6757 (2020). https://doi.org/10.3390/su12176757.
12. Kapusta, J., Munk, M., Drlik, M.: Cut-off time calculation for user session identification by reference length. In: 2012 6th International Conference on Application of Information and Communication Technologies (AICT). pp. 1–6. IEEE, Tbilisi, Georgia (2012). https://doi.org/10.1109/ICAICT.2012.6398500.
13. Agrawal, R., Srikant, R.: Fast Algorithms for Mining Association Rules in Large Databases. In: Proceedings of the 20th International Conference on Very Large Data Bases. pp. 487–499. , Santiago de Chile (1994).
14. Cochran, W.G.: THE COMPARISON OF PERCENTAGES IN MATCHED SAMPLES. Biometrika. 37, 256–266 (1950). https://doi.org/10.1093/biomet/37.3-4.256.
15. Conover, W.J., Conover, W.J.: Practical nonparametric statistics. Wiley, New York, NY Weinheim (1999).
16. Patil, K.D.: Cochran's Q Test: Exact Distribution. Journal of the American Statistical Association. 70, 186–189 (1975). https://doi.org/10.1080/01621459.1975.10480285.

The Impact and Use of AI in Modern University Education

Kristyna Kilianova, Petra Kockova, Marta Slepankova, and Katerina Kostolanyova

1 Introduction

Artificial intelligence (AI) is one of the most significant technological advancements of our time, fundamentally changing the way many industries, including education, operate. Its use in educational institutions is becoming increasingly common, and its benefits and impacts are the subject of much research and debate. AI applications offer a multitude of possibilities that can not only streamline educational processes but also bring about profound changes in the way we learn and teach others.

AI in education can be divided into several key areas that highlight its benefits and potential for effective use in higher education environments. These include the personalization of learning, the automation of administrative tasks, the improved accessibility of learning, the increased efficiency of teaching, the provision of continuous support and feedback through virtual assistants, the development of skills, and the support for researchers.

1.1 The Advantages of Utilizing AI in the University Environment

One of the most significant benefits of AI in education is the ability to personalize learning. Traditional education systems often adopt a one-size-fits-all approach, which may not be optimal for the individual needs of students. Per-

K. Kilianova (✉) · P. Kockova · M. Slepankova · K. Kostolanyova
University of Ostrava, Ostrava, Czech Republic
e-mail: kristyna.kilianova@osu.cz; petra.kockova@osu.cz; marta.slepankova.s01@osu.cz; katerina.kostolanyova@osu.cz

M. Turčáni (ed.), *15th International Scientific Conference on Distance Learning in Applied Informatics*, EAI/Springer Innovations in Communication and Computing,
https://doi.org/10.1007/978-3-031-81261-3_10

sonalized AI-driven learning represents a transformative approach that adapts educational content and interactions to the needs, preferences, and pace of individual students [1].

Zavalevskyi [2] notes that AI tools can greatly assist students with special educational needs through personalized learning programs and adaptive technologies that consider their individual needs. For example, tools such as Google Text-to-Speech allow students to convert text to spoken word, and AI-powered translators such as DeepL can help students with language barriers understand learning materials and develop communication skills in a foreign language. Assistive technologies for the deaf, such as Ava or Microsoft Translator, can provide captioning and translation from spoken word to text, allowing students to follow lectures and participate in discussions. Mobile apps such as Seeing AI or Google Lens can help students with visual impairments by using their smartphone's camera to recognize objects, text, and people in their environment. It is essential to note, however, that while some visually impaired students can use mobile apps in much the same way as regular users, just taking advantage of higher brightness, contrast, etc., for example, blind students benefit from special accessibility features that provide audio descriptions of visual elements and allow them to more easily navigate digital technology and the way they live in the modern age. Tools such as MindMeister use AI to organize thoughts and create visual maps, which can be useful for those with organizational or memory problems.

The potential of AI to automate administrative tasks has also been the subject of several studies. For example, Pupe [3] proposed a web architecture model that integrates machine learning for process automation in an undergraduate center. Joo [4] presented a data mining-based system for automating administrative tasks and analyzing student activities in public education in Korea. Gunawan [5] implemented robotic process automation to improve the administrative process at a private university, significantly reducing time and costs. Taken together, these studies demonstrate the potential of automation to improve efficiency and productivity in educational administration, especially in the areas of application processing, verification, digitization and archiving of student documents, analysis of classroom availability, and time preferences of students and instructors, leading to optimal scheduling of lectures and seminars.

The introduction of AI in these areas not only saves time for teachers and university staff but also improves the functioning of the entire education system. This is particularly evident in the field of nontechnical adult education, where AI can improve the understanding of complex technical concepts [6].

1.2 The Impact of AI on University Education

Every technology has its negatives. For example, Guilherme [7] argues that this change in the relationship between teachers and students could lead to the replacement of human teachers by artificial intelligence, which could disrupt educational processes and undermine the personal interaction that is crucial for students' emotional and social development. Interactions with teachers and classmates play a crucial role in the educational process, and limiting them can have a negative impact on students' overall development.

The more pressing issue may be the potential misuse of AI or the ethical and security issues associated with it. As Ivanov [8] notes, the use of AI in higher education raises concerns about potential misuse, including distortion of datasets and algorithms, plagiarism, micromanagement of students and staff, and privacy concerns. Mackay [9] argues that the increasing reliance on online learning has served to exacerbate these problems by lowering the barrier to breaches of academic integrity. Furthermore, the emergence of AI-enabled generative chatbots has further complicated the situation, as students have easy access to content-generating systems that they can present as their own [10]. The use of AI in essay generation, assignment solving, and test answering has the potential to lead to a significant increase in cases of academic dishonesty. This is because AI can be used to produce academic work that may not be subject to the same level of scrutiny as work produced by humans. In addition, AI systems that provide instant answers and solutions can reduce the need for students to think creatively and critically. Over-reliance on AI can limit the generation of original ideas and limit students' ability to solve problems independently and approach tasks creatively.

In conclusion, AI has the potential to fundamentally transform education systems around the world. The significance of AI in education is not limited to technological advancements; it also encompasses the potential to enhance accessibility, quality, and efficiency in education. While AI does present certain challenges, such as ethical considerations, data privacy and security, or its potential misuse, the benefits are considerable and could result in significant improvements in the educational process.

1.3 Follow-Up to Pilot Research

It is widely acknowledged that university students occupy a unique position in this transformation, as they stand at the threshold of careers that will be inextricably linked to the use of AI. Concurrently, they are the primary source of new ideas, and AI becomes not only a means to enhance the efficiency of studies but also a tool that provides novel avenues for research and innovation [11].

The principal objective of the pilot study, which spanned the entirety of 2023 and was divided into two phases, was to ascertain the potential opportunities afforded by AI to students, their expectations and concerns, and to observe any shifts in their perspectives over time. The study comprised a comprehensive survey involving 109 respondents from various universities. The findings were presented at international conferences, namely Edulearn23 and Inted24, and revealed the following.

A total of 88.1% of students demonstrated familiarity with the principles of AI and its applications, representing a notable increase in understanding compared to previous research. Additionally, there was a notable increase in hands-on experience with AI-enabled applications, with virtual assistants and chat applications such as Apple's Siri, Google Assistant, and ChatGPT being the most common. In terms of attitudes, 58.2% of students were found to have a positive or rather positive attitude towards the use of AI in everyday life and education, with only 3.7% of students having a strongly negative view on the issue. Furthermore, students expressed a range of views on the ethical implications of using AI in academic writing. Approximately 56% of students believe that the use of AI should depend on the specific situation and purpose. Forty percent consider the use of AI ethically acceptable if it respects copyright and citation rules. In contrast, 10% of students consider it ethically unacceptable. The results indicated that only 31.8% of respondents had practical experience in academic writing. Furthermore, 50% of students stated that they had never used AI in academic writing, while 18.2% planned to do so. However, 77.3% of students believed that AI could improve the quality and efficiency of work by automating repetitive tasks. Additionally, there was a strong interest in the potential for AI to support people with limited mobility and in crisis situations. Furthermore, students perceive AI as a crucial tool for supporting education, healthcare, and the advancement of new technologies.

Building on previous research, we have decided to continue and deepen this research as artificial intelligence (AI) becomes increasingly integrated into our lives. AI is becoming an integral part of many areas, from everyday life to the professional sphere. With the growing influence of AI, it is imperative that university students learn how to work effectively with this technology, understand its capabilities and limitations, and become more concerned with safety and ethical issues. The continuation of this study with an expanded sample of respondents will allow us to gain valuable insights into how students perceive AI in their future careers. The aim of this follow-up study is to obtain their attitudes, opinions, and practical experiences to ensure comparable results. This information can assist in the development of educational programs and strategies that provide university students with the necessary skills and knowledge to successfully integrate AI into their professional lives. To better process the information we gathered, we asked the following research questions:

- What are the attitudes of university students towards AI in education?
- What specific experiences and skills related to AI have they acquired during their studies?
- How do students expect AI to affect their future career paths, and how are they preparing for these changes?

2 Methodology

The research was based on a previous pilot study, which was divided into two phases. The first phase ran from January to June 2023, and the results were published at the international conference Edulearn23. The second phase of the pilot study commenced in July 2023 and concluded in November 2023. The results were then presented at the international conference Inted24.

The findings of the pilot study informed the construction of a more robust electronic questionnaire, designed to provide the most comprehensive view of students' views on AI in the university environment and their future careers. The research instrument comprised 26 questions, including closed-ended, semi-closed, open-ended, and Likert-scale assessment questions. Data collection was conducted between December 2023 and May 2024. The research sample included 378 respondents. The data were subjected to quantitative analysis with a view to identifying key trends and differences in student views on AI. The results of this analysis are presented in the following section.

3 Results

Our study analyzed the opinions and attitudes of 378 students studying at different universities in the Czech Republic. The research sample consisted mainly of students from the University of Ostrava (23.81%), Palacký University Olomouc (16.67%), and Charles University (12.17%). Students from the University of Hradec Kralove accounted for 11.90%, and the University of West Bohemia for 7.41%. Other universities, such as the Technical University of Ostrava and Masaryk University, had a smaller but still significant representation, with 5.29% and 5.82% of respondents, respectively. The universities with the lowest representation were Tomas Bata University in Zlín (4.76%), Jan Evangelista Purkyně University in Ústí nad Labem (4.76%), the University of Pardubice (4.50%), Silesian University in Opava (2.38%), and Brno Technical University or Prague University of Economics and Business, with 0.26% of respondents. The age range of the research sample reflected the typical demographic cross-section of the university population, including students from the first years of undergraduate studies to students in postgraduate programs.

We used a questionnaire to investigate whether students knew and understood the concept of AI and its applications. Students rated their knowledge on a scale of 1 (very good) to 5 (poor). The highest percentage of respondents (25.40%) rated their knowledge of AI as commendable, with the second highest rating of good at 24.60%; 20.90% of students rated their knowledge as very good, 14.29% of the students rated their knowledge as sufficient, while 14.81% considered their knowledge to be insufficient. These data show that most respondents have at least an average or better understanding of AI and its applications, indicating a growing interest and awareness in this area (Fig. 1).

Familiarity with AI concepts is closely related to attitudes towards the use of AI in everyday life, including teaching at university. A total of 29.10% of respondents do not have a strong opinion about the use of AI in everyday life, while 24.60% of respondents have a positive opinion, 18.25% of respondents have a very positive

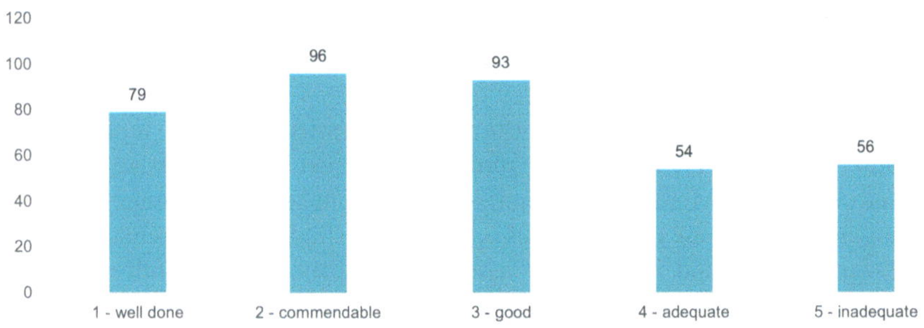

Fig. 1 Are you familiar with the concept of artificial intelligence (AI) and its applications?

attitude towards the use of AI, and 20.37% of respondents have mixed feelings about the use of AI. Only 7.67% of respondents do not identify with the use of AI in their daily lives.

The second graph shows the frequency of AI technology use in the learning process among the students surveyed. The results show that the largest proportion of respondents (33.60%) use these technologies frequently, and 33.33% use them very frequently; 19.58% of the respondents stated that they rarely use AI technologies, and the smallest number of respondents (13.49%) have never used any AI technology. These data suggest that most students (over 66%) use AI technologies regularly, indicating the increasing integration and importance of these technologies in the educational process. A smaller proportion of respondents use AI technologies less frequently or not at all, which may be due to various factors such as lack of awareness, access to technology, or personal preference (Fig. 2).

Awareness of applications that use AI principles to function is also related to the frequency of use. A total of 316 students, or 83.60% of respondents, said they were familiar with digital assistants such as Apple's Siri, Google Assistant, or Amazon Alexa; 82.28% are familiar with or use chat applications such as OpenAI's ChatGPT, 52.12% are familiar with image recognition applications such as Google Lens or Microsoft Seeing AI, and 32.80% are familiar with AI image creation applications such as DALL-E, DeepArt, Prisma, or Midjourney. AI tools for scientific writing, language translation, graphics creation, and others were the least represented.

Given that a large proportion of respondents are familiar with and use AI tools and applications, we were interested to know how much confidence students have in these tools and applications. A total of 85.45% of respondents have some confidence in AI applications but still maintain a level of caution and circumspection, while 12.17% of respondents said they have no confidence in these applications and are concerned about their accuracy and reliability. Only 2.38% of respondents have a great deal of confidence in AI applications and believe that they perform their functions correctly and accurately (Fig. 3).

This graph shows how respondents have encountered AI-enabled applications or technologies in their studies or teaching. The largest group of respondents, 35.45%,

Fig. 2 How often do you use AI technology (apps) in your learning process?

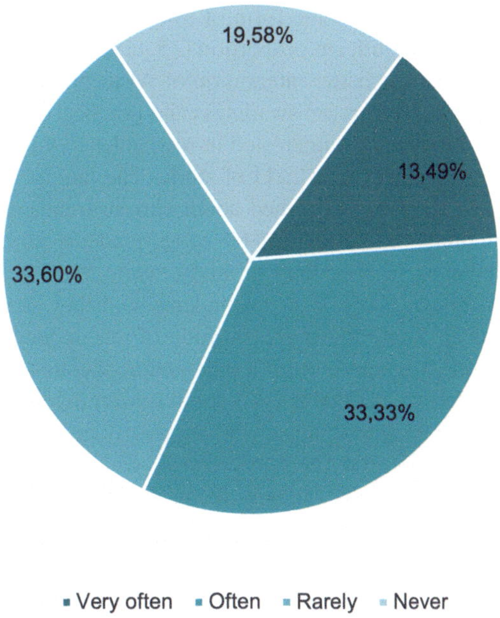

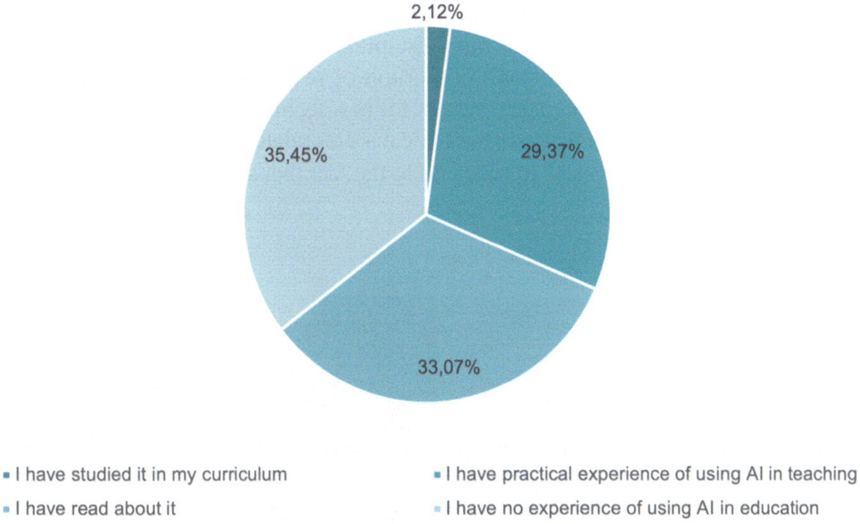

Fig. 3 How have you encountered applications or technologies using AI in your studies or learning?

said they had no experience of using AI in education; 33.07% of respondents said they had read or heard about AI and its use in education; 29.37% of respondents have practical experience of using AI in education, and only 2.12% of students are studying AI in their curriculum. This data suggests that although many respondents

(64.55%) have some form of awareness or experience with AI in education, there is still a significant proportion (35.45%) who have no experience. This points to the potential for better integration of AI technologies into the educational process and the need to increase awareness and practical use of AI among students and teachers.

We were also interested in respondents' knowledge of current trends or applications of AI in their field of study. The largest group of respondents, 37.30%, said they were well-informed about current trends and applications of AI in their field of study. A total of 28.04% of respondents said they were not very well-informed about AI in their field of study, while 15.87% considered themselves moderately informed; 14.55% of respondents said they were not interested in information in this area, and the smallest group of respondents, 4.23%, said they were very well-informed about current trends and applications of AI in their field of study. These results suggest that there is room for improvement in AI awareness among students, especially among those who are poorly informed or not interested in the field.

To reinforce their awareness, we asked respondents how education should prepare students for new technologies, especially those such as AI. As we can see from the attached graph, 37.84% of respondents agreed that education should include the provision of specific courses or training on new technologies, while 26.29% of respondents said that new technologies should be included in the curriculum of degree programs. A more balanced group of respondents, 21.28%, felt that independent study and research in new technologies should be encouraged, and 11.09% of students then stated that a multidisciplinary approach to new technologies should be encouraged. Only a small proportion of respondents (3.50%) were not sure how education should prepare students for new technologies (Fig. 4).

The use of AI in education also raises ethical issues related to its use, especially in the context of academic theses. In response to the fact that some Czech universities

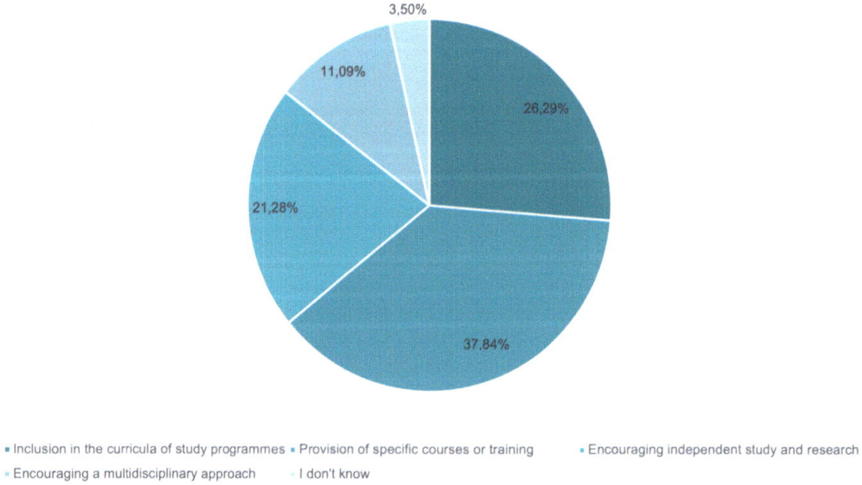

Fig. 4 How should education prepare students for new technologies like AI?

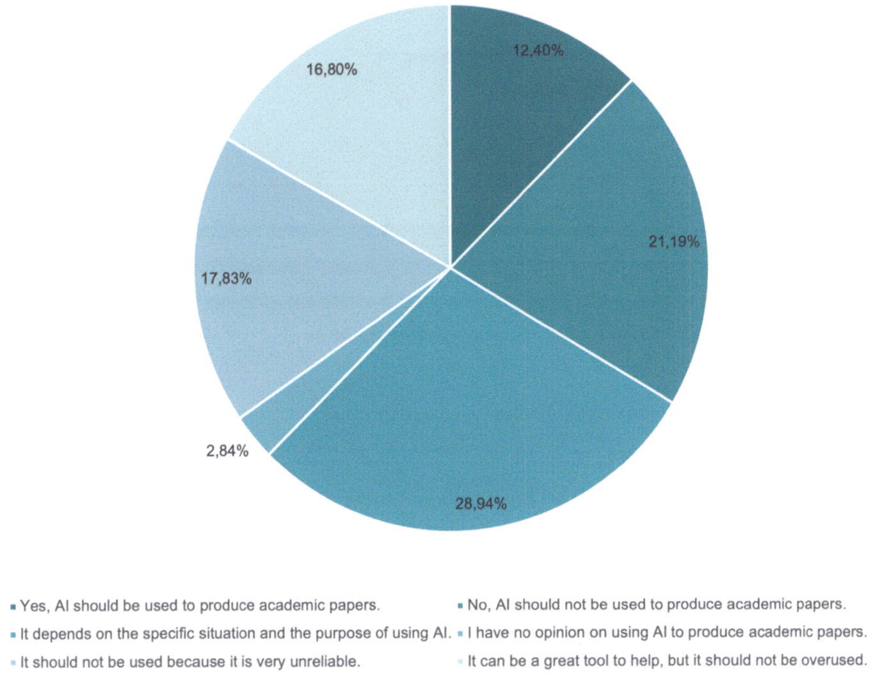

Fig. 5 What do you think about whether AI should be used to produce academic papers?

have decided not to support AI in the writing of these theses or have even canceled them, we were interested in how students themselves view this issue.

Figure 5 shows the respondents' views on the use of Artificial Intelligence (AI) for writing academic papers. A total of 12.40% of respondents said that AI should be used to write academic papers, while 21.19% of students disagreed with this opinion. The largest group of respondents (28.94%) said that it always depends on the specific situation and purpose of using AI. A total of 2.84% of respondents did not have an opinion on this issue; 17.83% of respondents said that AI should not be used because of its unreliability, while 16.80% of students think that AI can be a useful tool but should not be overused.

In the next question, we were interested in students' practical experience of using this technology. An overwhelming majority, 55.04%, said that they had never used AI to write an academic paper. A total of 10.85% of respondents said that they had not yet used AI but planned to do so in the future, while 34.11% of respondents had practical experience of writing academic papers using AI. We also asked these students about their experiences. For 25.00% of the students, it was a successful method that helped them produce academic papers, while for 48.00%, it was an unsuccessful method. For 17%, it was not their own experience, but the experience of classmates and friends. Of this 17%, 12% were not helped by AI, and 5% were helped to complete their work successfully. The remaining 10% of respondents gave

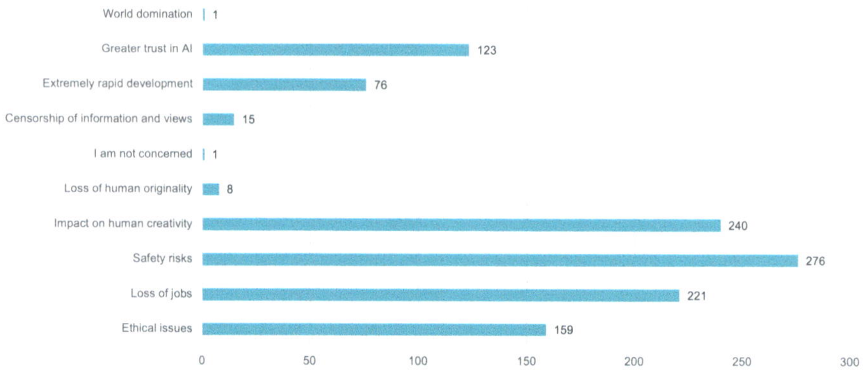

Fig. 6 About the development of AI, what are your concerns?

various reasons for when AI had helped them and when it had not, but these answers were not entirely specific to practice.

Regarding the ethical assessment of the use of AI to produce academic work, 47.6% of respondents agreed that it always depends on the specific circumstances and context of the use of AI. For 37.1% of respondents, the use of AI is ethically acceptable if AI is used in a way that respects copyright and citation rules. For 8.9% of respondents, the use of AI is ethically unacceptable because it violates copyright and citation rules, and 6.5% of students do not pay particular attention to ethical issues related to AI.

In addition to ethical issues (42.06%), the following graph shows other concerns that students have about AI: 276 students, or 73.02%, are worried about security risks, 58.47% are worried about job losses, 32.54% are worried that people will put more trust in AI, and 20.11% are worried about the extremely rapid development. In percentages, respondents fear censorship of information and opinions (3.97%), loss of human originality (2.12%), world domination by AI (0.26%), and one student said he had no concerns (Fig. 6).

These data show that respondents' main concerns focus on security risks, the impact on human creativity, job losses, and ethical issues. These results suggest that students' knowledge in these areas may need to be deepened.

So, what role should AI play in the future? The following graph shows that the greatest emphasis is placed on research and development of new technologies (86.24%), improving energy systems (77.51%), and helping people with reduced mobility (74.07%). Respondents also see an important role for AI in supporting education and research (74.87%), improving healthcare (69.58%), improving the quality and efficiency of work (62.17%), and in crisis situations (62.70%). The development of autonomous vehicles (45.24%) and the improvement of the environment (50.26%) are also important, although less so (Fig. 7).

Respondents also recognize that there are a few significant benefits to using AI in education. The biggest benefit is more effective exam preparation, cited by 20.11%

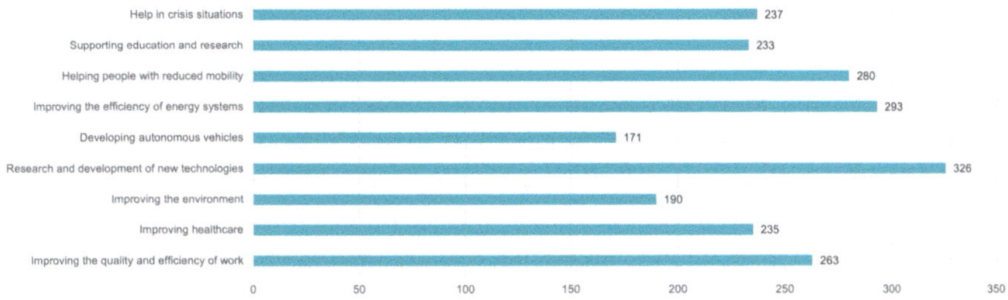

Fig. 7 What role do you think AI should play in the future?

of respondents. Another significant benefit is help with essays and academic work, cited by 16.67% of respondents. AI also significantly improves the quality and efficiency of work, cited by 14.81% of respondents. In addition, AI can help with lesson planning and provide immediate answers to students' questions, which helps to better organize and support the learning process (1.85%). Another benefit is the development of critical thinking, mentioned by 3.44% of respondents, and gaining knowledge about the capabilities, risks, and limitations of AI (1.59%). Finally, AI enables students to work with information and ask accurate questions, which was mentioned by 6.61% of respondents. Overall, AI in education brings many positive aspects that can significantly improve the learning process and enhance students' development.

The use of AI in education raises several concerns. The biggest is the overuse of AI by children and students, cited by 18.78% of respondents. The same proportion of respondents (11.11%) were concerned about the reckless use of AI without verifying information and encouraging laziness. Other significant concerns include the loss of the ability to critically evaluate information (5.03%) and the erosion of human creativity (10.85%). Respondents also mentioned that AI could lead to job loss (0.79%), encourage unethical behavior such as cheating (3.97%), and contribute to the misinterpretation of answers (2.38%). They are also concerned that the technology is complex and that users may not have sufficient technical skills (2.12%), which can lead to various problems. Respondents were also concerned about the misuse of personal data (0.53%), copyright infringement (1.06%), and loss of privacy (0.79%). Overall, these concerns point to the need for a careful and responsible approach to integrating AI into the educational process, minimizing negative impacts and maximizing the benefits of the technology.

Although the large number of respondents (378 students) provides a solid data base, the novelty of this research lies in its comparison with the previous two pilot studies. This research not only confirms earlier findings but also provides deeper insight into changes in students' attitudes and experiences with AI over time. The results indicate a growing awareness and acceptance of AI technologies among students and suggest significant trends in their expectations and readiness for future careers in an AI-driven environment.

4 Discussion

University students' attitudes towards artificial intelligence (AI) in education reveal a generally positive outlook, with some reservations. Many students are familiar with AI and its applications, reflecting a growing awareness and understanding of the technology. The data show that a significant proportion of students regularly use AI technologies in their studies, suggesting a trend towards the integration of AI in educational processes.

The experiences and skills related to AI that students have acquired during their studies are diverse. Most students have interacted with digital assistants and chatbots, indicating practical exposure to AI. However, there is a notable gap in more specialized AI applications, suggesting an opportunity for educational institutions to expand the scope of AI-related coursework and training.

Students' expectations of how AI will affect their future careers are mixed, but largely optimistic. Many students believe that AI will improve the quality and efficiency of their work, particularly by automating repetitive tasks. However, there are also significant concerns about the ethical implications, job security, and the potential for misuse of AI. These concerns highlight the need for comprehensive education on the ethical and practical aspects of AI, preparing students not only to use these technologies but also to responsibly address the challenges they present.

5 Conclusion

The study highlights that university students are generally positive about the integration of AI in education, recognizing its potential to enhance learning experiences and future career opportunities. However, their experience with AI is mostly limited to common applications such as digital assistants and chatbots, indicating a need for broader and deeper AI education.

Students expect AI to play a significant role in their future careers, improving efficiency and effectiveness. They are preparing for these changes by gaining hands-on experience with AI technologies and advocating for their inclusion in the curriculum. However, ethical concerns and the potential impact on job security are important considerations that need to be addressed. Our findings are consistent with Ivanov (2023) and Mackay (2023), who also highlight ethical concerns, but differ in the degree of optimism about the benefits of AI, as observed in our larger and more diverse sample. This suggests that while ethical and safety concerns are universal, the degree of optimism about the educational benefits of AI may depend on the local context and level of exposure.

References

1. Jian, M. J. K. O.: Personalized learning through AI. In: Advances in Engineering Innovation 5, pp. 16–19 (2023).
2. Zavalevskyi, Y., Kyrilenko, S., Kijan, O., Bessarab, N., Mosyakova, I.: The role of AI in individualizing learning and creating personalized programs. In: Amazonia Investiga 13 (73), pp. 200–208 (2024).
3. Pupe, J. W. O., Aguilar-Alonso, I.: Web System as Support to Automate Processes of the Administrative Area of the Pre-University Center. In: International Conference of Modern Trends in ICT Industry: Towards the Excellence in the ICT Industries (MTICTI), pp. 1–8. IEEE, Sana'a, Yemen (2021).
4. Joo, S., Kim, J., Kim, S.: Establishment of data mining-based public education administrative work automation system and student activity analysis. In: Journal of Autonomous Intelligence 6 (3), pp. 1–9 (2023).
5. Gunawan, A., Wijaya, M. I.: Robotic Process Automation to Enhance Education's Administration Process: Case of Attendance Checking and Reporting. In: 2023 International Conference on Information Management and Technology (ICIMTech), pp. 1–6. IEEE, Malang, Indonesia (2023).
6. Seya, K., Kobayashi, N., Shirasaka, S.: Effectiveness of Story-based Visual and Agile Teaching Method for Non-technical Adult Learners Who Want to Understand Artificial Intelligence. In: 2019 8th International Congress on Advanced Applied Informatics (IIAI-AAI), pp. 768–774. IEEE, Toyama, Japan (2019).
7. Guilherme, A.: AI and education: the importance of teacher and student relations. AI and Society 34 (1), pp. 47–54 (2019).
8. Ivanov, S.: The dark side of artificial intelligence in higher education. The Service Industries Journal 43 (11–12), pp. 1055–1082 (2023).
9. Mackay, S.: Factors Influencing Academic Integrity Violations. In: Proceedings of the 54th ACM Technical Symposium on Computer Science Education V. 2, pp. 1340. SIGCSE, Toronto, Canada (2023).
10. Oravec, J.: Artificial Intelligence Implications for Academic Cheating: Expanding the Dimensions of Responsible Human-AI Collaboration with ChatGPT and Bard. Journal of Interactive Learning Research 34 (2), pp. 213–237 (2023).
11. Kilianova, K., Kockova, P., Zavodna, M., Klubal, L., Kostolanyova, K.: Artificial intelligence through the eyes of university students for the second time. In: INTED2024 Proceedings, pp. 4705–4712. IATED, Valencia, Spain (2024).

AI in Multicultural Dialogue Between Teachers and Parents of Refugee and Immigrant Children in Slovenia

Helena Konšak, Metka Kordigel Aberšek, Boris Aberšek, and Maja Kerneža

1 Introduction

Every year, foreign children, refugees, migrants, immigrants, asylum seekers, etc., are enrolled in the Slovenian educational system. Analyses show that foreign children have difficulty following lessons and other activities in kindergartens and schools. Poorer integration into the broader social environment is due to inadequate knowledge of the Slovenian language, undeveloped strategies, and tools for integrating foreign children into the education system at all levels of schooling, while further exacerbating the insufficient involvement of children and their parents in the local and broader Slovenian environment [1].

A discussion on multiculturalism in education is essential [2]. The field of education in intercultural education plays a vital role. The concept of multicultural education as critical pedagogy, educational reform, and an ongoing process is established to provide equal opportunities for education and acceptance in society to all cultural groups. For educators, knowledge of the principles and characteristics of intercultural education, intercultural competencies, and multicultural dialogue is essential knowledge and skill in today's time.

Developing a multicultural teacher/educator as an essential process involves recognizing one's value judgments and theoretical foundations of understanding

H. Konšak
Primary school Bratov Polančičev, Maribor, Slovenia

M. K. Aberšek · M. Kerneža
Faculty of Education, University of Maribor, Maribor, Slovenia

B. Aberšek (✉)
Faculty of Natural Science and Mathematics, University of Maribor, Maribor, Slovenia
e-mail: boris.abersek@um.si

M. Turčáni (ed.), *15th International Scientific Conference on Distance Learning in Applied Informatics*, EAI/Springer Innovations in Communication and Computing,
https://doi.org/10.1007/978-3-031-81261-3_11

multicultural education, ultimately improving the quality of teaching, collaboration, and coexistence.

Statistical data show that the number of immigrants in Slovenia is increasing yearly, as nearly 78,000 foreign citizens immigrated to Slovenia between 2008 and 2012. In 2022, there were 36,316 (61% of them aged 25–64 years old). Among them, 31% had primary or lower education, 54% had secondary education, and 15% had higher education [3].

As a result of the immigration of foreign citizens to the Republic of Slovenia, there is also an increasing percentage of immigrant children in public schools and kindergartens. Regardless of the initial position of immigrant children, upon enrollment in the Slovenian educational system, their legal status is equalized with all other enrolled children. The problem that arises is how to enable effective integration for these children, ensure equal participation, and successfully overcome any difficulties that may arise due to the different language or cultural environment they come from, while maintaining the protected right to express their belonging to their nation or national community, to nurture and express their culture, and to use their language and script.

Collaboration with parents is one of the most promising paths that definitely leads to the effective integration of refugee children into a new social community. Family ties and the family, with its culture, customs, beliefs, and traditions, significantly influence a child's psychosocial development. Schools and teachers should care for the school's unity and the child's family and collaborate well with parents. Especially in today's multicultural Slovenia, collaborating well with parents to raise awareness, involve, consider, and provide them with information about their work is essential. Before engaging in conversations with parents of children of different nationalities, we must prepare well. Before the widespread use of AI translators, the valid recommendation was that communication should also involve translators or intermediaries [4]. This way, by actively involving parents, we achieve that they assist in achieving the goals of multicultural education [5]. The family environment is a child's first natural and supportive environment, where they acquire speech and language in everyday events and discover literacy. Of course, each parent responds differently to the teacher's invitation to collaborate, so the teacher needs to seek new ways and forms of collaboration that help in translation. When we raise awareness among parents about the importance of reading to their children, singing, and playing word games in their native language from a young age, we also show respect for their culture and national identity. By talking about this and giving it importance, we change the entire approach to books, reading, learning, communication, and play (which greatly encourages language development), while also conveying to them how important it is to support their child's integration into the school community and motivate them in the process of acquiring knowledge.

Only with parents can we empower children to develop reading literacy competency in the local language/Slovenian language more easily.

Immigrant children differ significantly from each other, not only in the language they speak at home but also in their emotional expression, group inclusion, culture, dress, etc. [4]. Language abilities and reading literacy competencies in children and their progress in achieving curriculum goals often depend not only on the child's language affiliation but also on the parents' education and awareness of the importance of schooling. "Most weak readers have less educated parents; they come from socially and economically disadvantaged families with no educational resources, including books. In most countries, immigrant status can significantly harm reading achievements, especially for students who speak a different language at home than the language of instruction at school. However, the influence of the domestic and family environment varies by country, indicating that in some educational systems, the strong connection between the environment and achievements can be reduced." [6].

Therefore, communication is the key to successfully integrating a refugee child. Both communication with the child and communication with their parents were the basis of the international project Intercultural Interaction and Understanding between Teachers and Parents (INTREF).

Being aware that intercultural collaboration with parents is essential for defining the curriculum for teachers' support of refugee children's integration and their school progress, and at the same time, knowing that refugee parents do not speak the Slovene language (many of them also do not speak English), we searched for a solution on how to overcome the language barrier. In such situations, traditional pedagogy would advise the collaboration of an interpreter, which is a good solution but not easy and costly to implement.

2 Research Focus

In the light of the INTREF project, which aims at the social inclusion of children of refugees and migrants through innovative learning tools and creative activities, the development of rhetorical strategies and pedagogical approaches for teachers in virtual learning environments was crucial [7]. We decided to implement AI and with it search for answers to the following research questions regarding interculturality, acceptance, and the individualization of refugee children in the Slovene school system:

RQ1: How do parents perceive teachers' knowledge of their child's culture, and how do they evaluate teachers' and domestic children's respect for cultural differences? Are there differences in opinions between parents and teachers?

RQ2: How do parents evaluate the acceptance of their child into the school community—of teachers and his classmates? Are there differences in opinions between parents and teachers?

RQ3: What is the opinion of parents regarding the question of individualization? Are there differences in opinions between parents and teachers on how much knowledge should be expected of the refugee child and how long the adaptation process should take? Are there differences in opinions between parents and teachers?

RQ4: Is AI a proper substitution for a human interpreter? Can it be successfully used for communication between teachers and parents in the educational process of refugee children?

3 Research Methodology

3.1 Sample of Research

The sample of this study was selected through a convenience sampling method, encompassing a total of 47 participants from a primary school situated in an urban setting. This sample comprised 30 primary-level teachers and 17 parents, representing a cross-section of the school's community.

3.2 Instrument and Procedures

To explore teachers' opinions on interculturality, acceptance, and individualization in the Slovene school, we formed three sets of questions. To get insight into the perception and evaluation of intercultural knowledge and respect, we formed four questions: "As their female teacher, the RF students respect me and always take me into account." "The parents of my RFs know that the knowledge acquired at school is important and they consistently support the lessons." "I am well aware of the customs of the cultural circle of RF students, and I strictly follow them in my lessons." "The school kitchen consistently offers meals in line with the RF child's cultural environment." The pilot implementation of this questionnaire showed that such questions cannot be used in communication with parents, so we paraphrased them into: "It is right that my child respects and follows the teacher strictly." "I know that the knowledge acquired at school is very important for my child and therefore I always encourage him to attend classes and study." "My child's teacher knows our customs very well and strictly follows them in his lessons." "The school kitchen consistently offers meals in accordance with our cultural environment" (see Table 1). Afterwards, the questionnaire for acceptance and individualization was formed, and in parallel, the same questions were paraphrased for parents (see Tables 2 and 3).

The next step was choosing the AI translation device. With machine translation tools, we considered the most commonly used and freely accessible applications. When using them to overcome language barriers in communication at school, children, parents, teachers, and other users must be familiar with them, so simple guidelines for their use would be valuable in the introduction process. The following translators made it to the shortlist:

Table 1 Comparative analysis of perceptions on interculturality among teachers and parents

Interculturality: teachers			Interculturality: parents		
	M	SD	M	SD	
As their female teacher, the RF students respect me and always take me into account.	4.28	0.830	*4.76*	*0.437*	*It is right that my child respects and follows the teacher strictly.*
The parents of my RFs know that the knowledge acquired at school is important, and they consistently support the lessons.	3.72	0.841	*4.88*	*0.332*	*I know that the knowledge acquired at school is very important for my child, and therefore I always encourage him to attend classes and study.*
I am well aware of the customs of the cultural circle of RF students, and I strictly follow them in my lessons.	3.17	0.658	*4.12*	*1.54*	*My child's teacher knows our customs very well and strictly follows them in his lessons.*
The school kitchen consistently offers meals in line with the RF child's cultural environment.	4.10	0.976	*0.18*	*0.883*	*When my child is in trouble, the D students usually help him successfully.*

Table 2 Comparative analysis of perceptions on acceptance among teachers and parents

Acceptance: teachers			Acceptance: parents		
	M	SD	M	SD	
My RFs are very well integrated into the class community.	3.52	0.911	*4.41*	*0.795*	*My child likes to go to school and feels good in his class.*
My D students always include the RF students in their play.	3.62	0.820	*4.29*	*0.849*	*At home, my child talks about D's students, including him in their play.*
In my class, the RFs have friends among the D students.	3.59	0.867	*4.29*	*0.849*	*My child also has friends in his class among students D.*
As a rule, my D students help when they notice that RCs are having trouble.	3.86	0.743	*4.12*	*0.993*	*When my child is in trouble, the D students usually help him successfully.*

1. Google Translate

Google Translate is one of the most popular and widely used translation tools, supporting over 100 languages. It offers text translation, speech translation, image translation, and real-time video translation, making it very useful for various educational purposes. It allows listening to the pronunciation of translations, which is especially useful for language learning, correct pronunciation, and direct active

Table 3 Comparative analysis of perceptions on individualization among teachers and parents

Individualization: teachers			Individualization: parents		
	M	SD	M	SD	
In the school system of our country, in order to receive a positive grade, an RF student must demonstrate knowledge at the level of the minimum standards as defined by the official curriculum.	3.97	0.981	*4.53*	*0.514*	*My child needs to know as much as his classmates D in school to get a positive grade.*
In the school system of our country, an RF student has a sufficiently long period at his disposal, during which the teacher monitors his progress with formative methods.	3.10	1.448	*4.12*	*0.928*	*The period when the teacher monitors my child's progress without grades is long enough.*
I think it would be necessary for RF students to lower the minimum standards in all school subjects.	2.24	1.272	*2.88*	*1.728*	*I don't think my child should be asked as much as local children.*
An RF student should be placed in a class appropriate to his/her abilities and not simply based on his/her chronological age.	4.14	1.098	*2.35*	*1.455*	*It would be better if they put my child in a class based on his prior knowledge, rather than just looking at how old he is.*

communication. Google Translate enables the translation of entire documents, such as Word and PDFs, while preserving basic text formatting. It can be integrated into applications and websites, providing easy access to translation services. With the Word Lens feature, users can point their device's camera at text in a foreign language, and the tool will translate and display the text over the original. Google Translate allows text translation without an Internet connection for specific languages, which is useful when Internet access is unavailable.

2. Microsoft Translator

This tool provides text and speech translation in over 70 languages. It allows real-time conversation translation, making it an excellent tool for meetings between parents and teachers where a language barrier exists. Microsoft Translator can be integrated into Microsoft Education products, increasing its utility in the school environment. It is designed to translate text, speech, conversations, and even entire websites in real time. The tool is accessible online and integrated into various Microsoft products, such as Office (Word, etc.), Edge, Skype, and others. Microsoft Translator enables multi-person real-time conversations where participants speaking different languages can interact, with the tool translating between them in real time. The application allows translation without an Internet connection for some languages, which is particularly useful in situations with limited Internet access.

3. DeepL Translator

DeepL Translator is an advanced machine translation tool that has quickly gained popularity due to its ability to provide accurate and meaningful translations. DeepL supports fewer languages, but the list of languages is expanding yearly. It is committed to high standards of privacy and data security. It allows the translation of entire documents (e.g., Word, PowerPoint, PDF) while preserving the original formatting.

4. Papago Translator

Papago is specialized in East Asian languages, including Korean, Japanese, and Chinese, and supports a total of 13 languages. It is known for its accuracy in these selected languages and offers text, voice, and image translation, facilitating communication between users of different languages. Papago also provides a detailed dictionary and phrase base that allows users to understand language nuances and learn the correct use of words and expressions. "Papago" originates from Esperanto and means "parrot," a bird known for its ability to mimic human speech. The "conversation" function allows two people speaking different languages to communicate through the device, with Papago translating their conversation in real time.

We chose Google Translate among these translators because it is currently the most popular, easily accessible (even offline!), and allows translation in both written and verbal communication channels.

From then on, we had to decide how to collect data. We could send questionnaires with statements to teachers via email, but it did not seem sensible to do so for data collection from parents. Together with parents—with the support of AI translators—teachers filled out the questionnaires. Thus, data collection proceeded like a semi-structured interview. The teacher asked the parent a question from the questionnaire in their language, which, in this case, served similarly as guiding questions in a semi-structured interview. Then, the communication between the parent and the teacher about that question followed. This communication continued until the parent decided whether they agreed with the statement and to what extent, which the teacher then marked on a five-point Likert scale.

3.3 Data Analysis

Teachers' and parents' responses were anonymized to maintain the confidentiality of all participants. Data were coded and entered into the database for analysis. The analysis was performed in IBM SPSS Statistics 29 and was guided by the research questions. The data analysis involved employing descriptive statistics to compare the mean and standard deviation across responses from two groups: teachers and parents. This comparison was three thematic areas: interculturality, acceptance, and individualization. The objective was to uncover the patterns, similarities, and differences in perceptions and experiences regarding the integration of refugee children into the educational system and, at the same time, to explore the usability of AI in the process of communication between the domestic teacher and a refugee child's parent.

4 Research Results

The research results reveal significant insights into the perceptions of Slovene teachers and parents of refugee children in Slovene schools regarding interculturality, acceptance, and the process of individualization. The results are presented in the following tables.

Statistically significant differences were observed in most categories in the domain of interculturality in Table 1. Regarding respect for authority and teacher leadership, it appears that parents ($M = 4.76$, $SD = 0.437$) reported a higher average value of respect for authority and leadership compared to the teachers themselves ($M = 4.24$, $SD = 0.830$), which could indicate that teachers, on one side, and parents, on the other, experience the teacher's role in the educational process differently. Especially, parents respect and value their child's education and teacher ($M = 4.88$, $SD = 0.332$), significantly higher than teachers themselves ($M = 3.72$, $SD = 0.841$), which especially emphasizes the importance that parents place on formal education for children, supported by promoting educational values at home. The results show a high rating for the cultural integration of their child in the classroom ($M = 4.12$, $SD = 1.054$), indicating that parents particularly value the effort and awareness of teachers and classmates regarding the cultural integration of their child.

The analysis of perceptions of acceptance reveals significant insights into how parents and teachers see their child's/their students' feelings within the school environment. Parents ($M = 4.41$, $SD = 0.795$) reported a higher sense of belonging and enjoyment at school compared to the perceived level of the teacher regarding their students ($M = 3.52$, $SD = 0.911$), indicating that students and their parents view the school environment more positively as a space of belonging and positive experiences. When participants evaluated the extent to which refugee students form friendships with domestic students, it was found that parents again reported a high average value ($M = 4.29$, $SD = 0.849$), while teachers ($M = 3.59$, $SD = 0.867$) perceived a lower level of friendships between refugee and domestic students. As for the support of refugee students by domestic students, the differences in perceived support were not statistically significant, as both groups reported relatively high averages, with the highest average rating from parents ($M = 4.12$, $SD = 0.993$).

The adjustment and recognition of progress, consideration of adjusted academic standards, and placement of students based on their abilities were monitored under the title Individualization. When stakeholders evaluated the adjustment of academic expectations and standards in comparison with domestic students, parents to the greatest extent ($M = 4.53$, $SD = 0.981$) agreed that their children should know as much as domestic students, while teachers mainly demand the achievement of minimal knowledge standards ($M = 3.97$, $SD = 0.981$). Teachers moderately agree that refugee children have enough time for their progress to be monitored within formative assessment ($M = 3.10$, $SD = 1.448$), while parents agree that the period during which the teachers monitor their child's progress without grades is long enough ($M = 4.12$, $SD = 0.928$). Statistically significant differences among participants indicate a need for better monitoring and recognition of individual student progress. Regarding the adjustment of academic standards, teachers disagree that it is necessary to adjust the knowledge standards for refugee students in all subjects ($M = 2.24$, $SD = 1.272$), and parents moderately agree that their child should know as much as local children ($M = 2.88$, $SD = 1.728$). On the question of student placement in class based on abilities, there are significant differences showing discrepancies in opinions regarding the appropriateness and effectiveness of such placement. Parents agree that refugee children should be placed in a class appropriate to their abilities and not based on their chronological age ($M = 4.14$, $SD = 1.098$), while teachers significantly disagree with this ($M = 2.35$, $SD = 1.455$).

5 Discussion and Conclusion

Besides traditional rhetorical skills, the development of digital literacy and the ability to effectively use technological tools have become key competencies for teachers. This not only improves accessibility and understanding between teachers and students from different cultural environments but also allows for a more personalized and adaptable approach to learning. As Chipalo [8] points out, parental involvement in children's education is vital for a child's learning outcome and for his socialization in the new social environment. Also, Kovacevic, Klimek, and Drower [9] agree that parental involvement in school education is beneficial for the academic success and emotional well-being of refugee children. Therefore, they developed a so-called "Field Model" for social justice and parent partnerships in multicultural education for encouraging the collaboration between parents and teachers. The model also has the same aim as the *Inclusive parent engagement practices*, which includes lessons learned from a school community Collaborative Supporting Newcomer Refugee Families by Georgis, Gokiert, Ford, and Mulki [10]. The findings of all this, and several other research, identify a language barrier as the main obstacle for communication between parents and teachers. For the solution of this problem, the presence of the translator was suggested. But at the same time, it was pointed out that such a solution would, in many communities, many schools, not be available. Mainly because of two reasons: persons who would speak both languages, the language of school education, teachers' language, and students' and parents' mother tongue, are not at the disposal. And on the other hand, hiring professional translators would be too expensive for regular help.

At the beginning of the third decade of the twenty-first century, the circumstances had changed: firstly, many places on our planet had become too dangerous to live in, because of climate changes and wars. Consequently, more and more people search for a better, safer place to live, among them many families. And one of the first problems they want to solve in the place of their new settlement is, where can my child/where can my children go to school? In such circumstances, understanding rhetoric has become even more important for teachers than it was in previous times. Enabling teachers to effectively communicate with various groups, including the parents of refugee children, is thereby contributing to greater intercultural understanding and collaboration [7]. The second essential change is commonly available AI, in the context of communication with a refugee child's parents, their capacity to translate immediately with practically no expenses. This means the essential problem, exposed in previous decades, the problem with the absence of translators' help, is now, with the help of AI, solved.

In our project, the AI made it possible to get insight into the problem areas of a refugee child's integration. AI made possible to collaborate with parents, who did not speak the teachers' language and could not use the English language. Such collaboration was especially vital in teaching refugee children, where teachers and parents operate in an environment rich with diverse cultures and languages. Using AI made in the first phase of the INTREF project (*Intercultural interaction and understanding between teachers and parents*) possible and successful.

The AI made it possible to compare the responses of two groups: teachers and parents in three thematic areas: interculturality, acceptance, and individualization. AI made it possible to uncover the patterns, similarities, and differences in perceptions and experiences regarding the integration of refugee children into the educational system. On the basis of information we have got with the help of AI in the communication process with parents, we could define the content of educational modules for teachers. To create the circumstances for a more successful integration of refugee children in the society of their new home and to integrate parents into the school environment, which would, with no doubt, lead to a greater academic success of their children.

References

1. Novak, M.: Smernice za izobraževanje otrok tujcev v vrtcih in šolah. Zavod Republike Slovenije za šolstvo. Ljubljana (2009).
2. Ule, M.: Predsodki in diskriminacija. Ljubljana: Znanstveno in publicistično središče. Zbirka ALFA. Ljubljana (1999).
3. Statistični urad, Republika Slovenija. https://www.stat.si/StatWeb/Field/Index/17/98. Last accessed 2024/04/30.
4. Kristan, A.: Izzivi multikulturnega izobraževanja (Challenges of Multicultural Education). Vzgoja in izobraževanje 1/2016, (47), Ljubljana, (2016).

5. Smolnikar, B.: Multikulturalizem v teoriji in praksi vrtca. Diplomsko delo. Univerza v Mariboru: Pedagoška fakulteta, Oddelek za predšolsko vzgojo. Maribor (2010).
6. Lakota, B. A.: Poučevanje branja v Evropi: okoliščine, politike in prakse. Ministrstvo za šolstvo in šport. Ljubljana (2011).
7. Kerneža, M., Zemljak, D., Kordigel Aberšek, M., Aberšek, B., Legvart, P., Konšak, H., Sari, H., Laurusone, I., Hanuliakova, I., Keskin, Y. M., Yiğit, E., Faruk Ayin, F. and Kılıç, M.: Education of refugee children in Slovenia: In-depth analysis within the INTREF project framework, PCE, Maribor, (2024).
8. Chipalo, E.: Parental Involvement in Children's Education in the United States: Evidence from Annual Survey of Refugees. (2024). https://doi.org/10.1080/08878730.2024.2328021
9. Kovacevic, Z., Klimek, B., Drower, I.: Refugee Children and Parental Involvement in School Education: A Field Model. In: Social Justice and Parent Partnerships in Multicultural Education Contexts. (2018). https://doi.org/10.4018/978-1-5225-3943-8.ch008
10. Georgis, R., Gokiert, R. J., Ford, D. M. & Mulki, A: Creating Inclusive Parent Engagement Practices. Lessons Learned from a School Community Collaborative Supporting Newcomer Refugee Families. Multicultural Education, 21, (3–4), 23-27 (2014).

Exploring the Use of GeoGebra Software in Teaching Stereometry at Primary Schools: An Action Research Study

Lilla Korenova, Angelika Schmid, and Antonin Demel

1 Introduction

The integration of digital technologies in education has become increasingly important in recent years, particularly in the field of mathematics. The use of software like GeoGebra offers significant potential to enhance teaching methods and improve student engagement and understanding. This potential is especially relevant in the context of primary education, where innovative teaching approaches can have a significant impact on students' long-term mathematical proficiency.

GeoGebra, a dynamic mathematics software that combines geometry, algebra, and calculus, provides a versatile tool for teaching complex mathematical concepts in an interactive and visually engaging manner. Its application in the classroom allows for the visualization and manipulation of mathematical objects, helping students develop a deeper understanding of the subject matter. Studies have shown that digital tools like GeoGebra can significantly enhance students' motivation and engagement [1, 2].

Constructivist teaching methods, such as the Hejny method, have been widely adopted in many schools across the Czech and Slovak Republics over the past two decades. These methods emphasize active learning and the construction of knowledge through experience and reflection, aligning well with the capabilities offered by digital technologies. The Hejny method focuses on developing students' mathematical thinking and problem-solving skills through a series of carefully designed tasks and environments that encourage exploration and discovery [3].

L. Korenova (✉)
Faculty of Education, Comenius University Bratislava, Bratislava, Slovakia
e-mail: korenova@fedu.uniba.sk

A. Schmid · A. Demel
Faculty of Education, University of Ostrava, Ostrava, Czech Republic

M. Turčáni (ed.), *15th International Scientific Conference on Distance Learning in Applied Informatics*, EAI/Springer Innovations in Communication and Computing,
https://doi.org/10.1007/978-3-031-81261-3_12

Today's primary school students, belonging to Generation Alpha, are digital natives who have grown up with technology as an integral part of their lives. This generational familiarity with digital environments makes the integration of technologies like GeoGebra into the classroom both a natural fit and a powerful tool for enhancing learning outcomes. Research indicates that Generation Alpha students benefit greatly from interactive and technology-enhanced learning experiences, which cater to their unique learning styles and preferences [4, 5].

Moreover, the use of Augmented Reality (AR) in educational settings has shown to be a highly motivating factor for students. AR can create immersive learning experiences that capture students' interest and make abstract concepts more tangible [6]. In mathematics education, AR can help visualize complex geometrical shapes and their properties, making learning more interactive and engaging [7].

This article presents the results of an action research study examining the implementation of an online GeoGebra Book in primary school mathematics teaching. The GeoGebra Book, developed using AR GeoGebra, was created as part of a diploma thesis by Antonín Demel [8]. The study focuses on the practical application of the GeoGebra Book in the classroom, utilizing qualitative research methods, including semi-structured interviews and focus groups, to address the research problem and questions.

The primary aim of this research is to explore how the combination of GeoGebra software and the Hejny method can enhance students' mathematical literacy and motivation. The study evaluates the implementation process, the challenges encountered, and the outcomes observed in terms of student engagement and learning. The findings offer valuable insights and recommendations for educators looking to integrate digital tools effectively into their teaching practices.

2 Issue of the Research

The integration of digital technologies in mathematics education has transformed teaching methods, improving student engagement and understanding. Tools like GeoGebra offer interactive and visual learning experiences, making abstract concepts more accessible. Hohenwarter and Preiner [1] note that GeoGebra enhances students' motivation by allowing them to interact with mathematical concepts, fostering a deeper understanding.

Digital technologies facilitate personalized learning, enabling students to learn at their own pace and according to their individual needs. This adaptability is crucial for addressing diverse learning styles and promoting deeper understanding [9]. Furthermore, digital tools support collaborative learning, allowing students to work together on mathematical problems and share their insights [10]. Technology in the classroom can create a more engaging and interactive learning environment that encourages collaboration and critical thinking [11, 12]. These collaborative experiences are essential for developing problem-solving skills and building mathematical confidence [13, 14]. GeoGebra, known for its versatility across educational levels, supports dynamic exploration of mathematical concepts, making learning more engaging [15].

Augmented Reality (AR) within GeoGebra further enriches education by overlaying digital content onto the real world, making abstract concepts tangible. AR has been shown to boost student motivation by providing immersive learning experiences [6]. In mathematics, AR aids in understanding spatial relationships and geometric properties through interactive 3D models [16].

Constructivist teaching methods, such as the Hejny method, emphasize active learning, where students build their understanding through experience and reflection. This approach contrasts with traditional methods that often focus on rote memorization and passive learning. Constructivism encourages students to explore, ask questions, and solve problems, fostering a deeper and more meaningful understanding of mathematical concepts. As Fosnot and Perry [17] note, constructivist teaching methods align well with the nature of mathematical inquiry, which involves exploration, hypothesis testing, and logical reasoning. Steffe and Gale [18] emphasize that such approaches promote autonomy and confidence, allowing students to take ownership of their learning.

The Hejny method, widely used in the Czech and Slovak Republics, develops mathematical thinking through tasks that encourage exploration. It has been successful in engaging students and improving their proficiency by leveraging their curiosity and promoting collaborative learning environments [3].

Generation Alpha, born into a digital world, benefits greatly from technology-enhanced learning experiences. Digital literacy, a crucial skill for this generation, is effectively developed through tools like GeoGebra [19]. Digital literacy involves both the use of tools and the critical thinking needed to evaluate digital information [20].

The benefits of integrating digital technologies into mathematics education, particularly through tools like GeoGebra and AR, have been highlighted. Constructivist methods, exemplified by the Hejny method, combine well with these technologies to offer interactive and engaging learning experiences that foster deep understanding and critical thinking. This review sets the stage for exploring how the combination of GeoGebra and the Hejny method can be implemented effectively in primary school mathematics education.

3 Methodology

3.1 Action Research Design

The methodology for this study is based on an action research design, which is particularly suitable for educational settings where the goal is to implement and evaluate new teaching methods or tools. Action research is a participatory and iterative process that involves identifying a problem, planning an intervention, implementing the intervention, and then reflecting on and evaluating the results to inform future practice [21]. This cyclical process allows for continuous improvement and adaptation based on feedback and findings from each cycle.

In this study, the action research design was used to investigate the implementation of the GeoGebra Book in primary school mathematics teaching. The primary aim was to explore how the combination of GeoGebra software and the Hejny method can enhance students' mathematical literacy and motivation. The iterative nature of action research enabled the researchers to make adjustments to the implementation process based on the observations and feedback collected during the study.

The action research was conducted in three main phases: planning, action, and reflection:

1. **Planning Phase:** This phase involved developing and customizing the GeoGebra Book to align with the curriculum and meet the specific needs of the students. The planning phase also included setting research objectives, selecting the sample, and designing data collection instruments.
2. **Action Phase:** In this phase, the GeoGebra Book was introduced into the classroom setting. Teachers were trained on how to effectively integrate the software into their lessons, and students began using the GeoGebra Book as part of their regular mathematics instruction. Observations were made to monitor how students interacted with the software and engaged with the mathematical content.
3. **Reflection Phase:** After implementation, data collected through observations, interviews, and focus groups were analyzed. The reflection phase focused on assessing the effectiveness of the intervention, understanding the challenges faced, and identifying areas for improvement. Feedback from participants was used to refine the GeoGebra Book and the implementation strategy for future cycles.

3.2 Data Collection Methods

To gather comprehensive data on the implementation and impact of the GeoGebra Book, qualitative research methods were employed, including semi-structured interviews and focus groups. These methods were well-suited for exploring participants' experiences, perceptions, and attitudes in depth [21].

3.3 Semi-structured Interviews

Semi-structured interviews were conducted with teachers and students to collect detailed information about their experiences using the GeoGebra Book. This method allowed for flexibility in questioning, enabling the interviewer to probe deeper into specific areas of interest based on the responses given. The interviews were guided by a set of predetermined questions but also allowed for spontaneous follow-up questions to explore unexpected topics that arose during the conversations.

The interviews focused on several key areas, including:

- Teachers' perceptions of the effectiveness of the GeoGebra Book in enhancing students' mathematical understanding and engagement.
- Students' experiences using the GeoGebra Book and how it influenced their motivation and learning.
- Challenges faced during the implementation of the GeoGebra Book and potential solutions.

3.4 *Focus Groups*

Focus groups were conducted with groups of students to gain insights into their collective experiences and perspectives. This method is particularly useful for exploring how students interact with each other and the learning material in a group setting. Focus groups allow participants to build on each other's responses, providing a richer and more nuanced understanding of their experiences [22].

The focus group discussions covered topics such as:

- Students' overall impressions of the GeoGebra Book and its usability.
- The impact of AR features on their engagement and understanding of mathematical concepts.
- Suggestions for improving the GeoGebra Book to better meet their learning needs.

3.5 *Participants and Sampling*

The participants in this study were drawn from primary schools in the Czech and Slovak Republics, where the Hejny method is widely implemented. The sample included both teachers and students who were involved in the mathematics classes where the GeoGebra Book was introduced. A purposive sampling strategy was used to select participants who could provide relevant and insightful information about the implementation and impact of the GeoGebra Book [23].

Teachers Teachers who participated in the study were those who had experience with the Hejny method and were willing to integrate the GeoGebra Book into their teaching. These teachers were selected based on their willingness to participate and their ability to provide valuable feedback on the effectiveness of the GeoGebra Book. Specifically, the study involved:

- Kateřina B., who, at the time of the research, was teaching at Masaryk Primary School and Kindergarten Bohumín (Masarykova základní škola a Mateřská škola Bohumín), Seifertova 601. She had just started her teaching career there.
- Lukáš Ž., a teacher at František Cardinal Tomášek Primary School Studénka (Základní škola Františka kardinála Tomáška Studénka), who has been teaching for 11 years and serves as the school prevention methodologist and deputy headmaster. He has been using GeoGebra since his university studies, albeit starting about 11 years earlier than Kateřina B.

Students The students who participated in the study were those enrolled in the mathematics classes where the GeoGebra Book was used. The sample included a diverse group of students to capture a wide range of experiences and perspectives. Students were selected to ensure a balanced representation of different skill levels, ensuring that the data collected reflected the experiences of both high-achieving and struggling students. The specific details of the student sample were as follows:

- Students from various grades, including sixth, seventh, eighth, and ninth grades, were involved.
- For instance, on January 10, 2024, action research was conducted during two teaching hours in the sixth grade at František Cardinal Tomášek Primary School Studénka (Základní škola Františka kardinála Tomáška Studénka). This was followed by further lessons on January 11, 2024, using borrowed iPads from the University of Ostrava.
- The seventh and eighth graders received lessons as part of their school curriculum on virtual models and their creation, while the ninth graders engaged in lessons under the broader topic of Digital Technologies.

This diverse and comprehensive sampling approach ensured that the study gathered rich, relevant data on the use of the GeoGebra Book in primary school mathematics education.

3.6 Iterations of the Action Research Cycle

Each cycle of the action research involved iterative processes designed to refine the intervention. The cycles were as follows:

1. Cycle 1
   - **Planning:** Initial development of the GeoGebra Book and preparation for classroom implementation. This included designing the GeoGebra Book based on the Hejny method and ensuring it aligned with the curriculum requirements for primary school mathematics.
   - **Action:** Introduction of the GeoGebra Book in a pilot classroom setting. The GeoGebra Book was implemented in one classroom where teachers integrated

it into their lessons, and students started using it as part of their mathematics instruction.
- **Observation:** Monitoring student engagement and collecting preliminary feedback. Observations focused on how students interacted with the GeoGebra Book and how it influenced their understanding and motivation.
- **Reflection:** Analyzing initial results and making necessary adjustments. Based on the feedback and observations, the GeoGebra Book was refined to address any issues and improve its effectiveness.

2. Cycle 2

   - **Planning:** Refinement of the GeoGebra Book based on feedback from Cycle 1. Adjustments were made to the content and structure of the book to better meet the needs of the students and align with the feedback received.
   - **Action:** Expanded implementation across multiple classrooms. The revised GeoGebra Book was introduced to additional classrooms, increasing the sample size and diversity of participants.
   - **Observation:** Collecting more comprehensive data through observations and interviews. More detailed data were gathered from a broader range of students and teachers to understand the impact of the GeoGebra Book more thoroughly.
   - **Reflection:** Further analysis and additional modifications to improve the intervention. The data were analyzed to identify patterns and trends, leading to further refinements of the GeoGebra Book.

3. Cycle 3

   - **Planning:** Final adjustments to the GeoGebra Book and strategies for implementation. The last set of refinements was made to ensure the book was as effective and user-friendly as possible.
   - **Action:** Full-scale deployment in participating schools. The final version of the GeoGebra Book was rolled out in all participating schools, with comprehensive support for teachers and students.
   - **Observation:** Systematic data collection through various methods. A combination of surveys, interviews, and classroom observations was used to gather extensive data on the implementation and impact of the GeoGebra Book.
   - **Reflection:** Comprehensive evaluation of the intervention's impact and preparation of final recommendations. The data were thoroughly analyzed to evaluate the success of the GeoGebra Book in enhancing mathematical literacy and motivation. The findings were used to prepare detailed recommendations for future use and further development of the GeoGebra Book.

These iterative cycles ensured that the GeoGebra Book was continuously improved based on real-world feedback and observations, making it a more effective tool for enhancing primary school mathematics education.

3.7 Data Analysis

The data analysis process involved several steps:

1. **Quantitative Analysis:** Survey data were analyzed using statistical software to identify patterns, correlations, and significant differences among participant responses. Descriptive statistics and inferential tests were employed as appropriate.
2. **Qualitative Analysis:** Interview transcripts and observational notes were analyzed using thematic analysis. Key themes and categories were identified through iterative coding and comparison. NVivo software was utilized to manage and analyze the qualitative data.
3. **Mixed-Methods Integration:** The findings from the quantitative and qualitative analyses were integrated to provide a comprehensive understanding of the research questions. Triangulation was used to cross-verify the results and enhance the validity of the conclusions [24].

The methodology employed in this study provides a robust framework for exploring the implementation and impact of the GeoGebra Book in primary school mathematics education. By using an action research design, incorporating AR GeoGebra, and employing qualitative data collection methods, the study aims to provide comprehensive insights into how digital tools can enhance mathematical learning and engagement. The findings from this study will contribute to the ongoing development and refinement of educational practices that integrate technology and constructivist teaching methods.

4 Implementation of the GeoGebra Book

4.1 Integration of GeoGebra with the Hejny Method

The integration of GeoGebra with the Hejny method represents a synthesis of advanced digital tools and established constructivist teaching practices. The Hejny method, known for its focus on students' active learning and discovery, complements the interactive and visual capabilities of GeoGebra. This combination aims to enhance students' mathematical understanding by providing a dynamic learning environment where abstract concepts are made tangible through interactive visualizations.

In practice, GeoGebra was used to create an online book containing a series of interactive applets that align with the Hejny method's principles. These applets allowed students to manipulate geometric shapes, explore their properties, and engage in problem-solving activities. The GeoGebra Book was designed to support the curriculum by incorporating tasks that encourage exploration, pattern recognition, and logical reasoning, all of which are central to the Hejny method. The integration process involved aligning GeoGebra activities with the existing lesson plans and ensuring that the digital content reinforced the constructivist learning objectives.

4.2 Pilot Verification Process

The pilot verification process was an essential phase of the action research, aimed at testing the practicality and effectiveness of the GeoGebra Book in a real classroom setting. This phase involved a small-scale implementation of the GeoGebra Book to gather initial feedback and make necessary adjustments before broader deployment. The pilot was conducted over a 6-week period in two primary schools that had already implemented the Hejny method in their mathematics curriculum.

During the pilot, students used the GeoGebra Book during their regular mathematics lessons. Teachers facilitated the sessions, guiding students through the interactive tasks and providing support as needed. Observations were made on how students interacted with the applets, the challenges they faced, and the overall impact on their engagement and understanding. Semi-structured interviews with teachers and focus groups with students were conducted to collect qualitative data on their experiences.

Implementation and Feedback

Teacher Involvement: Teachers played a crucial role in the implementation of the GeoGebra Book. For instance, Kateřina B. and Lukáš Ž. provided valuable insights into how the GeoGebra Book was received in the classroom. According to Kateřina B., "The GeoGebra Book offered a new way to visualize mathematical concepts, which significantly helped students who struggled with traditional methods." Lukáš Ž. added, "Integrating the GeoGebra Book with the Hejny method created a dynamic learning environment, enhancing both engagement and comprehension among students."

Student Interaction: Students were generally enthusiastic about using the GeoGebra Book. Observations indicated that students were highly engaged and found the interactive elements particularly beneficial. One student noted, "I liked how I could manipulate the shapes and see the changes immediately. It made learning more fun and easier to understand." However, some challenges were also noted, such as initial difficulties in navigating the software and understanding certain applets without additional guidance.

Challenges and Adjustments: Several challenges were identified during the pilot phase. Teachers observed that while students were excited to use the technology, there were initial hurdles in understanding how to interact with the applets effectively. To address these issues, the following adjustments were made:

- **Applet Modifications:** Based on student feedback, applets were made more intuitive. Simplified navigation and clearer instructions were added to help students understand the tasks better.

- **Instructional Support:** Additional instructions were provided within the GeoGebra Book to guide students through complex tasks. This included step-by-step tutorials and examples.
- **Task Difficulty:** The difficulty level of tasks was adjusted to better match the students' skill levels. More challenging problems were introduced gradually to ensure that all students could progress at their own pace.

Qualitative Data Collection

Semi-structured Interviews: Semi-structured interviews were conducted with both teachers and students. These interviews aimed to gather detailed feedback on the usability and impact of the GeoGebra Book. For example, one teacher mentioned, "The flexibility of the GeoGebra Book allowed me to adapt the lessons to suit different learning paces, which was very beneficial." Another teacher highlighted the motivational aspect, stating, "Students were more motivated to participate in the lessons when using the GeoGebra Book compared to traditional textbooks."

Focus Groups: Focus groups with students provided additional insights into their collective experiences. Students expressed appreciation for the interactive nature of the GeoGebra Book, with one student commenting, "Working with the GeoGebra Book in groups helped us learn from each other and understand the concepts better." However, some students also suggested improvements, such as, "It would be great if there were more games and puzzles related to the lessons."

Impact and Refinement

The feedback collected during the pilot phase was instrumental in refining the GeoGebra Book. The iterative process of action research allowed for continuous improvement based on real-world classroom experiences. Adjustments made to the GeoGebra Book ensured it became a more effective tool for enhancing mathematical literacy and engagement among primary school students.

The pilot verification process demonstrated the potential of the GeoGebra Book to transform mathematics education by integrating digital tools with innovative teaching methods. The success of this phase paved the way for broader implementation and further research into the long-term impact of such educational technologies.

4.3 Classroom Setup and Technology Integration

Effective implementation of the GeoGebra Book required careful planning of the classroom setup and integration of technology. Each classroom was equipped with the necessary digital devices, including tablets and computers, to ensure that all students could access the GeoGebra Book simultaneously. Ensuring reliable internet connectivity was also a priority, as the GeoGebra Book relied on online resources and real-time updates.

Teachers received training on using GeoGebra and integrating it into their teaching practices. This training covered the basic functionalities of the software, how

Fig. 1 Action research—student working with the model volume of a Pyramid in AR. (Source: Own work)

to navigate the GeoGebra Book, and strategies for facilitating student interactions with the applets. The goal was to ensure that teachers felt confident and capable of incorporating the digital tool into their lessons effectively.

In the classroom, technology integration involved using interactive whiteboards to display GeoGebra applets for whole-class discussions and demonstrations. Students worked on individual devices for hands-on activities, which allowed them to explore mathematical concepts at their own pace. This setup fostered a collaborative learning environment where students could easily share their findings and engage in group discussions (Fig. 1).

4.4 Challenges and Solutions

The implementation of the GeoGebra Book faced several challenges that required adaptive solutions to ensure success. One of the primary challenges was the varying levels of digital literacy among students and teachers. While Generation Alpha students are generally familiar with digital devices, some still required additional support to navigate the GeoGebra software effectively. To address this, introductory sessions were conducted to familiarize students with the basic functions of GeoGebra and the structure of the GeoGebra Book (Fig. 2).

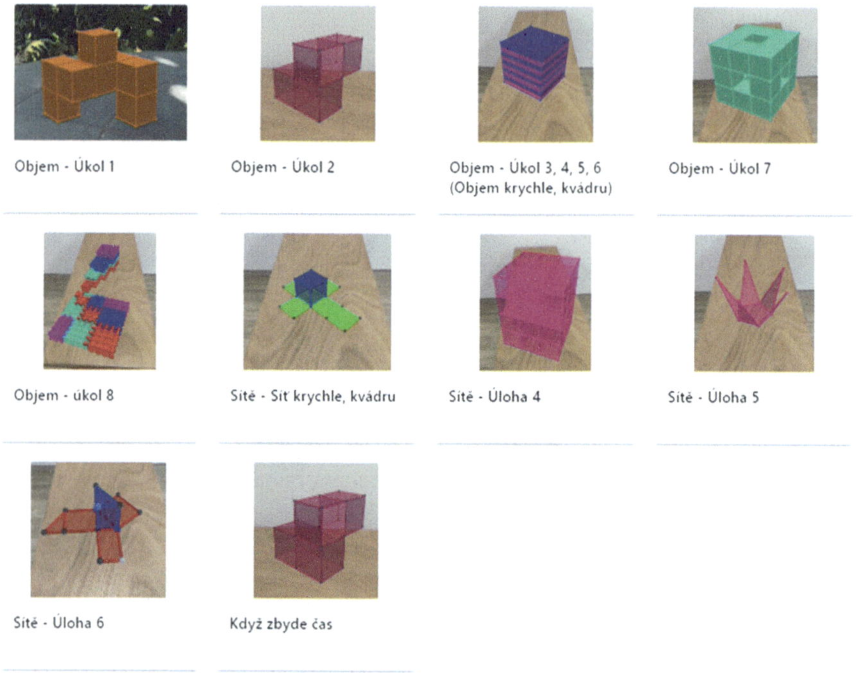

Fig. 2 Sample of applets from the GeoGebra Book. (Source: https://www.geogebra.org/m/rwcc6hvt)

Teachers also faced challenges in integrating the new tool into their established teaching routines. Some expressed concerns about balancing the use of digital tools with traditional teaching methods. Ongoing professional development and peer support were provided to help teachers feel more comfortable and confident in using GeoGebra. Additionally, the iterative nature of action research allowed for continuous feedback and adjustment, ensuring that the implementation process was responsive to the needs of both teachers and students.

Technical issues, such as connectivity problems and software glitches, occasionally disrupted the learning process. To mitigate these issues, technical support was made readily available, and backup plans were developed to ensure continuity of lessons even when technical difficulties arose. Ensuring that all devices were updated and compatible with the GeoGebra software also helped minimize disruptions.

The implementation of the GeoGebra Book within the framework of the Hejny method has shown promising results in enhancing student engagement and understanding of mathematical concepts. By addressing the challenges encountered during the pilot phase and making necessary adjustments, the integration of digital tools and constructivist teaching methods has the potential to significantly improve the quality of mathematics education in primary schools.

5 Results and Discussion

5.1 Analysis of Qualitative Data

The qualitative data collected through semi-structured interviews and focus groups were analyzed using thematic analysis. This approach enabled the identification of key themes and patterns that emerged from the participants' experiences.

1. **Enhanced Engagement and Motivation**: A prominent theme was the significant increase in student engagement and motivation. Both teachers and students noted that the interactive and visual nature of the GeoGebra Book made learning more enjoyable and stimulating. Students reported finding the tasks more interesting and being more willing to participate in classroom activities.
2. **Improved Understanding of Mathematical Concepts**: The use of interactive applets allowed students to visualize and manipulate geometric shapes, which contributed to a deeper understanding of mathematical concepts. Teachers observed that students were better able to grasp complex ideas and were more confident in applying their knowledge to solve problems.
3. **Challenges and Solutions**: Several challenges were identified during the implementation process, including technical issues and varying levels of digital literacy among students. Teachers also faced difficulties in balancing the use of digital tools with traditional teaching methods. To address these challenges, ongoing professional development and technical support were provided, and the GeoGebra Book was continuously refined based on feedback.

5.2 Impact on Students' Mathematical Literacy

The integration of the GeoGebra Book had a notable impact on students' mathematical literacy. Mathematical literacy, defined as the ability to apply mathematical knowledge and skills in various contexts, was enhanced through the use of interactive and exploratory learning activities provided by GeoGebra.

Students demonstrated improved problem-solving skills and a deeper understanding of geometric concepts. The ability to manipulate and visualize shapes in three dimensions helped students to develop spatial reasoning and to comprehend the properties of geometric figures more effectively. Teachers reported that students were more adept at identifying patterns, making conjectures, and testing their hypotheses, all of which are critical components of mathematical literacy.

Moreover, the constructivist approach of the Hejny method, supported by the GeoGebra Book, encouraged students to take an active role in their learning. This approach fostered a sense of curiosity and a willingness to explore mathematical ideas independently, further enhancing their mathematical literacy.

5.3 Student Engagement and Motivation

One of the most significant outcomes of the implementation was the increase in student engagement and motivation. The interactive nature of the GeoGebra Book captivated students' interest and made learning mathematics more appealing. Students were enthusiastic about using the applets and often expressed a desire to explore the tasks further outside of classroom hours.

Teachers observed that students were more attentive during lessons and participated more actively in discussions. The visual and hands-on aspects of the GeoGebra Book helped to maintain students' focus and facilitated a more dynamic and interactive classroom environment. This heightened engagement was particularly evident among students who had previously struggled with traditional teaching methods.

The use of AR features in the GeoGebra Book also played a crucial role in enhancing motivation. Students found the augmented reality elements to be exciting and engaging, which made learning more immersive. The ability to interact with 3D models and see the immediate impact of their manipulations helped to sustain their interest and enthusiasm for learning.

5.4 Interaction Between Students, Teachers, and Digital Environment

The implementation of the GeoGebra Book fostered a collaborative and interactive learning environment. The digital tool facilitated various forms of interaction between students, teachers, and the learning content.

Student–Student Interaction: The use of the GeoGebra Book encouraged students to work together on tasks and share their findings. Collaborative activities helped students to develop communication and teamwork skills, as they discussed their approaches and solutions with their peers. This peer-to-peer interaction was instrumental in creating a supportive learning community where students could learn from each other.

Teacher–Student Interaction: The GeoGebra Book enabled teachers to engage with students more effectively during lessons. Teachers used the interactive applets to demonstrate concepts and guide students through exploratory tasks. The immediate feedback provided by the software allowed teachers to address misconceptions and provide targeted support. This interactive approach helped to create a more dynamic and responsive teaching environment.

Student–Content Interaction: The most significant interaction facilitated by the GeoGebra Book was between students and the mathematical content. The ability to manipulate geometric shapes and visualize their properties in real-time helped students to develop a deeper understanding of the material. The interactive nature of the tasks encouraged students to explore and experiment, fostering a more active and engaged learning process.

In summary, the implementation of the GeoGebra Book within the framework of the Hejny method significantly enhanced student engagement, motivation, and mathematical literacy. The interactive and visual nature of the digital tool, combined with the constructivist teaching approach, created a dynamic and effective learning environment that supported students' development in mathematics. The insights gained from this qualitative research provide valuable guidance for future implementations and highlight the potential of digital tools in transforming mathematics education.

6 Recommendations

6.1 Best Practices for Using GeoGebra in Primary Schools

The successful implementation of GeoGebra in primary schools requires thoughtful planning and a strategic approach. Based on the findings from our study, we recommend the following best practices to maximize the benefits of using GeoGebra in mathematics education:

Align GeoGebra Activities with Curriculum Goals: Ensure that GeoGebra tasks are closely aligned with curriculum objectives and learning outcomes. This alignment helps to reinforce the concepts being taught and ensures that the digital tools are used effectively to enhance student learning.

Start with Simple Applets: Introduce students to GeoGebra using simple, intuitive applets that gradually increase in complexity. This approach helps students build confidence with the software and ensures they are not overwhelmed by too much information at once.

Integrate Interactive Demonstrations: Use GeoGebra for interactive demonstrations during lessons. Teachers can use the software to visually explain complex concepts, making abstract ideas more accessible to students. Interactive demonstrations also engage students and stimulate their interest in the subject matter.

Encourage Exploration and Experimentation: Allow students time to explore and experiment with GeoGebra on their own. Providing opportunities for independent exploration helps students develop problem-solving skills and fosters a deeper understanding of mathematical concepts.

Facilitate Collaborative Learning: Use GeoGebra to promote collaborative learning activities. Group tasks and projects that involve GeoGebra can encourage peer interaction and cooperative learning, which are essential components of the constructivist approach.

6.2 Strategies for Integrating Digital Tools with Constructivist Methods

Integrating digital tools like GeoGebra with constructivist teaching methods can enhance the learning experience by providing students with interactive and engaging ways to explore mathematical concepts. The following strategies can help achieve this integration effectively:

Create Constructivist Learning Environments: Design learning environments that support exploration, inquiry, and discovery. Use GeoGebra to create tasks that encourage students to investigate, hypothesize, and test their ideas within a supportive and interactive setting.

Use Real-World Problems: Develop GeoGebra activities that are based on real-world problems. This approach helps students see the relevance of mathematics in everyday life and enhances their motivation to learn.

Incorporate Scaffolding: Provide scaffolding to support students as they engage with GeoGebra tasks. Scaffolding can include hints, step-by-step guides, and differentiated instructions to cater to the varying needs and skill levels of students.

Foster Reflective Thinking: Encourage students to reflect on their learning process and the outcomes of their explorations with GeoGebra. Reflection activities, such as journals or group discussions, can help students consolidate their understanding and articulate their thought processes.

Integrate Technology Seamlessly: Ensure that the use of digital tools is seamless and integrated into the overall teaching strategy. Avoid using technology for its own sake; instead, focus on how it can enhance and complement the constructivist learning experience.

6.3 Teacher Training and Support

Effective implementation of GeoGebra and other digital tools requires adequate teacher training and ongoing support. Based on our findings, we recommend the following approaches to teacher training and support:

Professional Development Workshops: Offer comprehensive professional development workshops that cover the basics of using GeoGebra, its features, and best practices for integrating it with constructivist methods. These workshops should provide hands-on experience and practical tips for teachers.

Ongoing Support and Mentoring: Provide ongoing support through mentoring programs, where experienced GeoGebra users can offer guidance and assistance to their peers. Regular follow-up sessions can help address any challenges teachers may encounter and provide opportunities for continuous learning.

Create a Collaborative Community: Establish a collaborative community of practice where teachers can share their experiences, resources, and strategies for using GeoGebra. Online forums, social media groups, and regular meetups can facilitate this exchange of ideas and foster a supportive network.

Resource Libraries: Develop and maintain resource libraries that include lesson plans, GeoGebra applets, and instructional videos. These resources should be easily accessible and provide teachers with ready-to-use materials that can be adapted to their specific classroom needs.

Focus on Pedagogical Integration: Training should not only cover the technical aspects of using GeoGebra but also emphasize its pedagogical integration. Teachers should learn how to design lessons that effectively combine GeoGebra with constructivist teaching methods to maximize student learning outcomes.

6.4 Future Research Directions

While this study has provided valuable insights into the implementation of GeoGebra in primary school mathematics education, there are several areas where further research is needed:

Longitudinal Studies: Conduct longitudinal studies to examine the long-term impact of using GeoGebra on students' mathematical proficiency and attitudes toward mathematics. Such studies can provide a deeper understanding of how sustained use of digital tools influences learning outcomes over time.

Comparative Studies: Investigate the effectiveness of GeoGebra compared to other digital tools and traditional teaching methods. Comparative studies can help identify the unique benefits and potential limitations of GeoGebra, informing best practices for its use in diverse educational settings.

Student-Centered Research: Focus on student-centered research to explore how different student populations, including those with varying levels of mathematical ability and digital literacy, engage with GeoGebra. Understanding these differences can help tailor instructional approaches to better meet the needs of all students.

Teacher Experiences and Beliefs: Study the experiences and beliefs of teachers regarding the use of GeoGebra and other digital tools. Research in this area can shed light on the factors that influence teachers' willingness to adopt and integrate technology into their teaching practices.

Innovative Pedagogical Models: Explore innovative pedagogical models that integrate GeoGebra with other emerging technologies, such as virtual reality (VR) and artificial intelligence (AI). Researching these models can provide insights into how the next generation of educational technologies can further enhance mathematics teaching and learning.

By addressing these areas through future research, educators and policymakers can continue to refine and improve the integration of digital tools like GeoGebra in primary school mathematics education, ensuring that they effectively support student learning and engagement.

7 Conclusion

7.1 Summary of Key Findings

The integration of GeoGebra with the Hejny method in primary school mathematics education has yielded significant insights and positive outcomes. The key findings from this study are summarized as follows:

Enhanced Student Engagement and Motivation: The interactive and visual nature of the GeoGebra Book significantly increased student engagement and motivation. Students found the tasks more interesting and were more willing to participate in classroom activities, leading to a more dynamic learning environment. Teachers observed a noticeable increase in students' excitement about lessons and their participation levels.

Improved Understanding of Mathematical Concepts: The use of interactive applets allowed students to visualize and manipulate geometric shapes, contributing to a deeper understanding of mathematical concepts. This was evident in students' ability to grasp complex ideas and apply their knowledge to solve problems. Students reported a better understanding of shapes and their properties due to the ability to interact with them on the screen.

Effective Integration with Constructivist Methods: The combination of GeoGebra and the Hejny method created a supportive and interactive learning environment that fostered active learning, exploration, and discovery. This integration proved to be effective in enhancing students' mathematical literacy and problem-solving skills. Teachers noted that the constructivist approach, combined with GeoGebra, helped students develop a stronger conceptual understanding and improved their problem-solving abilities.

Overcoming Challenges: The study identified and addressed several challenges related to digital literacy and technical issues. Through ongoing professional development and support, teachers were able to integrate GeoGebra effectively into their teaching practices. Initial technical difficulties were quickly overcome with proper training and support, making the integration smooth and effective.

7.2 Implications for Mathematics Education

The findings from this study have several important implications for mathematics education:

Adopting Digital Tools: The positive impact of GeoGebra on student engagement and understanding highlights the potential benefits of incorporating digital tools into mathematics education. Schools should consider adopting similar technologies to enhance their teaching practices and improve student outcomes. Implementing digital tools like GeoGebra can revolutionize mathematics education by making it more interactive and engaging for students.

Supporting Constructivist Approaches: The successful integration of GeoGebra with the Hejny method underscores the value of constructivist teaching approaches. Educators should be encouraged to use methods that promote active learning, exploration, and collaboration, supported by digital tools that facilitate these processes. The constructivist approach, supported by GeoGebra, has been instrumental in enhancing students' critical thinking and collaboration skills.

Teacher Training and Professional Development: Effective implementation of digital tools requires comprehensive teacher training and ongoing professional development. Schools should invest in training programs that equip teachers with the necessary skills and confidence to integrate technology into their classrooms. Continuous professional development is crucial for teachers to stay updated with technological advancements and integrate them effectively into their teaching.

Addressing Digital Literacy: Ensuring that both teachers and students possess the digital literacy skills needed to use educational technologies effectively is crucial. Schools should provide resources and support to develop these skills, ensuring that all participants can fully benefit from digital tools. Digital literacy programs are essential to equip both teachers and students with the skills necessary to utilize digital tools effectively.

7.3 Final Thoughts

The integration of GeoGebra with the Hejny method in primary school mathematics education represents a promising advancement in teaching and learning practices. This study has demonstrated the potential of digital tools to enhance student engagement, understanding, and mathematical literacy. By creating interactive and visually stimulating learning environments, technologies like GeoGebra can transform the way mathematics is taught and learned.

Moving forward, it is essential to continue exploring innovative approaches to integrating digital tools with constructivist teaching methods. Further research should investigate the long-term impacts, compare different digital tools, and explore new pedagogical models that incorporate emerging technologies. In doing so, educators can continue to refine and improve their practices, ensuring that they meet the evolving needs of students in an increasingly digital world.

In conclusion, the successful implementation of the GeoGebra Book within the framework of the Hejny method provides valuable insights and practical strategies for enhancing mathematics education. The lessons learned from this study can guide educators, policymakers, and researchers in their efforts to integrate digital tools into teaching and learning, ultimately leading to more effective and engaging educational experiences for students.

Acknowledgments The research has been supported by the following projects:
SGS09/PdF/2024: "The use of new digital technologies in the teaching of geometry for future mathematics teachers".
KEGA 026UK-4/2022 "The Concept of Constructionism and Augmented Reality in STEM Education".

References

1. Hohenwarter, M., Preiner, J.: Dynamic mathematics with GeoGebra. Journal of online Mathematics and its applications 7(1), 2–12 (2007).
2. Tatar, E., Zengin, Y.: Conceptual understanding of definite integral with Geogebra. Computers in the Schools 33(2), 120–132 (2016).
3. Hejný, M., Kuřina, F.: *Dítě, škola a matematika.* 1st edn. Portál, Praha (2001).
4. Prensky, M.: Digital natives, digital immigrants. On the Horizon 9(5), 1–6 (2001).
5. Veen, W., Vrakking, B.: Homo zappiens: Growing up in a digital age. 1st edn. Continuum, London (2006).
6. Dunleavy, M., Dede, C., Mitchell, R.: Affordances and limitations of immersive participatory augmented reality simulations for teaching and learning. Journal of science Education and Technology 18, 7-22 (2009).
7. Babinská, M., Dillingerová, M., Korenova, L.: Augmented reality and future mathematics teachers. In: Augmented Reality in Educational Settings, pp. 236–263. Brill (2019).
8. Demel, A.: GeoGebra Book. https://www.geogebra.org/m/rwcc6hvt, last accessed 2024/08/07.
9. Smyrnova-Trybulska, E., Morze, N., Varchenko-Trotsenko, L.: Adaptive learning in university students' opinions: Cross-border research. Education and Information Technologies 27(5), 6787–6818 (2022).
10. Tran, D., Pytlik, M., Kostolányová, K.: Developing communication skills through augmented and virtual reality apps on smart devices. In: EDULEARN20 Proceedings, pp. 3387–3394. IATED (2020).
11. Dana-Picard, T., Hershkovitz, S., Lavicza, Z., Fenyvesi, K.: Introducing golden section in the mathematics class to develop critical thinking from the STEAM perspective. Southeast Asian Journal of STEM Education 2(1). 151–169 (2021).
12. Santos, V.: The Use of Technologies to Promote Critical Thinking in Pre-service Teachers. In: Analysis, Applications, and Computations: Proceedings of the 13th ISAAC Congress, pp. 163-174. Cham: Springer International Publishing, Ghent (2021).
13. Dagienė, V., Futschek, G., Stupurienė, G.: Creativity in solving short tasks for learning computational thinking. Constructivist Foundations 14(3), 382–396 (2019).
14. Drijvers, P., Boon, P., Van Reeuwijk, M.: Algebra and Technology. In: Drijvers, P. (ed.) Secondary Algebra Education, pp. 179–202. Sense Publishers (2011).
15. Hohenwarter, J., Hohenwarter, M., Lavicza, Z.: Introducing dynamic mathematics software to secondary school teachers: The case of GeoGebra. Journal of Computers in Mathematics and Science Teaching 28(2), 135–146 (2009).
16. Berger-Haladová, Z., Ferko, A.: Towards Augmented Reality Educational Authoring. In: Smyrnova-Trybulska, E. (ed.) E-Learning and STEM Education, pp. 587–608. Katowice-Cieszyn (2019).

17. Fosnot, C. T., Perry, R. S.: Constructivism: A psychological theory of learning. In: Fosnot, C. T. (ed.) Constructivism: Theory, Perspectives, and Practice, pp. 8–33. Teachers College Press (2005).
18. Steffe, L. P., Gale, J.: Constructivism in education. 1st edn. Lawrence Erlbaum Associates (1995).
19. Brůžková, N., Koldová, H.: Vybrané aktivity pro výuku geometrie s využitím programu GeoGebra v kontextu ‚malé" revize RVP ZV. 1st edn. Jihočeská univerzita v Českých Budějovicích, Pedagogická fakulta, České Budějovice (2023).
20. Binkley, M., Erstad, O., Herman, J., Raizen, S., Ripley, M., Miller-Ricci, M., Rumble, M.: Defining Twenty-First Century Skills. In: Griffin, P., McGaw, B., Care, E. (eds.) Assessment and Teaching of 21st Century Skills, pp. 17–66. Springer, Dordrecht (2012).
21. Creswell, J. W.: Educational Research: Planning, Conducting, and Evaluating Quantitative and Qualitative Research. 5th edn. Pearson (2015).
22. Krueger, R. A., Casey, M. A.: Focus Groups: A Practical Guide for Applied Research. 5th edn. SAGE Publications (2014).
23. Patton, M. Q.: Qualitative Research and Evaluation Methods. 4th edn. SAGE Publications (2015).
24. Gray, D. E.: Doing Research in the Real World. 5th edn. SAGE Publications (2021).

Increasing the Active Involvement of Students in Chemistry Teaching at Lower-Secondary School

Tomáš Krcil and Eva Milkova

1 Introduction

It has long been discussed that learning achievement in science is not improving in the Czech Republic; on the contrary, according to the Czech School Inspection of the Ministry of Education Youth and Sports, it is frequently worsening. Chemistry is still among the most difficult and least popular subjects taught in the lower secondary school. The nature of the subject itself, that is, the natural science orientation of the subject matter with a considerable degree of abstraction, supported by the application of mathematical rules, makes it difficult for learners of this age category [1].

According to a number of studies, the subject of chemistry does not achieve satisfactory outcomes not only in terms of students' academic performance but also in terms of students' interest in the subject [5]. Another study that examined students' interest in the subject was published by the Czech Statistical Office. Once again, chemistry occupied the lower ranks in this research.

On the other hand, according to a number of studies, computer science is highly popular, not only within the Czech environment. More advanced and modern technologies are penetrating into schools with the aim of bringing them closer to contemporary life and, at the same time, making the presentation of the subject matter more varied and better illustrated. In numerous publications, authors have investigated the impact of the use of Information and Communication Technology (ICT) in teaching on students' learning outcomes, and they prove, in the vast majority, that this impact is positive [3, 4, 6, 7, 9, 12, 13, 18].

T. Krcil (✉) · E. Milkova
University of Hradec Kralove, Hradec Kralove, Czech Republic
e-mail: tomas.krcil@uhk.cz; eva.milkova@uhk.cz

M. Turčáni (ed.), *15th International Scientific Conference on Distance Learning in Applied Informatics*, EAI/Springer Innovations in Communication and Computing,
https://doi.org/10.1007/978-3-031-81261-3_13

Learner activity is an intensive effort of a student that is based on and arises from their inner emotions, interests, or life needs. The student uses effort to acquire knowledge, habits, and attitudes or to get good grades [10]. The goal of any good teacher should be to not only transfer knowledge to students but also increase their interest in the subject and their active involvement in the learning process [11, 15].

Considering the above-mentioned facts, we conducted a 4-year research aimed at finding out the impact of including ICT tools in the teaching of chemistry in the eighth grade of the lower secondary school, not only in terms of students' academic results but also in terms of students' activity in teaching. Partial results of the first aspect of the research were published at the DiVAI 2022 conference [8]; the research on the influence of ICT on students' activity in teaching is the subject of this paper.

2 Materials and Methods

In the Czech Republic, inorganic chemistry is taught in the eighth grade of the lower secondary school. The submitted research addresses the question of whether the inclusion of appropriate ICT educational tools in chemistry teaching in this grade will increase students' activity in the classroom.

For this research, the ICT tools that are expected to better illustrate the subject matter and make the teaching more attractive were chosen for inclusion in chemistry lessons. Specifically, the following tools were used—an interactive whiteboard with appropriate interactive programs, sensors, and the iTRIANGLE educational kit.

2.1 Interactive Whiteboards

Interactive whiteboards are increasingly entering schools and replacing traditional whiteboards. The first company to enter the market with these boards was SMART Technologies, based in Calgary, Canada. The first ever interactive whiteboard appeared in 1991. Today, SMART Technologies is still one of the leading manufacturers of interactive whiteboards in the world [16]. The interactive whiteboard did not come to this country until the beginning of the twenty-first century, mainly due to financial constraints. The interactive whiteboards have numerous advantages and can be a valid teaching tool. However, they also have their disadvantages; especially, it is important to know how to use them properly; otherwise, this tool can be counterproductive, as pointed out by Dostál [2] in his study.

In this research, the interactive whiteboard was used mainly for projecting presentations and teaching using interactive programs—Chemie obecná a anorganická—euroDidact, Chemie 1—Zebra do školy, Projektui.cz a Studium-chemie.cz.

2.2 Sensors

Sensors are educational measurement systems. In the Czech Republic, the two most well-known companies that produce sensors are Vernier and PASCO, the latter of which was used in the research. Sensors are most often used in science subjects so that students can visualize individual processes more clearly. The sensors work in both Windows and iOS operating systems. They can be connected to a computer or tablet, where the data is then collected and evaluated. Aside from classroom use, they can also be used in nature outside the school. Teachers can use a range of ready-made tasks that can be found on the www.experimentujeme.cz server or they can create their own tasks, for example, with the help of students. PASCO sensors can be used for frontal teaching as well as for independent work or group work. PASCO has a wide range of sensors—pH sensor, gaseous oxygen sensor, gaseous carbon dioxide sensor, barometer, conductivity sensor, and others [14].

2.3 Educational kit iTRIANGLE

iTRIANGLE kit is a set of measurement systems used for discovering facts about the world. The kit allows students to understand different connections, gain new knowledge in the field of science, measure quantities, and carry out a variety of projects. The kit contains a large number of parts, tools, and methodologies, as well as sensors, actuators, and a MASTER unit for measuring and subsequent processing of the obtained data. The kit helps students to understand the material more vividly. In class, students most often work in groups, which develops their cooperation.

3 Research

The research was conducted as a quasi-experiment as the selection of research participants could not be random, and the choice of control and experimental groups was determined by the school years of teaching. Data collection was conducted in the eighth grade of lower secondary school during 2019/20 and 2020/21 on the control group of students, and in the following 2 years (2021/22 and 2022/23) on the experimental group of students. The research took place in both the first and second terms of a given school year.

The chemistry teaching of the control group was carried out in a standard way using a chemistry textbook, and the teaching of the experimental group was carried out using the above-mentioned ICT tools.

Teacher–student activity tracking was measured five times each term using the Flanders Interaction Analysis System (FIAS).

Table 1 Distribution of the control group

	Boys	Girls	Total
School year 2019/20	25	25	50
School year 2020/21	25	25	50
Total	50	50	100

Table 2 Distribution of the experimental group

	Boys	Girls	Total
School year 2021/22	25	25	50
School year 2022/23	25	25	50
Total	50	50	100

The research was conducted as part of the dissertation and in accordance with the Code of Ethics of the University of Hradec Kralove. In the beginning, students were informed about the research to be conducted in chemistry classes. All students included in the research agreed to participate.

3.1 Research Objectives

The aim of the research was to investigate and analyze the impact of information and communication technologies implemented in the teaching of inorganic chemistry in the eighth grade of the lower secondary school in terms of teacher–student activity ratio.

3.2 Research Sample

A total of 200 students in the eighth grade of elementary school in Humpolec participated in the research. In each school year, 50 students participated in the research. In the first 2 years, teaching followed a standard approach using a chemistry textbook, that is, the control group consisted of 100 students. The same number of students was in the experimental group, which was taught using ICT tools. The ratio of girls and boys was always balanced (see Tables 1 and 2).

All teaching was conducted by one teacher (the first author of the paper) with the same content taught throughout the research.

3.3 Limitation

The research involved only students of one elementary school. The results of the research are therefore valid only for this specific area. Reflections on the possible generalization of the results obtained in the research are discussed at the end of the article.

3.4 Data Analysis

To obtain the resulting data, the FIAS method was used. This system was developed in the 1970s and is a type of microanalytical research technique. It uses predetermined behavioral categories to describe the teacher's management of teaching or the teacher's interaction with students. The research is conducted by coding the current interaction into the appropriate category every 3 s.

Flanders' system originally had "only" ten parameters to monitor, which caused many scientists' considerable concern. Researchers were most concerned about whether a "mere" ten inventoried parameters were sufficient to describe such a complex process taking place in the classroom. For this reason, a number of researchers enriched the system and created their own categories, in terms of both students and teachers [17].

The resulting data can be processed and evaluated in different ways. In the research mentioned above, we counted the interaction index according to Svatoš [17] (see Table 3) and expressed the data graphically using the frequency of the observed categories. This allowed us to find out who was more active in the lesson and which category was most represented in the lesson.

For this research, the modified Flanders system according to Svatoš [17] was used. We will now list the categories:

Teacher's Activities

T1—The teacher presents information, explains the material, and expresses their opinion.

T2—The teacher asks questions and elicits answers.

T3—The teacher praises, encourages, and approves student's answers.

T4—The teacher accepts, corrects, answers, and develops student's idea.

T5—The teacher summarizes the subject matter.

T6—The teacher gives instructions and commands and organizes activities.

T7—The teacher criticizes, admonishes, and wants to change student's misbehavior.

Table 3 The calculation of the interaction index

Activity index	Title	Interaction index
Ta	Index of teacher's acceptance of the student	(T3 + T4)/C
Tv	Index of teacher's teaching activity	(T1 + T2 + T5)/C
Tr	Index of teacher's class management	(T6 + T7)/C
So	Index of student's support seeking	(S1 + S2)/C
Sa	Index of student's activity	(S3 + S4 + S8)/C
Sp	Index of student's assertion	(S5 + S6 + S7)/C
At	Index of teacher's activity	Ta + Tv + Tr
As	Index of student's activity	So+Sa + Sp
Ii	Overall interaction index	As/At

Student's Activities

S1—The student asks the teacher. Seeks help from the teacher.

S2—The student asks a classmate. Seeks help from a classmate.

S3—The student responds, explains, and states their opinion when called upon by the teacher.

S4—The student responds, explains, and states their opinion on their own initiative.

S5—The student directs their own activity, organizes, and presents.

S6—The student works in groups.

S7—The student discusses the topic (class discussion).

S8—The student works on their own.

Other

O1—Silence, confusion, and incomprehensible communication.

The calculation of the interaction index is shown in Table 3. The letter C denotes the total number of codes recorded in the whole lesson reduced by the number of codes in category O1.

4 Research Results

Both the control and the experimental groups were surveyed each term in five pre-selected lessons (i.e., the measurements were taken in the same lessons each term for both the control and the experimental groups). Since both the control and the experimental groups had been researched for 2 years, the measurements for each group in both the first and second terms of the school year were taken ten times.

4.1 Descriptive Results

Figure 1 shows the ratios of average teacher–student–silence frequencies based on the 10 surveys for the first half of the year. Figure 1a corresponds to the control group and Fig. 1b to the experimental group.

Figure 2 presents all categories with their frequencies in greater detail.

Tables 4 and 5 show the interaction indices in the first half of the year.

In the second half of the year, 10 measurements were conducted once again for each group. The ratios of the average teacher–student–silence frequencies are presented in Fig. 3.

Figure 4 shows all categories in more detail with their frequencies in the second half of the year.

Tables 6 and 7 show the interaction indices in the second half of the year.

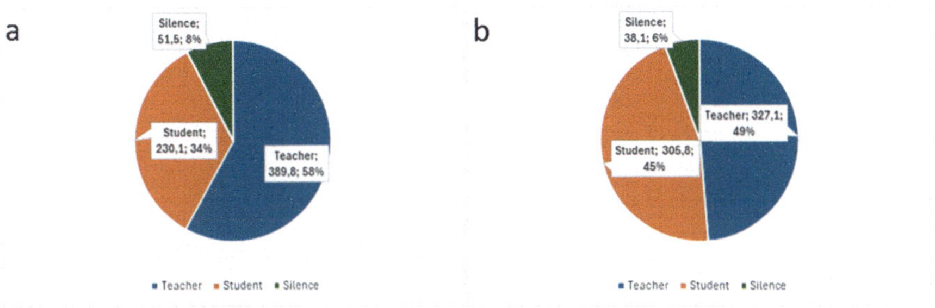

Fig. 1 Ratio of frequencies in the first term: (**a**) control group and (**b**) experimental group

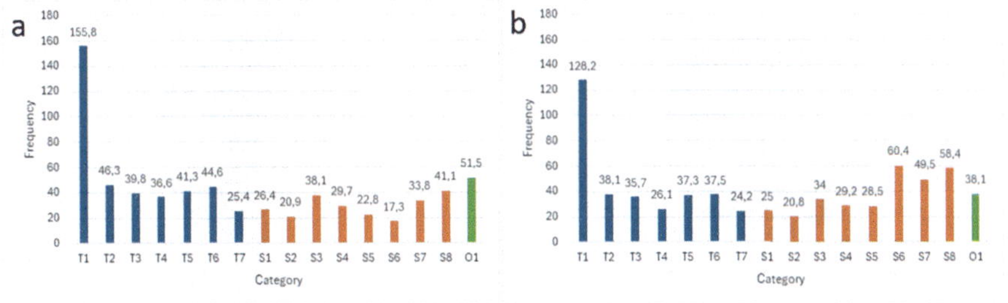

Fig. 2 Category distribution in the first half of the year: (**a**) control group and (**b**) experimental group

Table 4 Index of interaction for the control group in the first half of the year

Activity index	Title	Interaction index
Ta	Index of teacher's acceptance of the student	0.12
Tv	Index of teacher's teaching activity	0.39
Tr	Index of teacher's class management	0.11
So	Index of student's support seeking	0.08
Sa	Index of student's activity	0.18
Sp	Index of student's assertion	0.12
At	Index of teacher's activity	0.63
As	Index of student's activity	0.37
Ii	Overall interaction index	**0.59**

It is clear from the results that students who used ICT resources in the classroom were significantly more active. The tables and figures also show that there is a decline in student activity in the second half of the year for both groups, but the decline is higher for the control group.

Table 5 Index of interaction for the experimental group in the first half of the year

Activity index	Title	Interaction index
Ta	Index of teacher's acceptance of the student	0.10
Tv	Index of teacher's teaching activity	0.32
Tr	Index of teacher's class management	0.10
So	Index of student's support seeking	0.07
Sa	Index of student's activity	0.19
Sp	Index of student's assertion	0.22
At	Index of teacher's activity	0.52
As	Index of student's activity	0.48
Ii	Overall interaction index	**0.93**

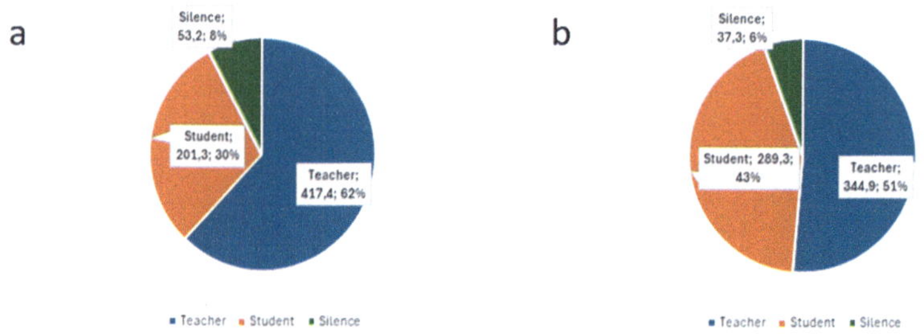

Fig. 3 Ratio of frequencies in the second term: (**a**) control group and (**b**) experimental group

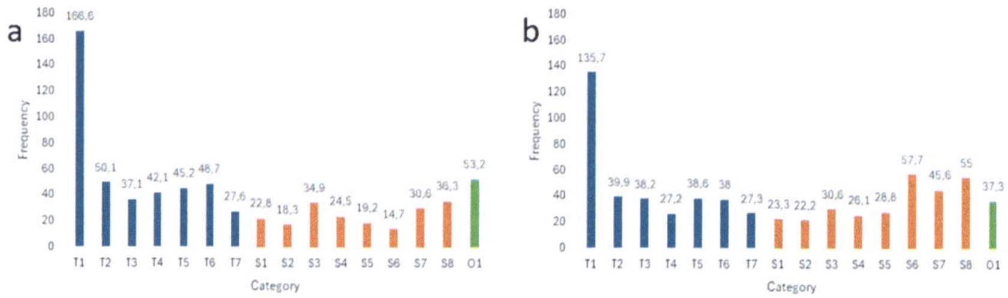

Fig. 4 Category distribution in the second half of the year: (**a**) control group and (**b**) experimental group

5 Discussion and Conclusion

The conclusions of the first part of the research described above, which investigated the impact of ICT on students' learning outcomes in the teaching of inorganic chemistry in the eighth grade of lower secondary school, were presented at the DiVAI 2022 conference and published in the article [8]. Statistical analysis of the data obtained in the research showed that students from the experimental group (using ICT tools) showed significantly better results in the half-year test than

Table 6 Index of interaction for the control group in the second half of the year

Activity index	Title	Interaction index
Ta	Index of teacher's acceptance of the student	0.13
Tv	Index of teacher's teaching activity	0.42
Tr	Index of teacher's class management	0.12
So	Index of student's support seeking	0.07
Sa	Index of student's activity	0.15
Sp	Index of student's assertion	0.10
At	Index of teacher's activity	0.67
As	Index of student's activity	0.33
Ii	Overall interaction index	**0.48**

Table 7 Index of interaction for the experimental group in the second half of the year

Activity index	Title	Interaction index
Ta	Index of teacher's acceptance of the student	0.10
Tv	Index of teacher's teaching activity	0.34
Tr	Index of teacher's class management	0.10
So	Index of student's support seeking	0.07
Sa	Index of student's activity	0.18
Sp	Index of student's assertion	0.21
At	Index of teacher's activity	0.54
As	Index of student's activity	0.46
Ii	Overall interaction index	**0.84**

students from the control group, both in the whole test and in its individual parts related to the ICT tools used.

Similar conclusions were reached when examining the effect of ICT on students' activity (see the previous chapter). The FIAS analysis showed that students' activity in class was higher in both terms in the experimental group and the interaction index was significantly higher in the experimental group than in the control group. All teacher categories, except for U3 (teacher praises, encourages, and approves students' responses), reached a mean value higher in the control group than in the experimental group. Conversely, all student categories, except for the Z3 category (student responds after teacher recall) were more frequent in the experimental group.

At the end of each school year, we always carried out a questionnaire survey in the observed groups to determine the popularity of chemistry lessons. Students recorded their subjective "level of liking" on a scale from 0 to 6 (0 = "extremely disliked," 6 = "very liked"). There was a significant difference between the control and experimental groups (see Table 8).

The higher interest in chemistry teaching among the experimental group can be attributed to the fact that the aforementioned educational ICT tools were used in teaching. Furthermore, the higher interest in education is also probably related to the higher activity during lessons of the students in the experimental group.

Table 8 Popularity among students

Group	Number of students	Average value
2019/20	50	3.42
2020/21	50	3.4
2021/22	50	4.08
2022/23	50	4.14

References

1. Bílek, M., Chroustová, K., Rychtera, J., Machková, V., Kolář, K., Štrofová, J., Sloup, R., Šmídl, M., Bártová, I.: Teachers' View on the Lower Secondary Chemistry Curriculum in the Czech Republic. In 3rd International Baltic Symposium on Science and Technology Education, BalticSTE2019. Scientia Socialis, Šiauliai (2019). ISBN 978-609-95513-6-4.
2. Dostál, J.: Interaktivní tabule—významný přínos pro vzdělání (2009), https://www.researchgate.net/publication/280805404_Interaktivni_tabule-_vyznamny_prinos_pro_vzdelavani, last accessed 2024/4/24.
3. Fernández-Gutiérrez, M., Gimenez, O., Calero, J.: Is the use of ICT in education leading to higher student outcomes? Analysis from the Spanish Autonomous Communities. Computers and Education 157 (2020).
4. Fussell, S. G., Truong, D.: Accepting virtual reality for dynamic learning: an extension of the technology acceptance model. Interactive Learning Environments (2021). https://doi.org/10.1080/10494820.2021.2009880.
5. Hőfer, G. et al.: Výuka fyziky v širších souvislostech—názory žáků. Západočeská univerzita, Plzeň (2005).
6. Hu, X. et al.: The relationship between ICT and student literacy in mathematics, reading, and science across 44 countries: A multilevel analysis. Computers and Education 125, 1–13 (2018).
7. Koğar, E. Y.: The investigation of the relationship between mathematics and science literacy and information and communication technology variables. International Electronic Journal of Elementary Education 11(3), 257–271 (2019).
8. Krcil, T., Milkova E.: The impact of ICT on Comprehensibility of the Chemistry Education. DiVAI 2022, vol. 14, pp 106–115. Wolters Kluwer, Alphen aan den Rijn (2022).
9. Kunina-Habenicht, O., Goldhammer, F.: ICT engagement: A new construct and its assessment in PISA 2015. Large-Scale Assessments in Education 8, 1–21 (2020).
10. Maňák, J.: Rozvoj aktivity, samostatnosti a tvořivosti žáků. Masarykova univerzita, Brno (1998).
11. Milkova E.: Multimedia Application – Effective Support of Education. In: Capay, M., Mesarosova, M., Palmarova, V. (eds.) DiVAI 2012, pp 13-21. Constantine Philosophe University, Nitra (2012).
12. Morris, T. H., Rohs, M.: The potential for digital technology to support self-directed learning in formal education of children: a scoping review. Interactive Learning Environments (2021). https://doi.org/10.1080/10494820.2020.1870501.
13. Nunvarova, J., Poulova, P., Prazak, P.: Digital Storytelling in Economics Subjects and its Effectiveness on Student Learning Outcomes by Gender and Different Economic Knowledge. Journal on Efficiency and Responsibility in Education and Science 16 (3), 208–219 (2023).
14. PASCO. PROFIMEDIA INTERACTIVE LEARNING: Univerzální edukační platforma pro experimentální výuku přírodních věd (2019), http://www.pasco.cz, last accessed 2024/4/24.
15. Sevcikova, A., Milkova, E., Moldoveanu, M., Konvicka, M.: Graph Theory: Enhancing Understanding of Mathematical Proofs Using Visual Tools. Sustainability 15(13) (2023).
16. SMART Technologies. SMART Board interactive whiteboard, https://www.smarttech.com/en/education/products/interactive-white-board/m700-series, last accessed 2024/4/24.

17. Svatoš, T.: Kategoriální systém podle Flanderse při evaluaci výuky v současné základní škole. 1 st edn., Pedagogika, Praha (2013).
18. Zhang, D., Liu, L.: How does ICT use infuence students' achievements in math and science over time? Evidence from PISA 2000 to 2012. Eurasia Journal of Mathematics, Science and Technology Education 12(9), 2431–2449 (2016).

Pre-service Preschool and Primary School Teachers' Position on Artificial Intelligence

Vincentas Lamanauskas and Rita Makarskaitė-Petkevičienė

1 Introduction

Recently, artificial intelligence (AI) has been attracting increasing interest, as well as some concern. Although it is not a particularly new technology in a general sense, however, the free availability of certain AI tools has given it some impetus. AI is becoming a more popular topic every year and causes a lot of controversy; however, even taking into account increasing attention, most people do not know what AI really is and that it has been a very important part of our daily life for a long time. It is obvious that constantly improving AI tools do not bypass the field of education today.

The application possibilities of AI do not bypass the field of education, moreover, those possibilities are constantly expanding. By applying AI, it is possible to create individualized training and various tasks adapted to educational needs, using interactive learning tools that help to better understand and master the educational material. One can observe the learners' progress and behavior during the education process, and this enables educators to identify problems and adjust teaching strategies to achieve the best possible results. As recently as a decade ago, researchers noted that the revolution in information technology (IT) and AI has encouraged (and continues to encourage) a wave of innovations, which promises to revitalize schools [1, 2]. Thus, one can agree with researchers who claim that AI-based technologies play and will play an increasingly important role in teaching

V. Lamanauskas (✉)
Vilnius University, Šiauliai, Lithuania
e-mail: vincentas.lamanauskas@sa.vu.lt

R. Makarskaitė-Petkevičienė
Vilnius University, Vilnius, Lithuania
e-mail: rita.makarskaite-petkeviciene@fsf.vu.lt

M. Turčáni (ed.), *15th International Scientific Conference on Distance Learning in Applied Informatics*, EAI/Springer Innovations in Communication and Computing,
https://doi.org/10.1007/978-3-031-81261-3_14

and providing education to students at various levels of the education system [5]. Although AI is an evolving technology, however, it can help solve challenges in the education sector, form conditions for using innovative teaching methods, help implement new teaching methods, and intensify progress in this field. The Education System cannot be left behind in this regard. For example, in Lithuanian schools, according to the updated general education programs, since September 2023, students have been taught about AI as well. Such a "revolutionary" change can also be compared to the emergence of the Internet when it was asked and discussed whether it was necessary to teach about the Internet and if so, how. It is likely that this dilemma has recently been associated namely with AI.

In recent years, there has been an abundance of scientific research related to the application of AI in education. Jho and Ha [16] explored generative AI productivity by extracting elements from the text. Not only the viability of AI for educational purposes was revealed, but also its shortcomings. Another recent study analyzed neural network and AI integration in image recognition for multi-object recognition. It was established that learning through AI robot recognition technology is more efficient than learning from the textbook in improving students' understanding of the concept of "cell division" and increasing their motivation to learn natural science [7]. A study conducted in Slovenia, in which 1215 teachers took part, examined the preparation of natural science teachers to assess the rapidly changing environment of AI in education. The results of the study showed that continuous professional development significantly improves the preparation of science teachers to assess using AI [19]. It is also clear that AI can increase the capacity of general education programs [12].

International surveys have shown that the attitude of university students toward AI and its inclusion in various science fields and studies is generally positive [10, 13, 25, 32]. The research also shows that it is very important to promote AI literacy among preschool and primary education teachers. The positive effect of AI knowledge acquisition and understanding of various aspects of AI literacy emphasizes the need for comprehensive AI training in teacher education programs [3]. A study conducted in Romania revealed that confidence in personal abilities to use AI and the perception of a greater number of benefits of AI have a positive and significant effect on the willingness to use AI in the educational process [27]. On the contrary, it is crucial for preschool and primary school teacher education policymakers to understand how to encourage future teachers to learn AI through effective teacher preparation programs [28].

Undoubtedly, it is important to analyze future teachers' position on AI, in order to better understand their needs, preparation, attitudes, and so on. Accordingly, it is necessary to improve university study programs so that future teachers acquire adequate AI literacy, that is, an appropriate skill set that enables them to firmly understand AI, teach/learn about AI, teach/learn how AI works, and teach/learn using AI. It is teachers who will have to prepare students for the rapidly changing world of AI with the uncertain requirements raised for future workforce skills.

Therefore, it is necessary to learn more about how future teachers accept AI [33]; it is especially important in the preparation of preschool and primary school teachers [18]. Despite the relatively large amount of research on the application of AI in education, there is still a shortage of research examining future preschool and primary school teachers' position (attitude, perception about AI, preparation to apply it, experiences, etc.) Thus, the present study will at least partially fill this gap.

Thus, the aim of the study was to find out future preschool and primary education teachers' position on AI. The following research questions were formulated:

- How familiar are preschool and primary education teachers with AI and how do they assess personal preparation in this area?
- How do future preschool and primary education teachers assess the impact of AI on education in the future?
- What are the fears of future preschool and primary education teachers about the use of AI in education?
- What AI tools and platforms do future preschool and primary education teachers use in their studies?

2 Research Methodology

2.1 Background

The conducted research is quantitative, informative, and descriptive in nature. A structured, non-standardized written survey was used. The purpose of such a study is to obtain empirical data that provides a more detailed picture of the researched phenomenon and its structural elements. The research does not aim to prove any causal relationships between variables but simply to assess the situation with respect to the characteristics of interest. As the researchers emphasize, such a study describes the characteristics of the studied population or the phenomenon. Applying this methodology, the focus is more on what the subject of the research is but not on why the research object is [24]. Moreover, in such a study, the description may focus on individual variables and on relationships between variables [22]. Descriptive research can also be used to purposefully collect significant data about prevailing conditions, practices, trends, and so on and then interpret them appropriately and accurately using statistical methods or even not using them at all [6].

It is assumed that such research can help to understand certain problems, identify important aspects and trends, and determine further research directions and/or actions. The study was conducted from March to April 2024 (during the spring semester).

Table 1 Distribution of respondents by year of study

Study year	1-year study	2-year study	3-year study	4-year study
N (%)	17 (15.2)	46 (41.1)	18 (16.1)	31 (27.7)

2.2 Sample

The research sample consisted of the university first to fourth-study-year students, future teachers of preschool and primary school education. All the respondents are women. From this point of view, possible differences between variables according to the gender factor are not analyzed. A total of 112 Vilnius University students participated in the study, but the sample consisted of the two structural units of the institution—the main unit and the stem unit, located in the northern region of Lithuania. In this way, the representativeness of the sample is partially increased, since students from different country regions study in both departments (Table 1).

This study uses a non-random, convenience sample. In convenience sampling, the selection of participants is not based on randomization or any specific criteria beyond their availability and willingness to participate in the study. Such a sample selection strategy was used because it is quite difficult to reach the entire population (part of the students study remotely); in addition, the population is relatively small, the study focuses only on a defined profile of university students, and the population itself is quite homogenous. On the contrary, the researchers assessed the availability of the respondents and their readiness to participate in the study. It was taken into account that probability sampling fails to be an adequate method when the population is hard to reach or hidden, and researchers have difficulty engaging and accessing these individuals [4]. The researchers took into account the methodological view that if researchers carefully conduct the research using a convenience sampling technique and controlling biases and uncertainty, it produces useful data [11].

Before conducting the survey, the respondents' consent to participate in the research was obtained. Participation was completely voluntary, and the necessary anonymity was maintained. The research instrument also included a motivational-explanatory part. The security of the data and personal information of research participants was fully ensured.

2.3 Research Instrument

A questionnaire developed by the researchers was used in the study. The questionnaire consisted of closed and open questions. In the written survey, everybody is given the same questions. Researchers hold a position that it is useful because it aims to gain a deeper understanding of the respondents' opinions or to assess the specific aspects, which may be difficult to cover with standardized questions. Hopefully, this will help uncover new ideas and trends. Particular attention was paid to the quality, clarity, and relevance of the questions in this study. After preparing the first version of the questionnaire, the content validity assessment was carried out with the help of two education experts, working in the field of research methodology. A pilot study was also conducted, in which ten students took part. The latter were not included in the main sample. After the pilot study, a minor adjustment of the research instrument was made.

Structure of the research instrument:

1. Demographic part (respondent gender, study year, study program).
2. Closed questions:

    2.1. Are you familiar with AI? (dichotomous scale—Yes; No).
    2.2. How familiar are you with AI? (rank scale—excellent, very good, average, a little, not at all).
    2.3. Do you think AI can improve the learning experience at university? (rank scale—yes, not sure, no).
    2.4. Do you think AI will have a big impact on education in the future? (rank scale—yes, not sure, no).
    2.5. Do you feel adequately prepared to use AI technologies in your future teaching career? (rank scale—yes, not sure, no).
    2.6. What AI skills and/or areas of knowledge do you think are important for future teachers to acquire? (Tick the most important) (nominal scale).
    2.7. Would you be interested in training or professional development on the use of AI in teaching/learning? (rank scale—yes, not sure, no).
    2.8. What are your biggest fears about the use of AI in education? (Tick those that cause the most irritation) (nominal scale).
    2.9. Have you noticed any impact of AI on your learning/studying performance? (rank scale—yes, not sure, no).
    2.10. Do you think AI can replace teachers in the future? (rank scale—yes, not sure, no).
    2.11. What AI tools or platforms do you use in your studies? (Mark which ones you use most often) (nominal scale).
    2.12. What skills do you think students can develop best using AI in their studies? (Tick no more than two) (nominal scale).
    2.13. Would you agree that AI monitors your academic progress? (rank scale—yes, not sure, no).

The presented questions were prepared exclusively for this research, were analyzed separately, and did not form a separate measurement scale. When questions are developed only for a specific study and are not intended to form a standard measurement scale, this indicates that the researchers have adapted their questions to collect information related to the specific objectives of the study. Such an approach allows for a flexible and nuanced approach to the research topic and can be particularly useful when investigating new or complex phenomena. Respondents had the opportunity to submit detailed comments to the closed questions. The researchers took the view that open-ended questions were important because they made it possible to learn unexpected information.

2.4 Data Analysis

An analysis of the collected data was performed using the main measures of descriptive statistics. Descriptive statistics, a type of quantitative data analysis, is used to describe or present data in an easily accessible, quantitative form [15]. The calculated absolute and relative frequencies are expressed as percentages. The aim is to summarize the data to reach a simple presentation as a result. According to Taherdoost [29], descriptive data analysis is recognized as the first type of data analysis. Spearman's rank correlation coefficient rho (ρ) is applied to determine possible relationships between variables, namely, to assess the strength of the relationship between variables, measured on the rank scale. This coefficient was chosen because it is a non-parametric measure of the strength of the relationship, which does not depend on the nature of the distribution of the variables [26]. Data analysis was performed using the SPSS statistical program.

3 Research Results

After analyzing the data, relevant results were obtained. First of all, it was established that all respondents are familiar with AI or similar AI-based tools in one way or another. Most students use AI tools in their studies. These can be a variety of AI-based tools or platforms. The results of the survey showed that the three platforms are used most frequently and to practically the same extent, for example, 32% use the Microsoft Teams platform, 30% the ChatGPT, and 25% Kahoot. Notably, 6% of respondents reported using Google Classroom and 7% used Quizlet.

After analyzing the respondents' assessment of the level of familiarity with AI, the following results were obtained (Table 2).

Table 2 Level of familiarity with AI

Level	*N*	%
Excellent	4	3.6
Very good	22	19.6
Average	59	52.7
A little bit	26	23.2
Not at all	1	0.9

Table 3 Respondents' position on AI (N (%))

Variable	Rating level		
	Yes	*Not sure*	*No*
AI improves the learning experience at the university	69 (61.6)	41 (36.6)	2 (1.8)
AI will have a significant impact on education in the future	104 (92.9)	8 (7.1)	0 (0.0)
Readiness to use AI technologies in the future teaching career	27 (24.1)	59 (52.7)	26 (23.2)
Interest in professional development about using AI for teaching/learning	89 (79.5)	23 (20.5)	0 (0.0)
AI impact on personal learning/study results	33 (29.5)	29 (25.9)	50 (44.6)
AI can replace teachers in the future	3 (2.7)	26 (23.2)	83 (74.1)
AI can track personal academic progress	29 (25.9)	58 (51.8)	25 (22.3)

As can be seen, most of the respondents are more or less familiar with AI. More than half of the survey participants consider their familiarity with AI to be average. Some characteristic comments of the respondents are presented here.

> It already has a lot of value for students because it's associated with easily achievable results, but as tools become available that detect the use of AI, then the transition to using it for educational purposes will follow. Such a powerful tool will change learning habits and inefficient methods currently used.

> In a few years, when powerful AIs become available for less or for free, students will be able to use a variety of AIs for different purposes and tasks. It is important to get acquainted with various artificial intelligences, and then it is only a matter of human creativity what to do with AI.

> In my opinion, I will definitely use AI in my work, but now I don't feel confident about my 'ability' to use it properly and effectively, if I found out in what ways and how AI could help me, then I would definitely not have anything against using it in my work.

After analyzing the collected data according to other applied variables, the obtained results are presented in Table 3.

As can be seen in Table 3, the respondents' position on the issue of AI is quite positive. The absolute majority of respondents agree that AI will be extremely important in the future and will have a significant impact on education. Such an approach is certainly adequate, considering the rapid technological progress in this field. It is understood that students adequately assess their understanding/competencies in AI technology. On the contrary, the results allow us to say that students do not feel a significant AI effect on personal academic results. Only one-third of the respondents positively assessed such an effect. Only a small minority believe that AI will replace teachers in the future. The vast majority of respondents do not agree with this. Accordingly, only a quarter of students would agree that AI could monitor their academic progress. A similar proportion of respondents believe that their preparation for using AI in their future work as a teacher is adequate. Here are some characteristic comments from the respondents.

> AI is inevitable, and its use among students will only increase. Therefore, it is important to familiarise yourself with it, learn how to use it and apply it to today's education.

> I think that teachers bring empathy to the classroom, which is extremely important. I don't think that AI can play this role.

> Unfortunately, I think it can have a negative impact on education because the further it goes, the more people use it, but often use it unethically/abusively.

> In my opinion, I will definitely use AI in my work, but now I don't feel confident about my 'ability' to use it properly and effectively, if I found in what ways and how AI could help me, then I would definitely not have anything against using it in my work.

> In my opinion, many people may have various thoughts about the 'possession' of AI, however, I don't think that AI can change the position of a teacher because a real person brings his experience, reacts to situations not based on something, for example, analytical research or similar, but real experiences of situations and knowledge acquired through experience.

It has been analyzed how the analyzed variables are related. Statistically significant correlations were found between some variables. The obtained results are presented in Table 4.

Table 4 shows that the more familiar students are with AI, the more prepared they feel to use AI technologies in their teaching career ($\rho = 0.393$, $p < 0.0001$). The level of familiarity is directly proportionally correlated with the possible impact on the study results ($\rho = 0.249, p = 0.008$). Also, a high, positive correlation coefficient is established between AI, which can improve the learning experience at university and readiness to use AI technologies in a teaching career ($\rho = 0.256$, $p = 0.007$). The impact of AI on the study results is also positively correlated with the potential impact of AI on education in the future ($\rho = 0.393$, $p < 0.002$).

It has been analyzed which skills and/or areas of knowledge are important for future teachers to acquire. The results are presented in Table 5.

Table 5 shows that ethical aspects are considered the most important in the field of AI. Meanwhile, the application of AI in the organization of lessons is not considered a very important area. This can be explained by the fact that students have little knowledge in this area and their understanding is limited at the current stage. However, students realize that it is quite important to understand AI and the principles of its work. This position is also illustrated by the respondents' presented comments.

> I think everything is very important. And you have to know a lot because the younger children are, the more receptive they are to innovations. That's why we older people need to learn and become more skilled so that we can be the best version of ourselves for the children.

> I think it's important to know how AI works, how to work with it so that this use is safe and effective. Also, it is necessary to be able to critically evaluate the received data and correctly interpret them.

> It is very important to be able to interpret, analyse and properly understand the information provided by AI because it may not always be appropriate.

Table 4 Spearman's rank correlation coefficients among variables

Variables	1	2	3	4	5	6	7	8
1. The level of familiarity with AI	1.000							
2. AI improves the study experience	**0.227**[a]	1.000						
3. The impact of AI on education in the future	0.111	−0.079	1.000					
4. Readiness to use AI technologies in a teacher's career	**0.393**[b]	**0.256**[b]	−0.047	1.000				
5. Professional development about the use of AI in education	0.097	0.018	−0.055	−0.026	1.000			
6. The impact of AI on study results	**0.249**[b]	**0.209**[a]	**0.285**[b]	0.071	0.107	1.000		
7. Position for teacher replacement by AI	−0.102	−0.084	0.163	−0.115	**0.188**[a]	0.111	1.000	
8. Monitoring academic progress with AI	0.139	−0.108	0.014	0.147	0.091	0.106	0.165	1.000

[a]Correlation is significant at the 0.05 level (two-tailed)
[b]Correlation is significant at the 0.01 level (two-tailed)

Table 5 AI skills and/or knowledge areas important for future teachers to acquire

Skills/areas of knowledge	%
Ethical aspects in the field of AI	40.4
Understanding of AI concepts and principles	31.5
Data analysis and interpretation	15.7
Collaboration with AI systems	10.0
Programming and coding skills	1.5
Application of AI in lessons	0.9

In my opinion, the most important skill is critical thinking. If you are looking for information with the help of artificial intelligence, it is very important to evaluate it critically, to choose what is correct, and what is perhaps not so good.

For the successful use of artificial intelligence, it is important to understand the principles of how this system works. You need to know how to properly formulate a question, to get the desired answer. Also, it is equally important to be able to select data critically. Artificial

Table 6 Skills best developed using AI in studies

Skills	%
Critical thinking	33.3
Systematic thinking	25.8
Ability to adapt/adaptability	21.5
Problem solving	8.6
Logical thinking	6.5
Creativity	4.3

Table 7 Fears about the use of AI in education

Fears	%
Privacy and data security	38.6
Ethical aspects	23.8
Potential damage	17.0
Lack of human interaction	13.0
Reduction of teacher jobs	4.5
Bias and discrimination	3.1

intelligence can offer many answers to questions and many solutions, but it is important for the future teacher to select the most relevant material that would be useful for the students.

Students' position regarding the skills that can be best developed/educated using AI has been analyzed. The results are presented in Table 6.

Table 6 shows that critical thinking is best developed using AI. Logical thinking, creativity, and problem-solving are the least developed.

I think it is precisely critical thinking because it is necessary to critically evaluate the received information and present it to coursemates, teachers, or students.

Processing the received information requires critical thinking.

I think that it is necessary to critically evaluate the information presented and not to believe it blindly; of course, not everyone has this already mentioned quality, that's why lessons, lectures, and seminars should be devoted to it.

I thought for a very long time whether to mark creativity only because artificial intelligence very often already presents all the ideas, however, I think that it would develop the creativity of some students because not everyone tends to 'copy down' all the ideas, sometimes a person just needs a starting point, and when he gets at least one minimal idea – he creates even more diverse ones.

Critical thinking because it is necessary to select which information is correct, which is not, which is suitable for use, and which is not suitable. Problem solving because when a problem arises, answers and solutions can be found with the help of AI.

It has also been analyzed what fears future teachers have about the use of AI in education. The results are presented in Table 7.

As seen in Table 7, privacy and security are the top concerns. In the second position, there is a concern about the ethical aspects of using AI. On the contrary, the respondents do not associate the development of AI technologies with the reduction of teachers' jobs, nor do they think that bias and discrimination cause anxiety in this respect.

> I think so because it can help the educator cope with the amount of information and plan time and educational activities. However, it is also important to be able to select information, apply it and evaluate it critically. In addition, AI can also have a negative effect, for example, if educational participants (teachers, students, etc.) use it 'blindly, trustingly.' I think that excessive use of AI can harm the development of abilities, competencies, and achievements of both teachers and students.

> It can already be noticed that both schoolchildren and students of educational institutions actively use artificial intelligence doing homework, writing essays, etc., so this reduces their creativity development, they 'think' less, so in the future, it may have a negative impact on their idea diversity, originality or creativity, difficulties may arise to come up with answers to questions "here and now", to present your ideas, views, or understanding on different topics.

> As AI is relatively new at the moment, it's difficult to predict the impact, but given that accessibility is rather easy at the moment, I'd dare say that the impact will definitely be felt and it's important to consider that perhaps this tool should be included in teaching/learning after all as a legal means of education.

> Children will no longer be able to think for themselves and will turn to AI for all their problems.

As can be seen, the respondents' positions are diverse; however, the expressed fears are clear enough. It is obvious that the use of AI in education has great potential, but it can also have negative effects that are not fully clear at this stage. Therefore, it is important to implement AI in education carefully and to attentively evaluate its impact, in order to maximally use its benefits and minimize negative effects.

4 Discussion

The aim of the study was to find out future preschool and primary education teachers' position on AI. The obtained research results show that students are generally familiar with AI in one way or another and, in addition to that, use some instruments/tools both for personal use and directly in the study process. The study made it possible to look at future teachers' experiences, positions, evaluations, fears, and apprehensions after the emergence of AI tools in the study process. The level of students' familiarity with AI is rated as average or good. AI definitely improves the learning experience at university and provides certain new skills. Similar results were demonstrated by Kuleto et al.'s [21] study, stating that AI provides students with new skills and a collaborative learning environment, which has great importance for the near future. The conducted study showed that students are aware of the importance of AI and the inevitable development of technology-based AI and also emphasize the importance of professional development in the field of AI. Similar results are shown by other studies stating that AI not only promotes changes in teaching/learning but also transforms the process itself [31]. The obtained results of the empirical study allow us to say that AI provides an effective teaching/learning experience. Students believe that AI can help the teacher cope with the amount of information and plan the time and educational activities. However, they also point to the importance of being able to select, apply, and critically evaluate information. In addition, AI can also have a negative effect, for example, if educational participants (teachers, students, etc.) use it uncritically. For example, respondents note that AI does not always provide accurate and correct information; in such a case, it needs to be verified using other sources, and this develops students' critical and logical thinking. Research results also showed that students use AI tools quite widely. Previous studies [14] showed similar results that students, as "digital citizens," intensively use AI and its generated tools to improve learning outcomes.

The study also revealed some concerns about the use of AI in education. It was revealed that privacy and data security (38.6%) and ethical aspects (23.8%) were of greatest concern. Similar results were obtained in other conducted studies [9, 30]. The study revealed other concerns as well, such as diminishing human interaction. The use of AI may present some challenges to human interaction. For example, if we rely too much on technology, the personal level of interaction between teacher and student may decrease. This could affect emotional bonding, which is very important in raising children. In addition, potential risks include over-reliance on technology and the reduced ability of people to communicate and collaborate without their assistance. Researchers argue that AI cannot replace the essential human teaching elements, such as building relationships, providing emotional support, and promoting critical thinking skills [17].

An important aspect in the context of the development and penetration of AI in education is the role of the teacher. AI may affect the role of teachers; however, it will be an additional form of change rather than a complete change of teachers. Various AI tools can help teachers carry out their tasks more effectively, but the aspects of human interaction, motivation, and empathy that are important to the teacher are likely to remain unchanged. The study showed that the majority of respondents (74.1%) did not think that AI would change teachers in the future. Other researchers also claim that the teacher's work cannot be replaced by AI technology [8, 23], even though there are some doubts about possible changes in the future [20].

The study has several limitations. Due to the homogeneity of the sample, it was not possible to compare possible differences by gender variable. Research shows that it is important to consider the gender aspect when preparing teachers, as, generally, there is a high percentage of women ready to become teachers [33]. On the contrary, the study sample is limited, so generalizing conclusions cannot be uncritically interpreted. However, considering that the population of students in preschool and primary education programs is not large, the results obtained and the conclusions drawn on the basis of such a sample are quite adequate and reasonable.

5 Conclusions and Implications

The results of the study allow us to state that future preschool and primary education teachers' familiarity level with AI is average. The general position on AI is positive, despite the fact that students do not feel that AI has a significant impact on their personal academic achievements. Students rate their understanding and skills in the field of AI sufficiently adequately and agree that more comprehensive learning about AI is needed, especially during university studies. The majority of students generally agree that AI will have a significant impact on education in the future, although they do not agree that AI should monitor their personal academic progress.

Students have been found to be quite active in using AI-based tools. The most frequently used tools are Microsoft Teams, ChatGPT, and Kahoot. From the students' point of view, using AI devices/tools develops critical and systemic thinking the best, and creativity and logical thinking are developed the least. One of the most important areas of knowledge regarding AI is ethical aspects and understanding the concepts and principles of AI.

It has been established that the more familiar students are with AI, the more they see a positive effect on their study results. Likewise, the impact of AI on study results is positively correlated with the potential impact of AI on education. Expressed fears about the use of AI in education are most associated with privacy, data security, and ethical aspects, and least with potential teacher job losses, bias, and discrimination. The majority of students do not believe that AI can replace teachers in the future.

The study was conducted with a limited sample, including only students from preschool and primary education study programs, which was its aim. Thus, additional research may warrant the involvement of large sample sizes from diverse teacher education programs to evaluate the generalizability of the results. Nevertheless, this research is likely to contribute to the new research based on AI-based education and teacher preparation.

References

1. Aberšek, B. (2005). Modern learning environments in combination with intelligent expert system. *Journal of Science Education*, *6*(Special issue), 13–14. https://www.proquest.com/openview/05b43312584cac1b2ca65e00a2431fae/1?pq-origsite=gscholar&cbl=28899
2. Aberšek, B., & Kordigel Aberšek, M. (2010). Information communication technology and e-learning contra teacher. *Problems of Education in the 21st Century*, *24*(1), 8–18. https://www.scientiasocialis.lt/pec/node/434
3. Ayanwale, M. A., Adelana, O. P., Molefi, R. R., Adeeko, O., & Ishola, A. M. (2024). Examining artificial intelligence literacy among pre-service teachers for future classrooms. *Computers and Education Open*, *6*, Article 100179. https://doi.org/10.1016/j.caeo.2024.100179

4. Berndt, A. E. (2020). Sampling methods. *Journal of Human Lactation, 36*(2), 224–226. https://doi.org/10.1177/0890334420906850
5. Bregant, J., & Aberšek, B. (2011). Artificial intelligence versus human talents in learning process. *Problems of Education in the 21st Century, 37*(1), 38–47. https://doi.org/10.33225/pec/11.37.38
6. Calderon, J. (2006). *Methods of research and thesis writing* (2nd Ed.). National Bookstore.
7. Chen, P.-Y., & Liu, Y.-C. (2024). Impact of AI robot image recognition technology on improving students' conceptual understanding of cell division and science learning motivation. *Journal of Baltic Science Education, 23*(2), 208–220. https://doi.org/10.33225/jbse/24.23.208
8. Fitria, T. N. (2023). The use of artificial intelligence in education (AIED): Can AI Replace the Teacher's Role? *Epigram, 20*(2), 165–187. https://doi.org/10.32722/epi.v20i2.5711
9. Gao, C. A., Howard, F. M., Markov, N. S., Dyer, E. C., Ramesh, S., Luo, Y., & Pearson, A. T. (2023). Comparing scientific abstracts generated by ChatGPT to original abstracts using an artificial intelligence output detector, plagiarism detector, and blinded human reviewers. *npj Digital Medicine, 6*(75). https://doi.org/10.1038/s41746-023-00819-6
10. Ghotbi, N., Ho, M. T., & Mantello, P. (2022). Attitude of college students towards ethical issues of artificial intelligence in an international university in Japan. *AI & Society, 37*(1), 283–290. https://doi.org/10.1007/s00146-021-01168-2
11. Golzar, J., Noor, S., & Tajik, O. (2022). Convenience sampling. *International Journal of Education and Language Studies, 1*(2), 72–77.
12. Gulson, K. N., Murphie, A., Taylor, S., & Sellar, S. (2018). *Education, work and Australian society in an AI world. A review of research literature and policy recommendations.* (Research Report). Gonski Institute for Education, UNSW.
13. Haseski, H. I. (2019). What do Turkish pre-service teachers think about artificial intelligence? *International Journal of Computer Science Education in Schools, 3*(2), 3–23. https://doi.org/10.21585/ijcses.v3i2.55
14. Ijaz, K., Bogdanovych, A., & Trescak, T. (2017). Virtual worlds vs books and videos in history education. *Interactive Learning Environments, 25*, 904–929. https://doi.org/10.1080/10494820.2016.1225099
15. James, D., & Simister, N. (2017). *Quantitative analysis.* INTRAC. https://www.intrac.org/wpcms/wp-content/uploads/2017/01/Quantitative-analysis.pdf
16. Jho, H., & Ha, M. (2024). Towards effective argumentation: Design and implementation of a generative AI-based evaluation and feedback system. *Journal of Baltic Science Education, 23*(2), 280–291. https://doi.org/10.33225/jbse/24.23.280
17. Kasneci, E., Sessler, K., Küchemann, S., Bannert, M., Dementieva, D., Fischer, F., Gasser, U., Groh, G., Günnemann, S., Hüllermeier, E., Krusche, S., Kutyniok, G., Michaeli, T., Nerdel, C., Pfeffer, J., Poquet, O., Sailer, M., Schmidt, A., Seidel, T., Stadler, M., et al. (2023). ChatGPT for good? On opportunities and challenges of large language models for education. *Learning and Individual Differences, 103*, Article 102274. https://doi.org/10.1016/j.lindif.2023.102274
18. Kaya, E., Yesilyurt, E., Mjavanadze, E., & Deniz, H. (2024). Artificial intelligence education in teacher preparation programs: Initial strategies and insights in AI education for pre-service elementary teachers. In J. Cohen, & G. Solano (Eds.), *Proceedings of Society for Information Technology & Teacher Education International Conference* (pp. 2092–2100). Association for the Advancement of Computing in Education (AACE). https://www.learntechlib.org/primary/p/224264/
19. Kerneža, M., & Zemljak, D. (2023). Science teachers' approach to contemporary assessment with a reading literacy emphasis. *Journal of Baltic Science Education, 22*(5), 851–864. https://doi.org/10.33225/jbse/23.22.851
20. Kolchenko, V. (2018). Can modern AI replace teachers? Not so fast! Artificial Intelligence and adaptive learning: Personalized education in the AI age. *HAPS Educator, 22*(3), 249–252. https://doi.org/10.21692/haps.2018.032
21. Kuleto, V., Ilić, M., Dumangiu, M., Ranković, M., Martins, O. M. D., Păun, D., & Mihoreanu, L. (2021). Exploring opportunities and challenges of artificial intelligence and machine learning in higher education institutions. *Sustainability, 13*, Article 10424. https://doi.org/10.3390/su131810424

22. Lans, W., & van der Voordt, T. J. M. (2002). Ways to study – Descriptive research. In T. M. de Jong & D. J. M. van der Voordt (Eds.), *Ways to study architectural, urban and technical design* (Chapter 6, pp. 53–60). DUP Science, Delft.
23. Louis, M., & ElAzab, M. (2023). Will AI replace teacher? *International Journal of Internet Education, 22*(2), 9–21. https://doi.org/10.21608/ijie.2023.312491
24. Manjunatha, N. (2019). Descriptive research. *Journal of Emerging Technologies and Innovative Research (JETIR)*, 6(6), 863–867.
25. Pokrivcakova, S. (2023). Pre-service teachers' attitudes towards artificial intelligence and its integration into EFL teaching and learning. *Journal of Language and Cultural Education, 11*(3), 100–114. https://doi.org/10.2478/jolace-2023-0031
26. Pukėnas, K. (2009). *Kokybinių duomenų analizė SPSS programa* [Qualitative data analysis using SPSS]. LKKA.
27. Sămărescu, N., Bumbac, R., Zamfiroiu, A., & Iorgulescu, C. M. (2024). Artificial Intelligence in education: Next-Gen teacher perspectives. *Amfiteatru Economic, 26*(65), 145–161. https://doi.org/10.24818/EA/2024/65/145
28. Sanusi, I. T., Ayanwale, M. A., & Tolorunleke, A. E. (2024). Investigating pre-service teachers' artificial intelligence perception from the perspective of planned behavior theory. Computers and Education: Artificial Intelligence, *6*, Article 100202. https://doi.org/10.1016/j.caeai.2024.100202
29. Taherdoost, H. (2020). Different types of data analysis; data analysis methods and techniques in research projects authors. *International Journal of Academic Research in Management (IJARM), 9*(1), 1–9. https://ideas.repec.org/p/hal/journl/hal-03741837.html
30. Tunjera, N., & Chigona, A. (2023). Investigating effective ways to use artificial intelligence in teacher education. In Johnston, S. J., & Singh, S. (Eds.), *Proceedings of the 22nd European Conference on e-Learning, ECEL 2023* (pp. 331–340). Academic Conferences International. https://papers.academic-conferences.org/index.php/ecel/article/view/1625/1765
31. Upadhyaya, P., & Kaur, G. (2023). Smart multi-linguistic health awareness system using RASA model. In *2023 International Conference on Sustainable Computing and Smart Systems (ICSCSS)*, Coimbatore, India (pp. 922–927). IEEE. https://doi.org/10.1109/ICSCSS57650.2023.10169275
32. Williams, C. J. (2015). An investigation of K-12 teachers' attitudes toward computer technology use in schools. *Journal of Business & Economic Policy, 2*(1), 71–87.
33. Zhang, C., Schießl, J., Plößl, L., Hofmann, F., & Gläser-Zikuda, M. (2023). Acceptance of artificial intelligence among pre-service teachers: A multigroup analysis. *International Journal of Educational Technology in Higher Education, 20*, Article 49. https://doi.org/10.1186/s41239-023-00420-7

Mining and Farming of Cryptocurrencies: Comparison of Economic and Environmental Demands

Ondřej Malý, Marie Hubálovská, and Štěpán Hubálovský

1 Cryptocurrencies

Cryptocurrencies are a type of money created electronically. According to the General Financial Directorate [1], it is an intangible, movable, and fungible object. One of the most important roles is played by their decentralization.

The most widely known cryptocurrencies are Bitcoin (BTC), Bitcoin Cash (BCH), Monero (XMR), as well as Ethereum (ETH), popular among miners with graphics card mining machines, Solana (SOL), Polkadot (DOT), and many others. Among the flourishing ones, we find Dogecoin (DOGE), Shiba Inu (SHIB), and many others.

Currently, according to CoinMarketCap [2], there are over 25,000 different cryptocurrencies, but this number is constantly increasing. Among the new ones, we can find primarily various imitations based on existing, better-known, and more sophisticated cryptocurrencies, for example, based on Bitcoin or Ethereum, but also cryptocurrencies built on completely new foundations.

1.1 Features of Cryptocurrencies, Pros and Cons, and Risks

As such, cryptocurrencies have been associated with decentralization since the very beginning. It is widely known that their functioning is ensured through peer-to-peer networks that communicate directly between the individual end devices of users

O. Malý · M. Hubálovská · Š. Hubálovský (✉)
University of Hradec Kralove, Hradec Kralove, Czech Republic
e-mail: stepan.hubalovsky@uhk.cz

M. Turčáni (ed.), *15th International Scientific Conference on Distance Learning in Applied Informatics*, EAI/Springer Innovations in Communication and Computing,
https://doi.org/10.1007/978-3-031-81261-3_15

of a particular cryptocurrency network. Through them, all user communication and transaction confirmation (mining, farming, validation) is carried out without any intermediaries. As a result, there is no central authority, such as a central bank, in the cryptocurrency environment and thus cannot be controlled by governments or other institutions. At the same time, there is an element of transparency as all transactions are traceable in the network, but the older the payment, the harder it is to trace. However, as a negative feature, we can mention the impossibility of canceling the payment, as there is no banking institution present in a supervisory and intermediary capacity.

Due to the use of peer-to-peer networks, cryptocurrencies are global and it is therefore possible to send money to the other side of the world in a moment. However, some countries have already banned cryptocurrencies. In addition, sending cryptocurrency between accounts is free, or for a small fee that makes the transaction verified faster. Since we only need an electronic device (computer, laptop, tablet, mobile, etc.) with an Internet connection (or bitcoinmat) to send cryptocurrencies, a cryptocurrency wallet, this makes them very widely available. On the contrary, the list of services like shops or restaurants where they are recognized as payment is very limited until now.

Another feature is pseudo-anonymity. This is both an advantage, as no sensitive data is associated with the payments, and a disadvantage, as cryptocurrencies are unfortunately also used for illegal trading, corruption, and financing of illegal activities. For example, it is one of the main payment platforms in the invisible part of the World Wide Web called the Darknet, which is known for drug, arms, or illegal activities.

An important element of cryptocurrencies is their encryption using asymmetric cryptography, another name for public key cryptography, a group of cryptographic methods that work on the principle of using different keys for encryption, which is done using a public key and a private key is used to decrypt the content. Because of its level of security, asymmetric cryptography is also used for authentication with electronic signatures, digital certificates, or Hypertext Transfer Protocol Secure (https) for secure network communication.

As an investment asset, cryptocurrencies have the potential for high appreciation and can therefore provide a high return on investment. However, logically, high appreciation goes hand in hand with a high risk of loss of value, as their value is based solely on supply and demand, as with other investment assets—stocks, foreign currencies, commodities, and so on. For example, all it takes is for a famous person to post an article or comment on a social network for the price of a cryptocurrency to drop by several tens of percent.

1.2 Cryptocurrency Procurement and Payments

Cryptocurrencies can be acquired in several ways. They can be purchased at a currency exchange, like foreign currency, at a predetermined rate.

It is also possible to buy or sell for cash in a machine called a Bitcoin machine, which at first glance is similar to a conventional deposit machine and an ATM.

Cryptocurrencies can of course be purchased on the Internet.

It is also possible to trade or earn cryptocurrencies on these exchanges by verifying cryptocurrency payments, so-called mining or validating or farming.

Cryptocurrency Mining—Proof of Work System.

Cryptocurrency mining is the process of confirming and verifying transactions on a blockchain using a blockchain. It was described by Satoshi Nakamoto [3] in the accompanying documentation for the Bitcoin cryptocurrency network. In order for a transaction to be valid, it needs to meet the conditions, which are: noticeable movement in the user's wallet, correct electronic signature, and others.

Due to decentralization, there is no central computer on the network to provide transaction verification, so it is provided by a peer-to-peer network of user computers, called miners, who verify payments by the amount of work they do to verify them—creating a block of transactions. To avoid network domination, the difficulty of creating a block is artificially increased by the need to solve a certain problem involving computationally intensive activity and an element of chance, but everything, on the contrary, must be easily verifiable.

The incentive to use miners' devices to verify payments is to reward them for creating a block of transactions that needs to be calculated.

1.3 Validation—Proof of Stake System

The Proof of Stake system was created in response to the Proof of Work (POW) system, according to the Binance Academy portal in 2011 [4]. The reason for the creation of this system was the desire to reduce the high-power consumption as in the POW system.

In addition, no blocks are created for cryptocurrencies that use Proof of Stake because they were all created at once and their number never changes in the blockchain. This means that there is no reward for mining a new block in Proof of Stake, as there is with Bitcoin, Ethereum, or other cryptocurrencies that use POW, but with Proof of Stake, only transaction fees are charged.

Here, transaction validation, or block validation, is done by having a user called a validator selected by the system. To become a validator at all, a user needs to "stake" a certain amount of a given cryptocurrency within his or her node as collateral (the amount of cryptocurrency in the stake is usually larger than the transaction fee), which guarantees fair dealing and has more credibility. The larger the amount of cryptocurrency a validator pledges, the more likely it is for the validator to obtain a block of transactions for verification. The system-selected validator then verifies the validity of all transactions in the block. If everything is in order, it seals the block and adds it to the blockchain. As a reward, it then receives cryptocurrency equal to the transaction fee. If the validator is no longer interested in validation, the system will issue his cryptocurrency in pledge along with the rewards. There is a security delay between termination and the subsequent issuance of cryptocurrencies, in case some form of cheating is discovered.

Due to its nature, the system is suitable for investors, but it is not suitable for new cryptocurrencies starting out and is not yet as secure as POW, as it has not yet gone through the load phase and a hole in the system may appear. The advantage, as can be seen from the principle of operation, is to deter attacks by economic force. Cryptocurrencies using this system are, for example, the upcoming Ethereum 2.0 (ETH2), Polkadot (DOT), or, for example, Solana (SOL).

The undeniable advantage of the Proof of Stake system is that the process is not as energy-intensive as POW, which determines the amount of electricity consumed and does not require expensive and powerful hardware. It may therefore be one of the greener options for cryptocurrency mining. This may result in an increase in the popularity of cryptocurrencies running on the Proof of Stake system and thus increase their value.

The disadvantage for validators is that the rewards are only in the amount of the transaction confirmation fee, which has to be small because, otherwise, it would not be worthwhile to transact in that cryptocurrency.

1.4 Farming—Proof of Space and Time

After 3 years of development, a cryptocurrency called the Chia Network (XCH) was launched in May 2021, with its developers, led by Bram Cohen [5], aiming to compete with Bitcoin, but in a much greener way.

A new term has evolved from this, with the term "cryptocurrency farming" being used instead of "cryptocurrency mining," as well as the name of the cryptocurrency itself, Chia, which refers to the simultaneously very popular seeds of the desert plant Salvia Hispanica, which are called the same and are a big trend at the moment.

The developers also came up with an algorithm for farming called Proof of Space and Time (POST).

The POST Algorithm Consists of Two Functions

The Proof of Space function is the use of empty space (but at least 100 GB), primarily on a hard disk drive (HDD) or possibly a solid-state drive (SSD), where large data files are written, so-called "plots," which contain "imaginary chia seeds" (another name for hashes).

The Proof of Time feature ensures security and prevents falsification of Proof of Space calculations. It also ensures that the process is greener and that there is no need for parallel scaling of computing power.

The principle of farming: The user, called the farmer, determines the maximum size of the beds containing imaginary chia seeds and gives permission to generate and store them on disk. A bed of imaginary chia seeds is actually a predefined computation for the blockchain. When creating a bed, a 100 GB file, up to 1200 GB of data needs to be written to disk to create the bed, from which the final

100 GB bed is then calculated. In this way, the farmer prepares his flower bed. The program then sends prompts to the farmer every 18 s that contain a transfer transaction, and for each prompt, the farmer (or his cryptofarm) scans his "beds" to find a hash—an imaginary chia seed corresponding to the desired transaction block. If the user has the given imaginary chia seed, he then receives a certain amount (2 XCH at a time), and the block of payments is sent to the blockchain. In principle, it is clear that the probability of receiving a reward increases with a larger number of beds. The problem here is that if the Chia Network becomes a popular currency, it may cause a shortage of SSDs or HDDs.

Since the POST algorithm involves large amounts of data, SSDs are used, which perform I/O operations much faster than HDDs. Unfortunately, standard SSDs have a lifetime of about 300–600 TB of writes, so an SSD hard drive can create about 600 beds (this is equivalent to 60 TB of imaginary chia seeds/beds stored on the drive). HDDs are also used for farming, which performs I/O operations slower, but for the same purchase price, the farmer has several times more memory than if he/she were to purchase an SSD, so he/she can create more beds.

Starting in the second half of 2021, a new plotting method is available that reduces the amount of write I/O from about 1200 to about 300 GB for a 100 GB bay, with the rest of it taking place in RAM. The resulting plot can be stored on a standard HDD.

2 Comparison of POW Mining and POST Farming

How it was already mentioned in the abstract part, the article has two goals:

1. The first goal is to describe the most well-known principle of payment verification by the POW **mining** system as well as to describe the new principle of payment verification by the POST **farming** system.
2. Another goal is to determine, through an experiment, whether farming cryptocurrencies using the POST system is a greener option for verifying payments than mining using the POW system.

To confirm whether it is more environmentally friendly to mine cryptocurrencies using the POW system or to farm them using the POST system, a home experiment was conducted by an undergraduate student.

The student designed and built two computers: one computer that mines the cryptocurrency Ethereum using powerful gaming graphics cards and the other computer that farms the cryptocurrency Chia using 3.5 in. HDDs, which appears to be a potential greener option for cryptocurrency mining.

The experiment aimed to compare cryptocurrency mining and farming with respect to electricity consumption and annual earnings. The experiment lasted 1 year.

The level of environmental friendliness of the payment verification method will be based on the annual amount of electricity consumed by both machines, so that if one machine has a smaller annual earning, it will be theoretically scaled up and recalculated so that both machines have the same annual earning value.

Among the possible pitfalls of the experiment, we can include the volatility of currency exchange rates, while it is generally known that the volatility of cryptocurrencies is even higher than that of normal currencies—furthermore, the movements of electricity prices (CZK/1 kWh), physical variables, and the gradual wear and tear of electronic equipment.

2.1 Ethereum Currency Mining

Hardware

- ASUS motherboard
- Intel Core 2 Duo dual-core
- processor
- RAM
- Mustang 2 GB DDR3
- 2x Zalman 1000 W computer power supply
- 5x AMD Sapphire Radeon R9 390 graphics card
- PCI-express 3.0 adapter for 4 graphics cards.

The price of the mining machine was 33,600 CZK.

Software

The operating system used in the mining machine, due to the motherboard used, was Linux OS, specifically in its CentOS 6.0 and Hiveon OS distributions.

A connection to a mining pool was used to mine the cryptocurrency Ethereum. The environment nanopool.org was chosen for the experiment.

Ethereum Currency Mining Progress

The period from the second quarter of 2021 to the first quarter of 2022 has been selected as the critical period for determining the profitability of mining.

During the period, the machine mined the cryptocurrency Ethereum, which was exchanged after each quarter at the exchange into the Euro currency. The exchanged value between the euro and the Ethereum cryptocurrency includes the exchange fees.

In total, 1.6 ETH were mined during the experimental period (1 year), converted into 4121 EUR, equivalent to 104,014 CZK.

The electricity price was fixed at 4 CZK/kWh during the experiment period.

The average hourly electricity consumption of the mining machine was 800 Wh/h, which is equivalent to 19.2 kWh/day (76.8 CZK/day) and 7008 kWh/year (28,032 CZK/year).

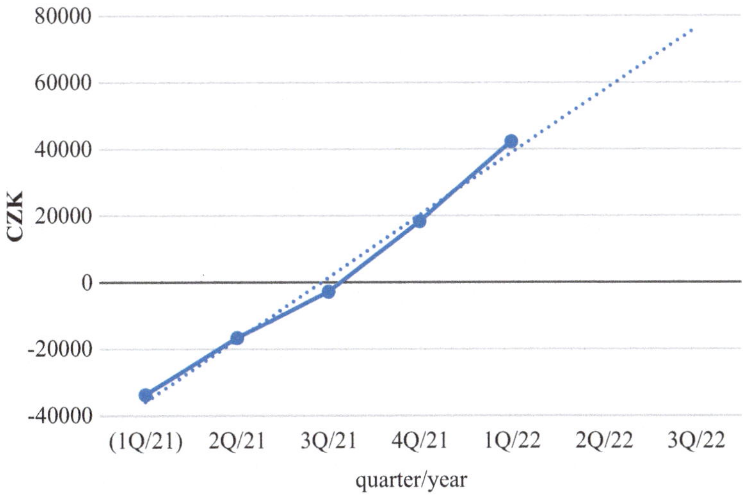

Fig. 1 Graph of *return* on investment in a cryptocurrency mining machine mining Ethereum

If we deduct the cost of the mining machine, which amounted to CZK 33,600, and the cost of electricity, which amounted to CZK 28,032, from the CZK 104,014 extracted, we arrive at CZK 42,382 in annual earnings.

We calculate the return on investment, which is the annual ratio of return to money invested, from the net annual earnings of CZK 42,382 and the cost of purchasing the mining machine and electricity, which together amount to CZK 61,632. From this, we get the result of the ratio, 0.688, which we then multiply by 100 to obtain the percentage value. The return on investment is thus 68% after the first year.

The graph in Fig. 1 shows that the cost of the mining machine has been paid for after only two quarters of mining. The deviations from the trend are due to exchange rate movements.

2.2 *Farming Chia Currency*

Hardware

- Baseboard
- B450M Pro4-F
- AMD Ryzen 53,600 processor
- 32 GB Kingston RAM
- 10x Toshiba Enterprise 3.5 in. HDD with 14 TB capacity

The price of the farming machine was 83,800 CZK.

Software

The operating system used was the aforementioned Debian OS. For farming itself, the official Chia currency program in the Linux Ubuntu/Debian variant was installed from GIT (distributed version control system).

Chia Currency Farming Process

As a critical period for determining the profitability, we take the period of Chia cryptocurrency farming from the beginning of January to the end of December 2022, when a debugged, stable cryptofarm was used for the experiment.

Over the course of the cryptofarm experiment, 7 hits were found and 20.2 XCH were farmed, corresponding to an average monthly earning of 1.69 XCH. In total, 20.2 XCH were farmed during the experiment period (1 year), exchanged for 792 EUR, equivalent to 19,221 CZK.

The electricity price was fixed at 4 CZK/kWh during the experiment period.

The average hourly electricity consumption of the cryptofarm was 120 Wh/h, which corresponds to an electricity consumption of 2.88 kWh/day (11.52 CZK/day). The annual consumption thus corresponds to 1051.2 kWh/year (4204 CZK/year).

If we subtract from the 19,221 CZK the costs of purchasing the cryptofarm, which amounted to 83,800 CZK, and the costs of electricity worth 6739 CZK, we, unfortunately, arrive at a net profit of −61,783 CZK after the first year.

The investment thus has a loss ratio of 78.78% after the first year.

The graph in Fig. 2 shows that the cost of the farm machine would be paid for after 4 years.

Comparison of Processes in Terms of Environmental Friendliness.

To compare the environmental friendliness of the two processes, it is necessary to recalculate some parameters. We do not take into account the origin of the electricity from the grid, whether it comes from fossil fuels or renewable sources.

The machine mining the cryptocurrency Ethereum in the experiment operated at a consumption of 800 Wh/h, extracting 104,014 CZK in a year.

For this purpose, we theoretically enlarge the Chia cryptofarm used in the experiment—specifically, 5.41 times to make the annual yield the same, which means expanding the capacity of the disks (by 44 pieces) and other components. Thus, the total purchase price of this cryptofarm would be 443,600 CZK instead of the original 83,800 CZK, which is 5.3 times more.

The power consumption, thus considering the consumption of 1 HDD hard disk of 5 W/h, would increase from the original consumption of the whole cryptofarm of 120 Wh/h to an average of 340 Wh/h by adding 44 more hard disks. This corresponds to a consumption of 2978 kWh/year.

Comparing the average power consumption of the Ethereum mining machine, which was 7008 kW per year, it can be concluded that farming the cryptocurrency Chia is roundly 2.4 times less energy intensive than mining the cryptocurrency Ethereum using a mining machine.

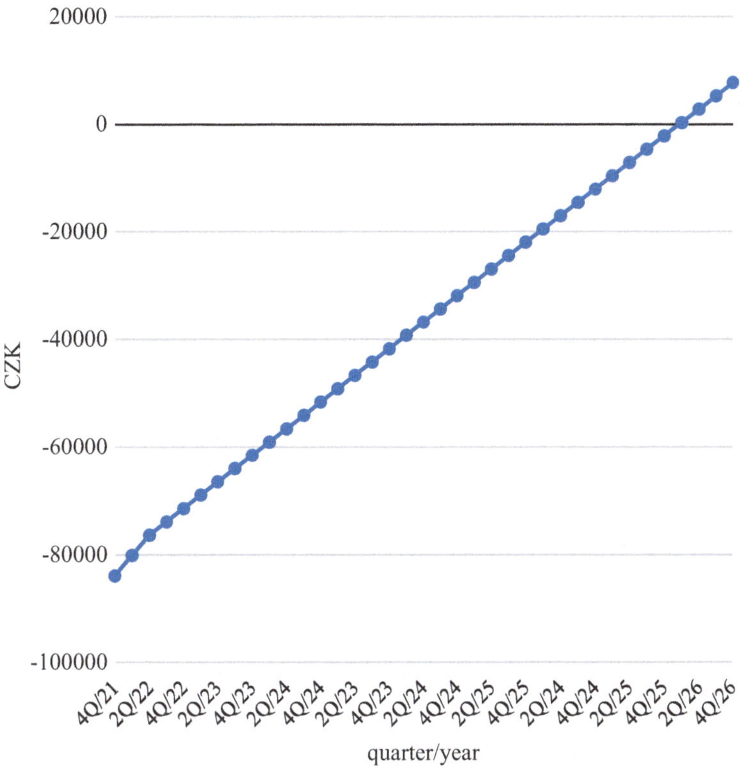

Fig. 2 Graph of return on investment in Chia cryptofarm

3 Conclusion

Ethereum cryptocurrency mining: a significant advantage of Ethereum cryptocurrency mining is a relatively fast return on invested funds (in our case, about 6 months).

Another advantage is the lower initial investment in the necessary hardware.

As a disadvantage, we must consider the high consumption of electricity, which, in addition to the computational power, is further converted into waste heat (waste heat can also be an advantage if the mining machine can be used as a heat source).

Cryptocurrency farming Chia: Here, the biggest advantage can be considered as several times less power consumption. Furthermore, there is almost no waste heat generated (in our experiment, passive cooling was sufficient, and even then the hard drives did not overheat.)

Thus, farming is significantly less energy intensive (2.3 times).

The disadvantage is a longer return on invested funds (profitability starts only in about the fourth year of farming, if taxes are not counted, otherwise even longer).

Furthermore, the disadvantage can be considered as a noticeably lower profitability, which includes a more financially demanding initial investment.

We cannot also forget the time spent fencing the drives, which can be months or years before the HDD hard drives are ready for actual farming. Further limiting is the lifespan of HDDs, which is estimated at 4–5 years.

Thanks to this, we can declare that under these conditions, farming Chia cryptocurrency is practically not worthwhile.

Future research will deal with a more detailed and long-term comparison of cryptocurrency farming and mining. The calculation will include costs that are missing in the above experiment (long-term development of energy prices, wage costs in the case of business, insurance, etc.).

Acknowledgments The article was supported by Specific Research Project of FacEdu UHK No 2126/2024.

References

1. Opluštilová, L. Information on the tax assessment of transactions with cryptocurrencies (e.g. Bitcoin). B.m.: General Financial Directorate. https://www.financnisprava.cz/assets/cs/prilohy/d-seznam-dani/Info_kryptomeny_GFR.pdf, last accessed 2022/03/30.
2. Chez, B. CoinMarketCap.com. CoinMarketCap. https://coinmarketcap.com/cs/, last accessed 2024/02/02.
3. Nakamoto, S. Bitcoin: A peer-to-peer electronic cash system 21260. B.m.: Decentralized business review. 2008. https://bitcoin.org/bitcoin.pdf, last accessed 2024/02/02.
4. Finance Academy. What is Proof of Stake (PoS)? Binance Academy. https://academy.binance.com/cs/articles/proof-of-stake-explained, last accessed 2024/02/02.
5. Cohen, B. Introduction to Chia. Chia network, Inc. https://docs.chia.net/introduction, last accessed 2024/02/02.

Virtual Reality in (Physics) Education: Impact on Critical Spots

Pavel Masopust, Jiří Kohout, Dominik Šilhánek, Libor Šmíd, and Miroslav Randa

1 Introduction

Virtual reality (VR) headsets have emerged as an innovative and highly effective educational tool that can revolutionize the way students learn and interact with educational content. By providing an immersive, interactive, and engaging learning environment, VR headsets have the potential to transform traditional educational methods, improve learning outcomes, and inspire a new generation of students.

1.1 Basics of VR Applications in Education

An Immersive Learning Experience VR headsets offer an unprecedented level of immersion in educational content, allowing students to become active participants rather than passive recipients of information. This level of immersion creates a sense of presence that helps students feel like a real part of the learning experience. Consequently, this can lead to a deeper understanding and longer-term retention of the material being taught.

Interactivity and Collaboration VR headsets allow students to interact with virtual objects, environments, and even other students or instructors in real time. This interactivity encourages collaboration and teamwork, which are essential skills in today's increasingly connected world. Students can participate in group projects

P. Masopust (✉) · J. Kohout · D. Šilhánek · L. Šmíd · M. Randa
University of West Bohemia, Pilsen, Czech Republic
e-mail: pmasop@kmt.zcu.cz

M. Turčáni (ed.), *15th International Scientific Conference on Distance Learning in Applied Informatics*, EAI/Springer Innovations in Communication and Computing,
https://doi.org/10.1007/978-3-031-81261-3_16

or problem-solving tasks, enhancing their communication skills and promoting the development of critical thinking.

Customizable Learning Environment With VR headsets, educators can create learning environments tailored to the needs and preferences of individual students. This flexibility allows for personalized learning paths that can adapt to different learning styles and paces. As a result, students who may struggle in traditional classrooms can thrive in a more adaptable and supportive virtual environment.

Increased Engagement and Motivation Using VR headsets can lead to increased student motivation. The novelty of the technology can spark curiosity, while the immersive nature of VR can make learning more enjoyable and less like an obligation. When students are more motivated to learn, they are more likely to absorb and retain information effectively. Additionally, the concept of gamification and the use of gamification principles in VR can further enhance motivation and engagement.

Experiential Learning VR headsets can bring abstract concepts to life and provide students with hands-on experiences that would be difficult or impossible to replicate in a traditional classroom. For example, students can virtually explore ancient civilizations, conduct science experiments in a simulated lab, or practice medical procedures on virtual patients. These experiences can help students better understand complex topics and make learning more relevant to their everyday lives.

Despite the many benefits, it is important to note that the widespread adoption of VR headsets in education also presents challenges. These include initial equipment costs, potential technical issues, privacy concerns, and the need for professional development for educators. However, as the technology continues to evolve and become more accessible, the benefits of using VR headsets in education are expected to far outweigh the challenges.

1.2 Current Progress of VR Education Research

Research on VR in education has attracted considerable attention in the literature since the 1990s, with numerous empirical studies evaluating the effectiveness of this approach. Several review studies and meta-analyses discuss current trends in this area based on empirical research findings, published in prestigious journals such as *Computers & Education* and the *Journal of Computer Education*.

VR is dynamically evolving, as is the definition of the term. Mikropoulos and Natsis [1] summarized the findings of 53 empirical studies on the use of VR in education, mostly in science and mathematics. From a technical perspective, interaction typically occurs through a keyboard, mouse, and stereoscopic glasses enabling spatial visual perception of a 2D object. From a didactic point of view, the authors highlight the issue of educational content processed in VR. Although a significant majority of the studies follow the view of Johnson et al. [2] that the chosen content and corresponding learning goal should be relevant, challenging, and have a clear potential for positive impact through VR, only a small portion of the empirical research based the choice of content on previous didactic research on learners' misconceptions and a sophisticated theoretical model. Such studies typically targeted older pupils and students. For example, the effectiveness of a VR teaching module for improving the understanding of quantum physics fundamentals was demonstrated, allowing first-year university students to experience being "inside a quantum atom" and better understand the physical laws of the microworld (Kontogeorgiou, Bellou, & Mikropoulos [3]).

The effectiveness of VR in education was addressed in a meta-analysis by Merchant et al. [4], based on 69 empirical studies focusing on educational games, simulations, and virtual worlds. It was shown that these approaches were effective in terms of learning outcomes compared to traditional methods. However, high effectiveness was observed in simulation studies where the authors used their own research tools (over 80% of cases), whereas the results were statistically inconclusive when using previously developed standardized tests. This raises the question of whether the reported effectiveness of VR was largely due to tailored measurement tools.

In recent years, attention has shifted toward the educational use of VR with head-mounted displays (HMDs). Early research includes a study by Kozhevnikova et al. [5], which focused on the conceptual understanding of relative motion in physics, previously identified as a problematic topic. It was shown that simulations using HMDs significantly improved performance (compared to computer simulations) on 2D tasks in a standardized conceptual test, while no difference was found for 1D tasks.

Jensen and Konradsen [6] analyzed 21 empirical studies on the effectiveness of HMDs in terms of various goals (cognitive, psychomotor, and affective). They found that studies typically focus on lower cognitive goals in terms of Bloom's taxonomy, such as memorization of facts. They identified a significant barrier to the wider deployment of VR in education as the high cost of developing modules and simulations, resulting in content being driven by VR developers (who do not typically focus on school education) rather than the needs of the education sector. Additionally, teachers or educators have limited opportunities to influence the creation of VR modules and are primarily users mastering the technology. The authors suggest that the essential question is not whether VR will continue to be used in education, but rather how and for which topics it will be utilized.

The most recent review study focusing on the use of HMDs in education was produced in 2020 (Hamilton et al. [7]). The rapid advancements in the field of VR are evidenced by the publication of numerous innovative empirical articles in the 2 years since the previous review, warranting the creation of another review study on the same topic. The authors analyzed 29 relevant publications from different perspectives, including content, educational objectives, and methodologies used. They found that VR was effective in about half of the studies focusing on cognitive objectives, particularly for complex and conceptual problems requiring a higher degree of spatial imagination and visualization skills. However, they also highlighted issues with the tools used to measure the effectiveness of VR interventions.

Another important factor related to the use of VR in education is the issue of negative side effects, specifically cybersickness or VR sickness. This topic has been addressed in a review study by Tian [8]. Researchers from the University of West Bohemia in Pilsen, where VR has been a focus for many years, have also explored this issue in their empirical research. Polcar and Hořejší [9] found that symptoms of cyber-sickness were manifested in more than half of the participants in a study using the Oculus Rift HMD, which focused on engineering issues.

The above suggests that although VR is frequently deployed in education globally (with Hamilton et al. [7] noting that 96% of UK universities are currently using it in some form), there has been a lack of a coherent didactic concept for selecting topics, designing relevant modules and interventions, and methodologically evaluating their effectiveness. The authors of this chapter worked on a project focusing on such a concept.

2 Concepts of Use in Education

The effective use of VR as an immersive technology in education is linked to two basic sets of issues:

- What learning content should be chosen, and how should it be implemented in a VR environment?
- What methods should be used in VR learning (including technical, organizational, and safety aspects)?

In line with the current widespread trend in education, the second set of questions is coming to the front attention. However, we believe that from the perspective of the didactics of specific disciplines, transdisciplinary didactics, and in line with the content-oriented approach in education (Slavík and Janík [10]), the first set of questions is at least equally important. This is also because, for most learners, VR is still a relatively new technology, and its deployment in the classroom can play a significant motivational role, largely regardless of the content (in a sense, VR is content in itself for learners). However, as VR becomes more prominent in various areas, this "newness" is expected to gradually fade away, and students will need to be engaged by the educational content itself.

2.1 *Possible Applications of VR in School Teaching*

The possibilities of using VR in the classroom are vast and varied. While this chapter focuses on its use in physics, here are two examples of useful applications/apps:

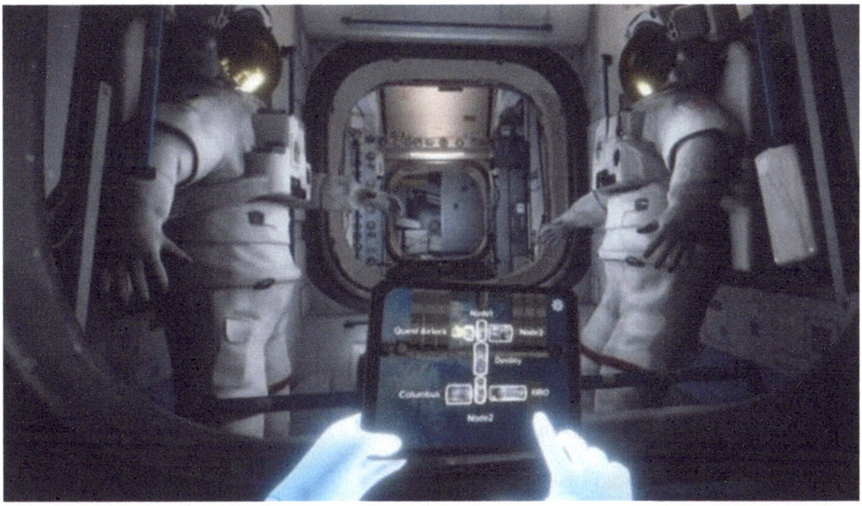

Fig. 1 International space station. (Source: direct screenshot from app in VR device)

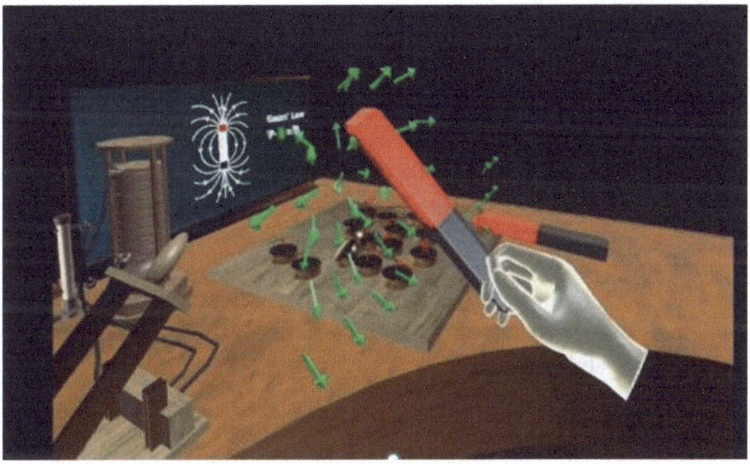

Fig. 2 Faraday's lab. (Source: direct screenshot from app in VR device)

International Space Station: accessible in store. This app lets users experience what it's like to be an astronaut on the International Space Station. They can interact with objects, navigate the station, and even go on a spacewalk (Fig. 1).

Faraday's lab: accessible in store. In this app, users will meet James Clerk Maxwell and Michael Faraday and can learn the basics of electromagnetism through interactive lessons and simulations (Fig. 2).

There are many possible applications of VR in education, including but not limited to: natural sciences and biology, language teaching, history, art, social and emotional learning, and education of people with special needs.

2.2 *Using VR to Address Critical Spots in the Curriculum*

The authors have been intensively studying critical spots of curriculum for a long time. The issue is described in more detail in their monography [11]. However, the criticality of the areas in question may have different causes, and it cannot be expected that VR will automatically help in resolving every critical spot. At the same time, the causes of criticality may be domain-specific. For example, in physics, the following causes are reported by Kohout [11]:

- Students' mathematical insufficiency.
- Problematic curricular continuity.
- Teacher's (dis)favorite topics.
- Insufficient school facilities.
- Attitude to didactic formalism.
- Excessive abstractness or complexity.
- Strong and difficult-to-overcome misconceptions of pupils' intuition.

For the first three points, VR cannot be expected to bring about a qualitative change. However, it can potentially address the problems of inadequate equipment in schools (students can experience experiments in VR instead of real life), excessive abstractness (immersion can reduce abstraction with appropriate content), and strong erroneous preconceptions (students can be directly convinced in VR that their intuition is failing, prompting a change in their thinking). VR can also help overcome didactic formalism, especially stolen cognition (the "parrot" metaphor), where students learn by memorizing without understanding. Conversely, if used incorrectly, VR can slide to the opposite extreme, concealed cognition (the "blissful fool" metaphor), where students are excited by the technology and playful context but fail to understand the subject matter because they cannot connect the VR experience with its underlying principles and the real world.

It should be emphasized that this risk exists even with well-crafted VR material (as it does with the best exhibits in science centers, etc.), and the only way to overcome it is through the quality work of a teacher or lecturer. They need to reflect on the VR experience with students, build on it, and connect it to the general laws of the field.

2.3 Concept of Use in Teaching

We conducted many classes in schools where we tested the concept of using VR. Here we briefly describe our attitude. The students were always completely immersed in the experience and the unusual situation after being placed in the VR space. It is not feasible to expect that an untrained user can put on VR headsets and immediately engage with an educational application. We found it useful to use a gradual adaptation method to VR, following a brief procedure:

- **First Introduction to VR:** Begin with the theory, a brief history, and an introduction to headsets. Provide safety and hygiene instructions.
- **First Steps Application:** Depending on the number of available VR headsets, distribute them among students. For the first encounter with VR, the "First Steps" app has proven to be effective. This free app for Oculus Quest and Quest 2 devices serves as an introduction to VR for new users. It is designed to help users become familiar with the VR environment and Oculus Touch controllers. The app consists of a series of interactive tutorials and mini-games that teach users how to navigate, interact with objects, and use the controllers effectively. Given the number of VR headsets, it is useful to divide students into groups and project the VR view onto another device (tablet, mobile phone) or a projector, so those without headsets can follow along. The "First Steps" app takes about 10 min to complete. Students who have already used the VR headsets can offer advice to those currently using them, complete a feedback questionnaire, or work on a related worksheet.
- **Educational Apps:** After the initial exposure to VR, introduce students to more advanced concepts and applications. Once they are accustomed to the VR environment, the immersion element remains strong but no longer interferes with the potential for "serious" educational use.

Screen Casting

During our visits to schools, we have often found that immersion is associated with a loss of contact with the surrounding environment. When a student puts on the VR headsets, the VR experience absorbs them, making it difficult for them to react to external instructions. If they get lost in the VR environment, either because they do not know how to operate the application or are unable to cope with the presented learning situation or task, it becomes challenging to assist them from the outside.

Instructions like "describe what you see and try to click on a control" only delay and disrupt the flow of the lesson. Therefore, it is almost necessary for the tutor or teacher to see what is currently displayed in the headset. Displaying an image on an external display is usually referred to as mirroring or casting. In Oculus Quest terminology, this is known as Cast. Oculus Quest supports three mirroring options:

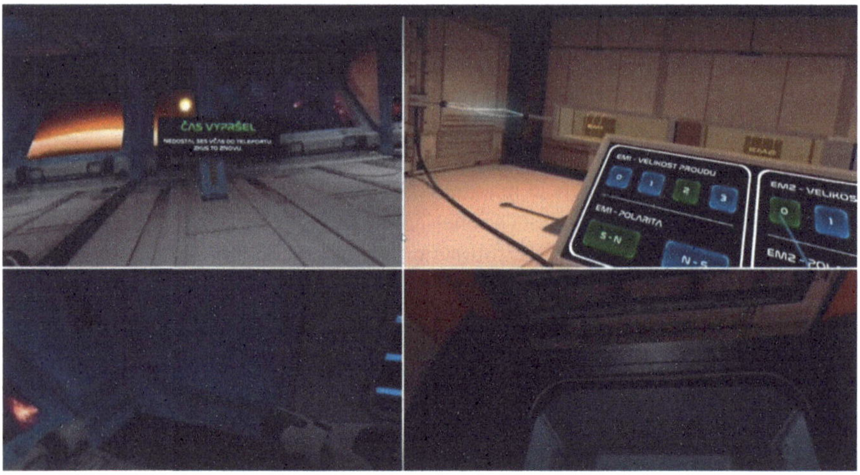

Fig. 3 Application OculusCaster. (Source: direct screenshot from PC app made by authors)

- **Display on a Mobile Device:** Each headset is associated with an Android or iOS device to which the headset image can be mirrored.
- **Mirroring to Chromecast:** Just as you can mirror to an app, you can mirror to a Chromecast device, a hardware device from Google that connects via HDMI to a compatible device (TV, projector, etc.).
- **Mirroring to a Web App:** There is a web application available at oculus.com/casting that allows screen mirroring.

For classroom use, we developed OculusCaster PC application. This is a Windows desktop application that allows image reception from the VR headsets. Figure 3 shows the OculusCaster application displaying four headset images simultaneously. The application can be set to receive images from up to six headsets, though multiple headsets are limited by the technical constraints of the Oculus infrastructure.

The application has been used in testing the developed applications and allows lecturers and teachers to instantly check what is happening in each headset. This makes it possible to immediately detect situations where the student is struggling, unsure of how to proceed, or does not know what to do.

For later analysis of the teaching unit, it is possible to record the PC screen, capturing four or six different approaches to solving the problem in one video. The data collected in this way is invaluable for the development of further applications and the creation of didactic and methodological principles for the use of VR in teaching.

3 Pedagogical Experiment with VR in the Context of Newton's Law of Action and Reaction

The law of action and reaction was identified as a critical spot in the curriculum of physics teaching at secondary schools in previous research conducted by the staff of the Physics Department of the Faculty of Education at the University of West Bohemia (ZČU). This research included identifying the criticality, justifying why this particular topic was critical, and designing experiments and tasks to address this issue.

A study of literature results, public comparative test results, and interviews with teachers identified the topic of action and reaction, that is, the content of Newton's Third Law, as a critical spot. This is echoed in international literature as well. For example, a study in *Physics Teacher* magazine [12] suggests reordering the teaching of Newton's laws so that the law of action and reaction is taught first. They report that this reordering has led to some improvement in students' understanding. Paul Hewitt, the founder of conceptual physics, also identified the law of action and reaction as the most problematic of all Newton's laws for understanding [13].

We designed a VR application targeting precisely Newton's Third Law.

The developed app (designed by authors and programmed by a professional VR studio) for the Oculus Quest headset places the student in a room on a spaceship, and their task is to get a key to the door.

The key is located on an EM1 electromagnet, but it is hidden behind an obstacle (a transparent plate) and cannot be reached directly. There are two windows in the barrier that could be used to remove the key, but the key needs to be moved there. In addition to the electromagnet with the key (EM1), there is another electromagnet (EM2). It has the same characteristics as the EM1, but is without the key, and near one of the windows, there is an ordinary magnet (M3) that can be placed in the barrier if necessary (see Fig. 4). The polarity and the magnitude (on a scale of 1–3) of the magnets can be adjusted.

The idea is that students will likely try to get EM1 to the window by manipulating EM2. They may attempt to set different currents in the two electromagnets to pull EM1 toward EM2, thinking that a higher current in EM2 will move EM1 while minimizing EM2's movement. However, using only EM1 and EM2, the problem is unsolvable because, according to the law of action and reaction, the forces between EM1 and EM2 are always equal, regardless of the current in either. Since EM2 is closer to the window, it always gets there first, and EM1 has no chance of reaching the window regardless of the parameters chosen.

The only way to get the key is to place the M3 magnet in the right-hand part of the barrier. By doing this, we can use the correct polarity to attract EM2 and then set the correct polarity of the currents to attract EM1 and EM2, allowing EM1 to reach the window (since EM2 will be attracted to M3 and will not block EM1).

The exact computation of this problem is quite challenging. Our focus was just on a schematic representation, showing that without M3, the velocities of EM1 and

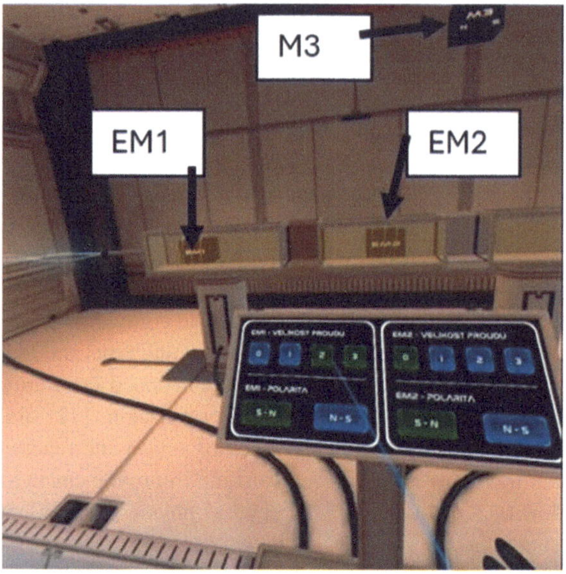

Fig. 4 Application "Key and electromagnet"

EM2 are always equal and proportional to the product of the magnitudes of the currents, causing EM1 and EM2 to meet halfway, to the left of the window.

3.1 Conduct and Results of the Pedagogical Experiment

To preliminarily assess the possible benefits of VR for overcoming the critical spots of the physics curriculum, a pedagogical experiment has been carried out. At one secondary technical school in Plzeň, four experimental classes (68 participants in the first year of study, i.e., typically 16 years old) and three control classes (56 participants of the same characteristics) were involved. All classes had the same physics teacher, and, in all classes, it was not the first lesson focused on the given topic in secondary school (the basic information about Newton's laws of motion had been provided several months before the research described). The experimental group first fulfilled the 25-item knowledge test described in [14], then experienced after 1 week one-lesson class with VR focused on Newton's laws of motion, and finally fulfilled the same test 1–2 weeks after the intervention. In the case of the control group, the same approach was used, but a classical lesson was provided instead of an experience with VR. The tests were fulfilled by pupils online using Google Forms; they had up to 45 min time for both pre-test and post-test.

The data were evaluated in terms of descriptive statistics for the total score of the knowledge test (1 point for the correct answer and 0 points for incorrect or no answer for each item, i.e., maximum of 25 points in the test). Additionally, the level of confidence of the answers to the individual group of items was evaluated on a 6-

point Likert scale (0 = just guessing, 5 = totally confident with the answer) and the perception of VR experience and its possible benefit for physics teaching by pupils were monitored based on several additional questions (see below) at the end of the post-test in the experimental group.

We found that the total score of the knowledge test slightly decreased from 12.94 ± 2.67 points (mean $\pm$ standard deviation) in the pre-test to 12.32 ± 2.78 points in the post-test in the case of the experimental group and from 12.31 ± 2.41 to 11.58 ± 2.59 points in the case of the control group. Based on the presented results, it is not possible to state that the intervention using VR contributed to a better understanding of the topic of Newton's laws of motion in comparison with the classical approach. Regarding the level of confidence, a very similar pattern was observed with only minimal change in the mean level of confidence from 2.48 to 2.51 points in the case of the control group and from 2.63 to 2.49 points in the case of the experimental group.

The additional questions revealed that the perception of VR by pupils was mainly positive with approximately 60% of participants believing strongly or partially that the VR experience contributed to a better understanding of the topic and 63% agreeing with the statement that VR has a significant potential to motivate pupils and help to keep attention during the lessons. Similarly, 93% believed that teaching with VR is more interesting in comparison with classical teaching and more than 80% agreed that VR is a valuable tool for physics teaching. Regarding possible improvements of VR for physics teaching, approximately half of the respondents would prefer to provide more diverse content and 30% would appraise more available sets and equipment for VR.

The presented results suggest that despite the positive perception of the VR experience by pupils, the pedagogical experiment using the modules developed was not successful in terms of improving the conceptual understanding of Newton's laws of motion. It may be partly due to the fact that it was the first experience with VR in an educational context for the overwhelming majority of participants. Thus, the physics content requiring intense thinking could be overshadowed by the practical issues of how to familiarize with the VR equipment provided and how to succeed purely technically in the quests that the pupils should fulfill during the VR experience. It would be probably beneficial to carry out before the intervention a lesson focusing on pupils getting familiar with the VR environment to make sure that during the intervention, all of them could be fully concentrated on the content and not be distracted by technical problems such as mounting of the headsets and so on. It would be thus beneficial to design the experiment in a different way and maybe focus in the first phase not only on (in terms of content) strongly challenging conceptual topics such as Newton's laws of motion where misconceptions are extremely common. It could be also beneficial to try the experiment on a gymnasium more focused on physics than secondary technical school.

4 Conclusion

The integration of VR in education presents a promising avenue for enhancing student engagement, motivation, and understanding of complex topics. Our exploration and experimentation with VR applications in school settings have revealed both the potential benefits and the challenges associated with this innovative technology.

The immersive and interactive nature of VR can transform traditional educational methods by offering students unique experiences that are difficult or impossible to replicate in a conventional classroom. For instance, VR field trips and interactive simulations can make abstract concepts tangible and foster a deeper understanding of subjects like physics and geography. Additionally, VR's ability to provide access to remote and global resources, along with its potential to reduce costs and environmental impact, underscores its value as an educational tool.

However, the deployment of VR in education is not without its hurdles. Our research highlights several critical spots that must be addressed to maximize the effectiveness of VR. These include ensuring students' familiarity with VR technology to avoid distractions during learning, developing a coherent didactic concept tailored to specific educational needs, and overcoming technical and organizational challenges. Moreover, the issue of cybersickness remains a concern, necessitating careful consideration of VR content and session duration.

An additional challenge lies in the limited availability of high-quality educational VR applications. Many existing VR applications are not produced by professional firms with a strong understanding of educational needs, resulting in content that may lack pedagogical value or fail to align with curriculum standards. This gap in professionally developed educational VR content makes it difficult for educators to find suitable tools that effectively support learning objectives.

Despite these challenges, the potential of VR to revolutionize education remains significant. The positive feedback from students regarding their VR experiences indicates a strong motivational impact and suggests that, with refined methodologies and continued development, VR can become a valuable asset in the educational landscape. It is crucial to continue exploring and addressing the practical and didactic aspects of VR integration to fully harness its potential for enhancing learning outcomes.

Acknowledgement This chapter was prepared with the support of SGS-2022-021 project provided by University of West Bohemia in Pilsen.

References

1. Mikropoulos, T. A., Natsis, A.: Educational virtual environments: A ten-year review of empirical research (1999–2009). *Computers & Education, 56*(3), 769–780 (2011)
2. Moher, T, Johnson, A, at al.: Bridging Strategies for VR-Based Learning, 536–543 (2019). https://doi.org/10.1145/302979.303153.

3. Kontogeorgiou, A., Bellou, J., Mikropoulos, T.: Being Inside the Quantum Atom. PsychNology Journal 6. 83–98 (2008)
4. Merchant, Z., Goetz, E., Cifuentes, L., Keeney-Kennicutt, W., Davis, T.: Effectiveness of virtual reality-based instruction on students' learning outcomes in K-12 and higher education: A meta-analysis. Computers & Education. 70. 29–40. (2014)
5. Kozhevnikov, M., Gurlitt, J.: Immersive and non-immersive virtual reality system to learn relative motion concepts. Proceedings of the 3rd Interdisciplinary Engineering Design Education Conference (2013), IEDEC 2013. 168–172. https://doi.org/10.1109/IEDEC.2013.6526781.
6. Jensen, L., Konradsen, F.: A review of the use of virtual reality head-mounted displays in education and training. Education and Information Technologies (2018). 23. 1–15. https://doi.org/10.1007/s10639-017-9676-0.
7. Hamilton, D., McKechnie, J., Edgerton, E., Wilson, C.: Immersive virtual reality as a pedagogical tool in education: a systematic literature review of quantitative learning outcomes and experimental design. Journal of Computers in Education (2020). 8. https://doi.org/10.1007/s40692-020-00169-2.
8. Tian, N., Lopes, P., Boulic, R.: A review of cybersickness in head-mounted displays: raising attention to individual susceptibility. Virtual Reality 26, 1409–1441 (2022)
9. Polcar, J., Horejsi, P.: Knowledge acquisition and cyber sickness: A comparison of VR devices in virtual tours. MM Science Journal. 2015. 613–616. https://doi.org/10.17973/MMSJ.2015_06_201516.
10. Slavík, J., Janík, T.: Kvalita výuky: obsahově zaměřený přístup ke studiu procesů vyučování a učení. Pedagogika 3/2012. 2012. 262–286.
11. Kohout, J., Masopust, P. & Mollerová, M. & Feřt, L. et al: 2019. Kritická místa kurikula fyziky na 2. stupni základní školy. I. Západočeská univerzita v Plzni, Plzeň. ISBN 978-80-261-0933-4.
12. Hestenes, D., Wells, M., & Swackhamer, G.: Force concept inventory. The physics teacher, 30(3), 141–158 (1992).
13. Hewitt, P. G.: Newton's (Often Misunderstood) Third Law of Motion. The Science Teacher, 84(2), 12 (2017)
14. Masopust, P., Kohout, J.: Nové experimenty a učební úlohy k zákonu akce a reakce jako kritickému místu kurikula. In: Proceedings of Veletrh nápadů učitelů fyziky, Brno (2021)

Impact of Redundant Link on Message Delivery in IQRF Networks

Gergely Sebestyen and Jozsef Kopjak

1 Introduction

Wireless sensor networks (WSNs) face a common challenge in determining the stability of radio connections. Environmental factors can terminate wireless connections between devices, rendering the network inoperative due to the absence of redundant routes. In IQRF mesh sensor networks, which use flooding routing, devices can be reached via several alternative paths [1–6]. This redundancy enhances network robustness by providing multiple routes for message delivery, thereby mitigating the impact of individual connection failures. Flooding routing involves all nodes in the network repeating the message rather than using the optimal route. While this approach increases robustness and redundancy, it also consumes more resources. In point-to-point communication, flooding routing is not optimal, but it becomes efficient for data collection in networks like IQMESH [11]. The stability of a network is significantly influenced by the number of redundant connections it possesses; the more redundant connections, the more robust and stable the network. However, determining which nodes can communicate with each other in a Time Division Multiple Access (TDMA) network using flooding routing requires a comprehensive network mapping solution. In IQMESH-like networks, routing addresses are allocated using a breadth-first search algorithm, but this method does not provide information about all existing network connections.

This chapter investigates the IQRF message delivery ratio for broadcast messages in networks utilizing flooding routing with TDMA media access. Previous research has shown that redundant paths enhance the robustness of these networks. However, after network setup and node discovery, IQRF networks do not display

G. Sebestyen (✉) · J. Kopjak
Kandó Kálmán Faculty of Electrical Engineering,Óbuda University, Budapest, Hungary
e-mail: sebestyen.gergely@kvk.uni-obuda.hu; kopjak.jozsef@kvk.uni-obuda.hu

M. Turčáni (ed.), *15th International Scientific Conference on Distance Learning in Applied Informatics*, EAI/Springer Innovations in Communication and Computing,
https://doi.org/10.1007/978-3-031-81261-3_17

all connections between nodes. To address this issue, we previously introduced network mapping methods that reveal all connections through the network map's adjacency matrix. Using these matrices, we tested how the number of redundant connections affects the packet delivery ratio, providing insights into optimizing network robustness and improving message delivery in IQRF networks [10, 12, 13].

2 Messaging in IQRF Networks

The topology of wireless networks determines the physical arrangement of network nodes, or devices, and the mode of communication between them. IQMESH-directed flooding routing leverages TDMA media access properties. Following the allocated time slots, nodes repeat the received message in their own time slot, enabling multi-hop message transmission across the network. While flooding routing is not optimal for point-to-point communication, using data collected from the network, as seen in IQMESH, can be efficient. If a wireless device is connected to the network through only one routing link, network communication can fail if this connection is lost. The more redundant connections present in the network, the stronger and more stable it is considered. Flooding routing, although slower, provides excellent robustness and reliability in densely routed networks. Instead of the optimal route, all network nodes repeat the message, resulting in a robust and redundant routing mechanism that uses more resources for message transmission between two points than the minimal required. Using flooding routing in a TDMA network results in a robust and redundant routing protocol. Figure 1 illustrates the message routing phases in an IQRF network that employs flooding routing. The process begins with the network coordinator (C) sending a message. Each routing node in the network repeats the received message in its assigned time slot, determined by its Virtual Routing Number (VRN). This process continues through multiple hops until the message reaches the addressed node (N3). The use of VRN addresses ensures that each node repeats the message in the correct sequence, enabling reliable multi-hop communication across the network [7–9, 14]

Each device in the network is assigned a unique logical network address upon joining, which can be done manually or automatically. In a TDMA network, every routing node repeats network messages to forward them to the recipient within their own time slot. Improper sequencing of devices can disrupt network routing. In IQRF networks, nodes can be placed in random order, and logical addresses cannot be used to determine TDMA time slots for proper routing. The discovery process in IQRF networks establishes network routing, during which nodes receive a VRN. The VRN indicates the routing distance from the network coordinator and determines the dedicated time slot for each node during routing. Zones are also established during discovery, indicating how many hops it takes to reach a node in a specified zone. Each zone's first node has the lowest logical address, finding devices in the next zone.

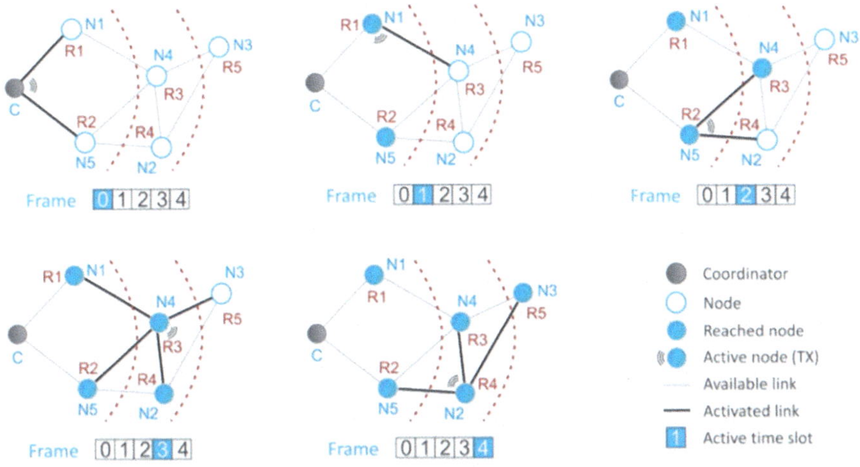

Fig. 1 Message routing phases in the IQRF network utilize flooding routing from the network coordinator (C) to node 3 (N3) using VRN addresses for routing [15]

The virtual routing sequence addresses this issue. In the IQHMESH network, the Discovered Full Mesh routing algorithm employs a Virtual Routing Structure [7–9, 14]. During discovery, the coordinator assigns unique VRN numbers to each routing device. The IQRF naming system's "discovery" procedure is based on a breadth-first search algorithm. During discovery, the coordinator identifies connected nodes and assigns them unique VRNs based on their route distance from the coordinator. The coordinator begins by finding its neighbors. Devices responding to search messages receive their VRN. All nodes with direct connections to the coordinator belong to Zone 0, making them directly accessible without routing. The coordinator then instructs all Zone 0 nodes to discover their first-hop neighbors, creating Zone 1, reachable by one routing hop from the coordinator, continuing until no more nodes are discovered [14]. After the final iteration, each discovered node has a unique VRN. Although discovery can identify all network devices, it creates a graph where the first node in the next zone has the lower logical address. Consequently, the network coordinator may not be aware of all network connections. An important fundamental property of the IQRF network routing mechanism is that routing nodes must receive the message before their assigned timeslot to forward it. In other words, if a node does not receive a message before its timeslot, it cannot forward it. In graph representation, this means that the in-degree of a node cannot be greater than its VRN number. This limitation impacts the availability of redundant paths and ultimately affects the message delivery efficiency. By ensuring that nodes receive messages in the correct order, the network can maintain robust and reliable communication, although this constraint may reduce the number of potential redundant paths available for message routing [4–6].

3 Simulation of Packet Delivery Ratio in IQRF Networks

In this study, we implemented a simulation to evaluate the packet delivery ratio in IQRF networks under various failure conditions. The simulation primarily focuses on broadcast message failures, which is designed to test how the presence of link failures affects the robustness and reliability of message delivery in the network. The simulation begins by creating a directed graph representing the IQRF network, with nodes and edges corresponding to network devices and their communication links. Each node in the network is assigned a VRN number, which dictates its time slot for message transmission in the protocol.

The simulation performs a broadcast operation from the network coordinator to all other nodes, repeating this process across each timeslot. During each timeslot, nodes that have received the message attempt to forward it to their neighbors. The simulation introduces a specified number of link failures randomly or consistently across timeslots, reflecting real-world scenarios where environmental factors or hardware issues may disrupt communication. The primary metric evaluated in the simulation is the packet delivery ratio, which measures the proportion of nodes that successfully receive the broadcast message despite the presence of link failures. By varying the number of link failures, the simulation assesses how the network's redundancy and stability factors influence its overall resilience.

The results of the simulation provide insights into the effectiveness of flooding routing in maintaining robust communication in IQRF networks. Higher redundancy in network connections generally leads to a higher packet delivery ratio, demonstrating the importance of multiple alternative paths for message transmission. The simulation helps in understanding the critical thresholds of link failures that an IQRF network can withstand while still ensuring reliable message delivery, guiding the design and optimization of more resilient WSNs.

3.1 Simulation Results

The simulations were designed to demonstrate the impact of network link failures on the message delivery ratio, which is defined as the proportion of nodes that receive a broadcast message. Given that IQRF network communication relies on flooding routing, every routing node repeats the broadcast message, allowing us to measure message reach across nodes. In our simulations, each node functions as a routing node. When a link failure occurs and no redundant link is available to relay the message, it is marked as a delivery error. We ran simulations for broadcast message delivery 10,000 times for each network configuration. Link failures were simulated by randomly disabling a link during each timeslot of the routing process, aiming to reflect more realistic link failure scenarios. Over the course of the simulations, we gradually increased the number of link failures from 1 to 6 per timeslot. The

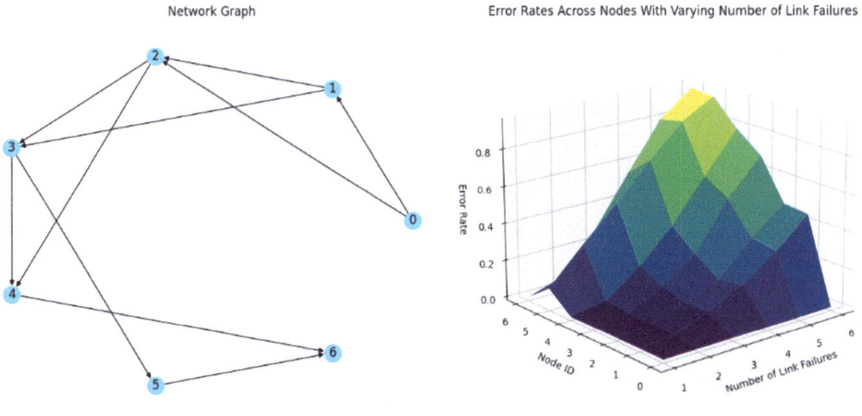

Fig. 2 Network 1—graph representation and the error rates result of the simulation

Table 1 Error rates of packet delivery for Network 1

Node No.	1 fail	2 fail	3 fail	4 fail	5 fail	6 fail
C	0.00%	0.00%	0.00%	0.00%	0.00%	0.00%
N1	0.00%	2.12%	8.54%	17.68%	32.02%	48.04%
N2	0.00%	2.12%	7.84%	16.40%	30.64%	47.44%
N3	0.00%	2.12%	11.11%	26.19%	46.87%	68.67%
N4	0.00%	4.70%	17.04%	35.24%	57.21%	77.20%
N5	10.75%	26.51%	46.66%	66.06%	81.62%	92.94%
N6	0.00%	6.74%	25.24%	53.48%	79.08%	94.04%

stability factor is defined by how many link failures the nodes can withstand without incurring delivery errors.

Simulation 1 The first simulation, depicted in Fig. 2, illustrates a network where most nodes have two redundant links, except for Node 5, which has only one link. Table 1 presents the error rates observed in this network configuration (Network 1). The results indicate that a single link failure does not disrupt network communications, except at Node 5. Due to its lack of a redundant link, Node 5 experiences a non-zero error rate even with a single link failure.

When two link failures occur, the impact extends to other nodes despite the presence of redundant links. This is because the failed links were randomly selected in each timeslot, occasionally resulting in both redundant links failing in successive timeslots, thereby preventing some nodes from receiving the broadcast message. As the number of link failures increases, the error rates rise significantly, underscoring the vulnerability of the network to multiple simultaneous link failures.

Simulation 2 In the next simulation, Network 2 is configured with all nodes having redundant links, including Node 5. The simulation results show that with one link

Table 2 Error rates of packet delivery for Network 2

Node No.	1 fail	2 fail	3 fail	4 fail	5 fail	6 fail
C	0.00%	0.00%	0.00%	0.00%	0.00%	0.00%
N1	0.00%	1.87%	6.18%	14.46%	26.50%	39.89%
N2	0.00%	1.87%	5.77%	13.46%	24.80%	38.93%
N3	0.00%	1.87%	6.84%	18.44%	35.18%	56.29%
N4	0.00%	1.87%	8.16%	22.15%	42.73%	63.90%
N5	0.00%	3.84%	14.58%	33.21%	55.97%	77.12%
N6	0.00%	3.68%	14.24%	34.94%	62.25%	84.58%

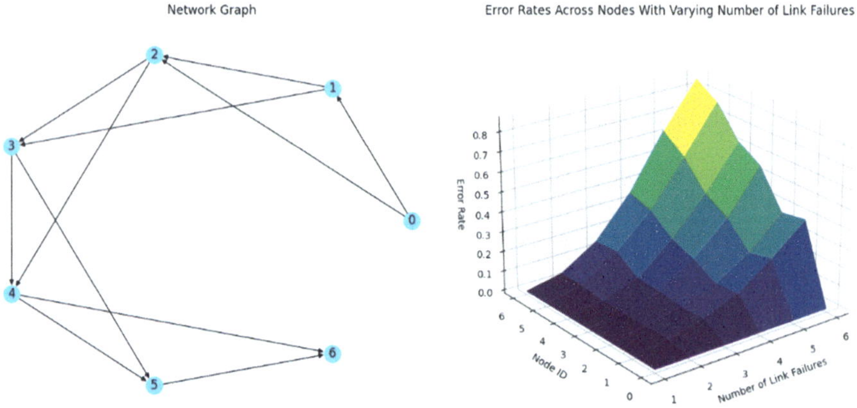

Fig. 3 Network 2—graph representation and the error rates result of the simulation

failure, there are no delivery errors. The redundant links effectively handle the single link failure, maintaining robust communication throughout the network.

However, when the network experiences two simultaneous random link failures in each timeslot, the delivery error rate remains low but not negligible. This indicates that while redundancy significantly enhances network resilience, it is not infallible. As the number of link failures increases further, the delivery error rate rises substantially, demonstrating the limits of redundancy in protecting against multiple concurrent failures (Table 2, Fig. 3).

Simulation 3 In the subsequent simulation, Network 2 was reconfigured to ensure that all nodes, including Node 5, had redundant links, as illustrated in Fig. 4. The results, shown in Table 3, indicated a significant reduction in error rates. This improvement is due to the increased number of links within the network, which decreases the probability of any single link being selected for failure. Node 5, which previously exhibited higher error rates due to its lack of redundancy, now demonstrates much lower delivery error rates, even at higher numbers of link failures. This is because the three redundant connections from its predecessors provide alternative paths for message delivery, reducing the impact of link failures.

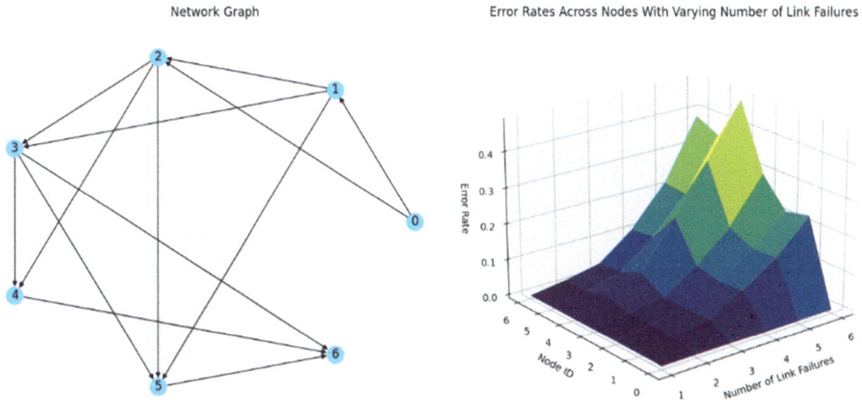

Fig. 4 Network 3—graph representation and the error rates result of the simulation

Table 3 Error rates of packet delivery for Network 3

Node No.	1 fail	2 fail	3 fail	4 fail	5 fail	6 fail
C	0.00%	0.00%	0.00%	0.00%	0.00%	0.00%
N1	0.00%	1.09%	3.53%	7.89%	14.45%	22.66%
N2	0.00%	1.09%	3.53%	7.82%	14.14%	22.48%
N3	0.00%	1.09%	3.53%	8.66%	17.54%	29.37%
N4	0.00%	2.32%	9.13%	19.72%	33.31%	49.53%
N5	0.00%	1.09%	4.62%	11.37%	22.42%	34.76%
N6	0.00%	1.09%	3.85%	10.20%	21.44%	37.88%

Thus, the network's overall resilience to random link failures is greatly enhanced, maintaining low error rates even under multiple simultaneous failures.

Simulation 4 The additional links were expected to increase the packet delivery ratio. In the next simulation, each node in the network was configured to have three redundant links from its predecessors, as shown in Fig. 5. This configuration resulted in a significant improvement in the network's performance, as illustrated in Table 4.

With up to two simultaneous link failures, the network maintained perfect communication, with all messages successfully reaching every node. When the number of simultaneous link failures increased to three, there were still some delivery errors, but the error rates were substantially lower compared to the network configuration with only two redundant links per node. This demonstrates that increasing redundancy to three links significantly enhances the network's resilience to link failures, ensuring higher reliability in message delivery even under adverse conditions.

Simulation 5 Modifying Network 4 by removing one of the redundant links from Node 5 resulted in higher delivery error rates when two simultaneous link failures occurred. The reduction in redundancy significantly impacted Node 5, leading to

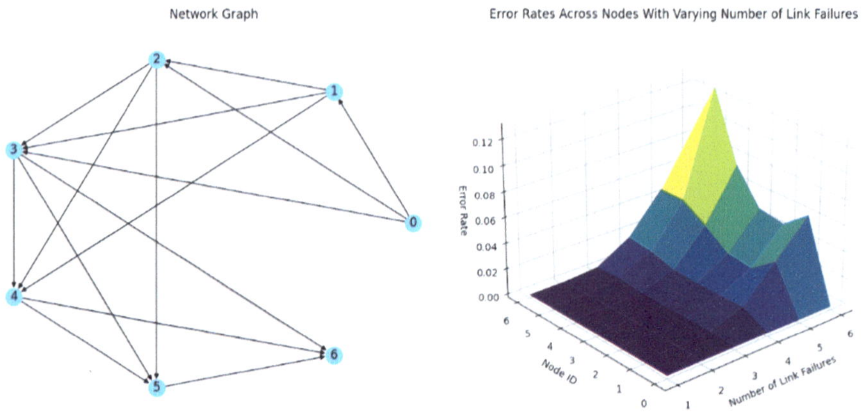

Fig. 5 Network 4—graph representation and the error rates result of the simulation

Table 4 Error rates of packet delivery for Network 4

Node No.	1 fail	2 fail	3 fail	4 fail	5 fail	6 fail
C	0.00%	0.00%	0.00%	0.00%	0.00%	0.00%
N1	0.00%	0.00%	0.24%	1.04%	3.30%	7.28%
N2	0.00%	0.00%	0.24%	0.87%	2.27%	5.46%
N3	0.00%	0.00%	0.24%	0.87%	2.25%	5.35%
N4	0.00%	0.00%	0.24%	1.08%	3.69%	8.81%
N5	0.00%	0.00%	0.50%	1.96%	5.51%	12.73%
N6	0.00%	0.00%	0.47%	1.74%	4.73%	10.89%

Table 5 Error rates of packet delivery for Network 5

Node No.	1 fail	2 fail	3 fail	4 fail	5 fail	6 fail
C	0.00%	0.00%	0.00%	0.00%	0.00%	0.00%
N1	0.00%	0.00%	0.33%	1.30%	4.27%	9.95%
N2	0.00%	0.00%	0.33%	1.39%	4.21%	9.82%
N3	0.00%	0.00%	0.33%	1.18%	3.40%	8.73%
N4	0.00%	0.00%	0.33%	1.75%	5.89%	13.93%
N5	0.00%	1.21%	3.74%	8.68%	15.76%	27.90%
N6	0.00%	0.00%	0.68%	2.29%	7.37%	16.62%

increased delivery errors, and also affected Node 6. This demonstrates that even a single reduction in redundancy can have a notable effect on network performance, particularly for nodes that rely heavily on multiple paths for robust message delivery (Table 5, Fig. 6)

Simulation 6 In the final simulation discussed in this chapter, we increased the number of redundant links to four per node. The results demonstrated that up to three simultaneous failures resulted in no delivery errors. Even with higher numbers of link failures, the error rates were significantly lower compared to earlier

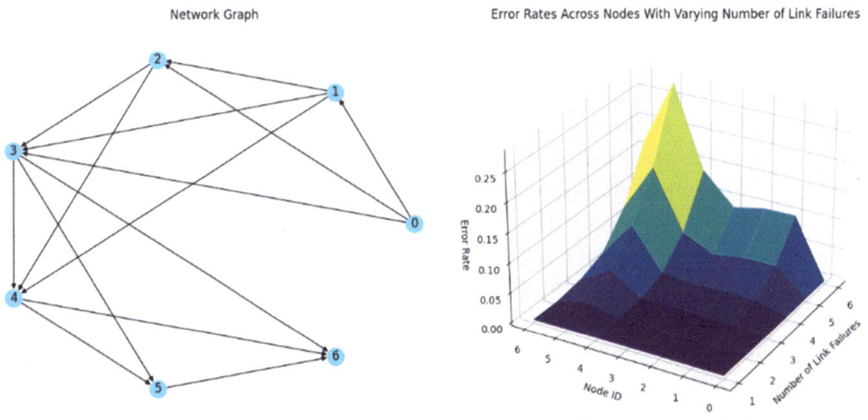

Fig. 6 Network 5—graph representation and the error rates result of the simulation

Table 6 Error rates of packet delivery for Network 6

Node No.	1 fail	2 fail	3 fail	4 fail	5 fail	6 fail
C	0.00%	0.00%	0.00%	0.00%	0.00%	0.00%
N1	0.00%	0.00%	0.00%	0.04%	0.15%	0.74%
N2	0.00%	0.00%	0.00%	0.04%	0.12%	0.78%
N3	0.00%	0.00%	0.00%	0.04%	0.14%	0.80%
N4	0.00%	0.00%	0.00%	0.04%	0.12%	0.83%
N5	0.00%	0.00%	0.00%	0.20%	0.57%	2.57%
N6	0.00%	0.00%	0.00%	0.05%	0.33%	1.32%

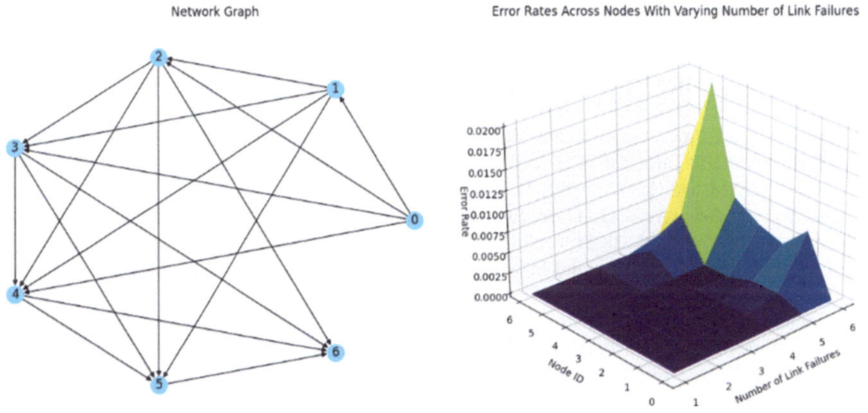

Fig. 7 Network 6—graph representation and the error rates result of the simulation

simulations. This highlights the substantial benefits of increased redundancy in improving network resilience and maintaining robust message delivery (Table 6, Fig. 7).

4 Conclusion

The simulation of packet delivery failures in IQRF networks highlighted the critical impact of redundant links on delivery error rates. One of the key findings from the simulations is the significant role that redundant links play in enhancing network robustness. By setting the minimum in-degree of nodes according to their VRN, the error rate can be effectively reduced to zero for up to three simultaneous link failures. This underscores the importance of ensuring adequate redundancy within the network to maintain reliable communication.

The simulations further demonstrated that increasing the number of redundant links significantly improves the network's resilience to link failures. Networks with higher redundancy exhibited much lower error rates, even under conditions of multiple simultaneous failures. This emphasizes that the strategic incorporation of redundant links is crucial for optimizing the robustness and reliability of IQRF networks, ensuring efficient message delivery even in adverse scenarios.

In conclusion, our findings provide valuable insights for the design and optimization of more resilient IQRF networks. By enhancing redundancy, network designers can significantly improve the robustness of communication systems, making them more reliable and capable of withstanding higher levels of link failures.

References

1. Pies, M.; Hájovský, R.; Velička, J. Wireless Measuring System for Monitoring the Condition of Devices Designed to Protect Line Structures. *Sensors***2020**, *20*, 2512, https://doi.org/10.3390/s20092512.
2. Kuchta, R.; Vrba, R.; Sulc, V. Smart Platform for Wireless Communication – Case Study; **2008**; p. 120; ISBN 978-0-7695-3106-9.
3. Kopják, J.; Sebestyén, G. Comparison of Data Collecting Methods in Wireless Mesh Sensor Networks. In Proceedings of the 2018 IEEE 16th World Symposium on Applied Machine Intelligence and Informatics (SAMI); February 2018; pp. 000155–000160., https://doi.org/10.1109/SAMI54271.2022.9780791
4. Kopjak, J.; Sebestyen, G. Deep Sleep Algorithms in Battery Powered TDMA Wireless Mesh Sensor Network. In Proceedings of the 2018 IEEE 18th International Symposium on Computational Intelligence and Informatics (CINTI); November 2018; pp. 000115–000118., https://doi.org/10.1109/CINTI.2018.8928209
5. Kopják, J.; Sebestyén, G. Energy Consumption Model of Sensor Nodes Using Merged Data Collecting Methods. In Proceedings of the 2022 IEEE 20th Jubilee World Symposium on Applied Machine Intelligence and Informatics (SAMI); March 2022; pp. 000027–000030., https://doi.org/10.1109/SAMI54271.2022.9780791
6. Sebestyén, G.; Kopjak, J. Network Mapping for TDMA Mesh Sensor Networks with Flooding Routing. (CANDO-EPE); November 2022; pp. 000153–000156., https://doi.org/10.1109/SAMI54271.2022.9780791
7. Sulc, V. Electronic Transceiver Module for Network Wireless Communication in Electric or Electronic Devices or Systems, a Method of Controlling It and a Method of Creating a Generic Network Communication Platform with Transceivers, US Patent 9336676B2, filed 2008-05-30

8. Sulc, V. System for wireless mesh network communication, US Patent 9743404B2, filed 2015-11-03
9. Sulc, V. Configuring a Wireless Mesh Network of Communication Devices with Packet Message Transmission, and Routing Packet Message Transmission in Such a Network, US Patent 20120163234A1, filed 2015-11-03
10. Gergely Sebestyen, Jozsef Kopjak, Estimating Battery Life in TDMA Mesh-based Wireless Sensor Networks, for Merged Data Collection Method. *Acta Polytechnica Hungarica***2024**, *21*
11. Sulc, V.; Kuchta, R.; Vrba, R. IQMESH Implementation in IQRF Wireless Communication Platform. In Proceedings of the 2009 Second International Conference on Advances in Mesh Networks; June **2009**; pp. 62–65., https://doi.org/10.1109/MESH.2010.17
12. Seflova, P.; Sulc, V.; Pos, J.; Spinar, R. IQRF Wireless Technology Utilizing IQMESH Protocol. In Proceedings of the 2012 35th International Conference on Telecommunications and Signal Processing (TSP); July **2012**; pp. 101–104.
13. DPA - IQRF Available online: https://www.iqrf.org/technology/dpa (accessed on 1 October 2023).
14. IQRF Tech s.r.o., IQRF OS User's Guide Version 4.03D for (DC)TR-7xD, 2018, Available online: https://www.iqrfalliance.org/data_files/news/user-guide-iqrf-os-403d-tr-7xd-181025.pdf (accessed on 1 Octber **2023**).
15. Obr1342_p240.Jpg Available online: https://cdn.sos.sk/novinky/obr/obr1342_p240.jpg (accessed on 19 December **2023**).

Mobile Citizen Science Applications for Teaching Biodiversity Within the Third Cycle of Elementary Education

Katarína Skokanová and Imrich Jakab

1 Introduction

Biodiversity, in the abbreviated form of the term biological diversity, encompasses diversity of genes within species, diversity of species within ecosystems, and diversity of ecosystems within the biosphere. The diversity at these three levels together creates biodiversity. In simplified terms, it is essentially the variety of life and its forms on Earth. Therefore, it is considered an invaluable natural heritage to be protected for future generations [1].

Species extinction has always been here, according to the Bratislava Regional Conservation Association (BROZ) [2], natural extinction represents a gradual but steady decline in the number of species. However, what we are currently experiencing is characterized as an indirect extinction, caused by human progress, causing loss or degradation of habitats. One of the greatest challenges to biodiversity preservation in today's world is climate change. Paradoxically, it is a two-way process. Climate change is one of the main drivers of biodiversity loss, but the destruction of ecosystems undermines nature's ability to regulate greenhouse gas emissions and protect against extreme weather, thus accelerating climate change and increasing vulnerability to it [3]. An effect of climate change is that native species are being replaced by invasive species with their consequences for reducing global biodiversity [4]. Lower biodiversity contributes to the global homogenization of the biosphere that will be a threat to food production and human health [5]. And so,

K. Skokanová (✉)
Katedra ekológie a environmentalistiky, Nitra, Slovakia
e-mail: katarina.skokanova@ukf.sk

I. Jakab
Springer Heidelberg, Heidelberg, Germany
e-mail: lncs@springer.com

M. Turčáni (ed.), *15th International Scientific Conference on Distance Learning in Applied Informatics*, EAI/Springer Innovations in Communication and Computing,
https://doi.org/10.1007/978-3-031-81261-3_18

whatever is done, or not done over the next few decades will determine how many animal and plant species, ecological communities and natural areas will survive [6].

The vast scope of biodiversity regarding the identification and characterization of plant and animal species is often taught in classrooms and from textbooks. According to the recent critical study "What Are You Taught in School?" [7]. Slovak students demonstrate weaker performance in global issues requiring critical thinking and awareness of interconnections compared to other OECD countries. There is also a weak connection between students' theoretical knowledge and real-life situations. However, biodiversity topics require broader contextual thinking. The protection of biodiversity requires active student involvement and physical activity in the field or nature. That is why biodiversity studies have great potential to become engaging and interactive through the use of information and communication technologies and outdoor activities, and it is increasingly important to consider this potential for learning and teaching opportunities in both formal and non-formal education [8].

One teaching approach that actively engages students in different fieldwork outside the school [9] and increases their awareness and understanding of environmental issues is citizen science (CS). This can be achieved through the use of CS smartphone applications that are offered by the platforms within their educational programs. First, students will learn much more about various plant and animal species when they see them with their own eyes, observe and examine them closely, and further relate them to other phenomena [9], but especially to environmental problems caused by human activity, resulting in a decline in biodiversity. Second, students will be much more motivated because they will contribute their bit to science, which is needed for change toward a better environment, and they will not feel like they are doing certain activities in school just because the teacher ordered them to do, but because it will have a greater cause.

The attention of this chapter has been focused on how the topic of biodiversity intersects with other subjects within the natural sciences. Our objective is to harness the potential of school environments or other educational institutions, by involving students, and potentially their family members, and put them in the role of citizen scientists for gathering various biodiversity-related data across Slovakia through CS programs.

2 Citizen Science

CS also called community-based monitoring [10], participatory sensing [11], or crowd-sourced science uses principles of open science and is a form of open collaboration in which volunteers, who are called citizen scientists, participate in the scientific process in various ways [8], mainly in a collection of various research data in relatively short periods of time, contributed by world-wide science projects [12]. Buytaert et al. [13] defined CS as "the participation of the general public (i.e., non-scientists) in the generation of new knowledge." Dawson et al. [14] add that it is "a means of engaging the public with science on the scale of individual experiments, creating a unique position of combining participation, monitoring, and social change." CS utilizes various platforms for the purposes of gathering or analyzing data, whether they are mobile applications with different tools or websites where data is published.

By involving students as citizen scientists, we can tap into an expansive and diverse pool of data contributors. Students can actively participate in scientific research, making significant contributions while gaining valuable educational experiences. If 1000 students participated in a single day during a Mapping Challenge or monitoring initiative, the amount of data collected would be much larger than using conventional research methods that do not have these possibilities.

However, the right credibility-building strategies have to be chosen while intending the project's vision [15, 16], participant recruitment and motivation [17], and the use of emerging technologies [17, 18]. This would not only contribute to monitoring of the spread or decline of certain plant and animal species, for example, due to climate change effects, but also to recording new unknown invasive species, so-called aliens [19]. And to be the first to photograph something, in a location where no one has ever seen it before, can be even more exciting.

CS projects have the potential to increase, in the first place, volunteers' knowledge of the field of study, such as ecology and conservation [20]; to engage in deliberation about conservation management actions [21]; and to promote positive attitudes toward nature [22]. Second, the outcomes of CS projects, including advancements in scientific research, have the potential to increase understanding of science or so-called "scientific literacy" for students [23]. This can be achieved through their own engagement in activities related to the procedures of working with the data or also contributing to decision-making about the direction of the project, at a higher level of participation.

Scientific investigations include many steps or activities in which the participants can be involved, and these are: choosing or defining questions for study; gathering information and resources; developing explanations (hypotheses); designing data collection methodologies (experiments, observations); collecting data; analyzing data; interpreting data and drawing conclusions; disseminating conclusions/ translate results into action; and discussing results and asking new questions that arose [8].

CS models involve a variety of activities and tasks at different levels of participation. From an educational perspective, these models differ by involving the public in the steps by altering the amount of control that participants have over the different steps that according to Bonney [8] can be divided into three major levels:

1. **Contributory projects**, which are generally designed by scientists and for which members of the public primarily contribute data;
2. **Collaborative projects,** which are generally designed by scientists and for which members of the public contribute data but also help to refine project design, analyze data, and/or disseminate findings;

3. **Co-Created projects,** which are designed by scientists and members of the public working together and for which at least some of the public participants are actively involved in most or all aspects of the research process.

Shirk [24] divides models of the CS projects into these five categories adding two more to the three above, according to what is the source of initiative for answering the research questions (outside the educational area, etc.):

4. **Contractual projects,** where communities ask professional researchers to conduct a specific scientific investigation and report on the results;
5. **Collegial contributions**, where non-credentialed individuals conduct research independently with varying degrees of expected recognition by institutionalized science and/or professionals.

In the field of CS, we can encounter organizations or institutions that offer not only different forms of engagement, such as training and workshops, community science events, and research partnerships with schools, universities, and municipalities, but also individual educational programs, with different levels of collaboration in project design, and all of these require different tools. While some projects use DIY (do-it-yourself) sensors or "citizen science kits," modern technologies offer all-in-one experiences through smartphone apps. These apps, with their online platforms, offer interactive features for more seamless data upload and collaboration.

Regarding the expansion of CS in Slovakia, CS is still in its early stages and has not that much developed systematic promotion so far. However, in June 2021, the Slovak Government approved the "National Strategy for Open Science in Slovakia" [25] starting with the first Open Science Action Plan 2021–2023. There are several Slovak CS projects that have been running for quite a few years already that may not identify themselves as such because the term was not yet widely used at the time of their inception. However, Slovak citizen scientists, collaborating with universities and NGOs, primarily engage in international projects facilitated by user-friendly platforms such as PlantNet, GLOBE Observer, Seek by iNaturalist, and Bioblitz.

3 Methodology

Since September 2023, there has been an education reform taken off, which brought a change in the State Educational Program [26] and that is currently being tested in 39 primary schools in Slovakia, to be implemented by 2026. As a result, the curriculum is no longer set for individual grades of primary education (ISCED 1) and lower secondary education (ISCED 2), but three cycles of primary education: the first cycle includes first to third grade, the second cycle consisting of fourth to fifth grade, and the third cycle consisting of sixth to ninth grade.

In the past, environmental education was included as one of the cross-curricular themes. Since September 2024, environmental literacy has been elevated as part of the new national curriculum, and environmental topics are directly included in the curriculum of educational fields and individual subjects. It covers environmental topics including ecosystem disturbances, biodiversity conservation, restoration, and impacts on ecosystems' biotic and abiotic components.

In this chapter, we have focused on the topic of biodiversity, more specifically where it intersects with topics in the educational field of "Human and Nature" (Biology, Physics, Chemistry), which can be tackled using the CS methods. In the following section, we present the selection of CS mobile applications/platforms suitable for lower secondary (formal) according to the following criteria:

1. Aligning with the biodiversity topics in the curriculum for the third cycle, specifically focusing on the following biodiversity topics, by Primack et al. [6]:
   - **What is biodiversity** (understanding species, ecosystem, genetic biodiversity).
   - **Biodiversity protection** (protection of species populations, protected areas).
   - **Importance and value of biodiversity** (economic value, protection of water and soil resources, climate regulation, recreation, education, consumer and production value, existential value).
   - **Threats to biodiversity** (habitat destruction and fragmentation, environmental degradation and pollution, global climate change, overexploitation of resources, invasive species, genetically modified organisms).
   - **Species extinction** (genetic variability loss, extinction rates, island extinction).
2. Applicable in the educational process for the third cycle for subjects such as Biology (B), Chemistry (CH), and Physics (PH).
3. Suitable for use in locations in Europe, with international applicability.
4. Appropriate for younger learners aged 10–15 years.
5. Functionality of mobile applications and seamless operation.

The actual process of selecting the apps was carried out according to the conditions mentioned above. However, the first condition that we looked at was to select applications in the given topics usable in Europe, including Slovakia, since research mostly pertains to various locations. First of all, we have looked at websites/platforms that gather individual projects/apps whether users can use them globally or in specific countries. For example, Zooniverse gathers CS projects mainly from the UK, Österreich Forscht from Austria, and so on. To gather Czech projects, Tomáš Bat'a University in Zlín and the Institute of Geonics in the Czech Republic have created a platform called Citizen Science.cz. In Slovakia, the Ministry of Education, Research, Development and Youth with the Centre of Scientific and Technical Information of the Slovak Republic has created a platform OpenScience.sk, where we have chosen three applications for this selection. Other platforms have also helped, such as Sci-Starter, which covers the USA, which groups thousands of projects from all over the world, where you can filter a specific location, and where we found three apps. Three other apps were found on the European platform Eu-citizen.science, which has equally good search capabilities. The next two apps were found on the Spotteron platform, which is also a brand of a series of own apps. Another four apps we already knew from our own use and use at partner institutions DAPHNE Institute and ERA Malta—the other two applications we have found in scientific papers [38, 39].

> National Geographic collects CS projects that are mostly about specific species and research questions related to them. And thus our other criterion, that we became aware of during the search, was that all apps should have a more general focus and be applicable to several species of animals or plants within a given topic. Another important verification of the applications was whether these applications were really designed and intended for data collection by users to be used for further research because there are many apps on the market that provide a database of data for identification or viewing additional information, but the user does not contribute any inputs. Among the last verification was the difficulty of the topic, the research for the specific age group, and the functionality of the application itself, which caused us to cut several apps from the list.

The results of the chapter will be stated in this unified structure:

- Curriculum theme + subject: The theme and topics are taken from the educational standards of the educational area "Man and Nature" for the third cycle [26] in the content standard, which is divided into a main curriculum theme, a subtopic and one additional subtopic from the section: Concepts, Relationships, Facts.
- Suitable apps: Here we present a selection of suitable CS mobile applications for the curriculum topics, a brief description of the apps' focus, and then how the above curriculum topics relate to biodiversity.
- Source: Websites where the specific apps are found and described also provide links for downloading the apps for Apple and Android operating systems.

4 Results

The list of recommended CS mobile applications/platforms focused on biodiversity topics overlapping with the curriculum of the third cycle [26]:

4.1 What Is Biodiversity

Curriculum theme (B): *Organism—Structures, organ systems of organisms and their functions—Body structure of plants and fungi.*

Suitable apps: **Mushroom Observer**: A globally used app for the identification of fungi and **Pl@ntNet**: a globally used app for the identification of plants, that observes morphological characteristics of plants and fungi and the diversity of their structures that are unique to each species and help determine their functions in ecosystems.

Source: https://scistarter.org/mushroom-observer, https://otvorenaveda.cvtisr.sk/en-gb/citizen-science-projects-in-slovakia/

Curriculum theme (B): *Organism—Life processes of plants and animals—Methods of reproduction*

Suitable app: **Spot-a-Bee**: European app for monitoring bee populations and their behaviors, to help understand genetic diversity and habitat preferences of pollinator species that determine variation and adaptation across species which is ultimately crucial for biodiversity conservation.

Source: https://www.spotteron.com/spotabee/info

Curriculum theme (B): *Evolution—Variability of organisms—Exploring the diversity of current and fossil organisms.*

Suitable app: **Fossil Discovery**: European application that is focused on the identification of fossils in images captured in the field, which helps to understand the evolution of biodiversity. Evolution and variation of organisms explore the diversity of current and fossil living things.

Source: https://apps.apple.com/us/app/fossil-discovery/id1583917783

Curriculum theme (B): *Organism—Structures, organ systems of organisms and their functions—Body structure of invertebrates and vertebrates.*

Suitable apps: **iNaturalist**: globally used app for identification and observation of animals; **eBird**: global app that serves as a database of bird observations, recording of sightings of wild birds. Comparing morphological traits of animals helps to understand their evolutionary development and adaptations to various environments.

Sources: https://www.inaturalist.org/, https://ebird.org/about/ebird-mobile/

Curriculum theme (B): *Heredity and Variability—Laws of inheritance—Crossing*

Suitable app: **INCREASE CSA**: home-based app for conservation of genetic resources on food legumes to develop novel varieties for healthier food production with the exchange of seeds via the app. Genetic principles are crucial for comprehending biodiversity at the molecular level and this app helps to understand genetic variation and inheritance patterns that give us insights into the diversity of traits within and between species.

Source: https://eu-citizen.science/project/341

4.2 *Biodiversity Protection*

Curriculum theme (CH): *Substances—Properties of substances—Waste separation*
Suitable app: **TrashOut**: a global app for reporting, monitoring, locating, and tracking illegal waste dumps, marking their GPS, and sharing information with local authorities. To preserve biodiversity also means the more separated waste the less waste in the environment and the less toxic substances that can harm or kill organisms in ecosystems. In addition, recycling reduces the need to extract new raw materials.
Source: https://www.trashout.ngo/#TrashMAp

4.3 *Importance/Value of Biodiversity*

Curriculum theme (B, CH, PH): *Organism—Life processes of plants and animals—Photosynthesis*
Suitable app: **PhotosynQ (PARSynQ)**: a global app for measuring photosynthetic parameters derived from absorbance and fluorescence measurements with a MultispeQ device that provides a detailed picture of the health of the plant used for plant breeding. The diversity of photosynthetic plants and microorganisms contributes to the resilience of ecosystems to environmental changes and pressures. Conversely, changes in biodiversity can affect photosynthetic processes in ecosystems and thus their stability. Source: https://scistarter.org/photosynq

4.4 *Threats to Biodiversity*

Curriculum theme (PH, B): *Interactions—Optical phenomena—Light pollution*
Suitable app: **Loss of the night**: a global app for recording the number of visible stars in the night sky at certain locations contributing to data on light pollution worldwide that is used for research on its impact on ecosystems and human health. Light pollution affects the life cycles of many animals and ecosystems. Artificial light can disorient various nocturnal species, such as birds, insects, and frogs, and affect their migration, reproduction, and behavior. Source: https://scistarter.org/globe-at-night

Curriculum theme (PH, B): *Interactions—Acoustic phenomena—Noise pollution* Suitable apps: **NoiseCapture**: a global app for measuring the noise level in surroundings using sensors on smartphones, **Hush Citylab**: global app for measuring the degree of silence in urban areas using sensors on smartphones, which can be used by policymakers for future planning. Animals such as birds, mammals, and fish can be affected by excessive noise from traffic, industry, or other human activities. This noise can affect their behavior, communication, migration, and even reproduction. Sources: https://noise-planet.org/noisecapture.html, https://scistarter.org/hush-city

Curriculum theme (CH, B): *The Earth—Exploring phenomena in the atmosphere and their impact on the environment—Individual and societal impact on climate change* Suitable app: **GLOBE Observer**: a global app for recording data in meteorology, hydrology, phenology to record seasonal events that are sensitive to changes in climate. Human activities causing CO_2 gases contribute to climate changes, such as increased temperature, and alter the timing of plant flowering which can affect the availability of food for pollinators. If plants and their pollinators are out of sync, it can reduce the reproductive success of both.

Source: https://daphne.sk/program-globe-sa-na-slovensku-osvedcil/

Curriculum theme (B): *Ecosystem—Consequences of disturbing the biotic and/or abiotic parts of an ecosystem—Environmental pollution.*

Suitable app: **EyeonWater**: European app for observation of water color that indicates the algae and organic content of the water that feeds organisms. Disturbance of abiotic components of the ecosystem, such as water pollution, alters the conditions for the survival of organisms and can lead to the extinction of sensitive species, thus significantly reducing biodiversity. Source: https://eu-citizen.science/project/103

Curriculum theme (CH): *Ecosystem—Consequences of disturbing the biotic and/or abiotic parts of an ecosystem—Environmental pollution*

Suitable app: **Dnes dýcham**: European app for measuring CO_2 concentrations, providing real-time data on air quality. Disturbance of abiotic components of the ecosystem, such as air pollution, alters the conditions for the survival of organisms and can lead to the extinction of sensitive species, thus significantly reducing biodiversity.

Source https://otvorenaveda.cvtisr.sk/en-gb/citizen-science-projects-in-slovakia/

Curriculum theme (B): *Ecosystem—Ecosystem interactions—Assessing the impact of invasive species on ecosystem relationships*

Suitable apps: **VISITOR**: Slovak app for monitoring of distribution, abundance, and ecology of non-native species; **Invasive Alien Species Europe**: European app for monitoring and managing invasive alien species in Europe. Invasive species displace native species and disrupt animal relationships in ecosystems, changing the structure and functioning of ecosystems and upsetting its balance, leading to biodiversity loss.

Sources: https://era.org.mt/citizen-science/ https://otvorenaveda.cvtisr.sk/en-gb/citizen-science-projects-in-slovakia/

5 Discussion

CS is a broad and complex concept that encompasses various forms of public participation in scientific research, where citizens, students, or other volunteers actively collaborate with scientists to address scientific questions. When school gets outdoors with CS activities, lessons become more efficient and up-to-date, also thanks to using smart technologies [12]. According to Doyle [27], CS has great potential for use in formal education, highlighting especially these benefits: fostering their curiosity, building a relationship toward their immediate environment, and providing them an unprecedented engagement with professional scientists. Moreover, CS projects help students develop presentation skills, which are essential for their future careers [9, 28].

In the chapter, we focused on the selection of apps applicable in teaching the topic of biodiversity for the third cycle in elementary schools. The results highlight 21 apps addressing various biodiversity topics, following the framework provided by Primack et al. [6]. Some CS projects along with their apps also offer educational programs involving their use. **Project Noah** [29, 30] offers an online curriculum called "Project Noah Nature School" for elementary and middle school students. This community formerly had a mobile app but is now web only. Probably, the most widespread CS project globally, founded by NASA, **the GLOBE project** [31, 32] provides educational programs with extensive materials, including teacher workshops, curriculum resources, and student field guides, that are delivered in Slovakia by the DAPHNE institute.

Some of the popular and widely used applications mentioned above are not yet specifically designed as educational tools, that is, **iNaturalist** [33, 34]**, TrashOut** [35]**, and Pl@ntNet** [36], hold significant potential for use in schools. Notably, one of the most widely used platforms in Europe and Slovakia, iNaturalist offers a "Teacher's Guide" to facilitate its integration into the classroom. Regarding the TrashOut app, a non-governmental organization INAK focused on climate action, has developed a teaching methodology as part of the international project Teaching Green [37], demonstrating how the app can be used in educational settings. Similarly, the non-governmental organization Živica has created an initiative for the Pl@ntNet app, offering the CS educational program "Online course: Let's go out with Pl@ntNet" for teachers/lecturers of primary and secondary schools.

6 Conclusions

The results of this chapter show a good range of mobile applications for integrating CS into the school curriculum. Some of the apps are also available in Slovak language versions (iNaturalist, TrashOut, Plantnet, eBird, and Globe Observer), although the rest of the apps' operating language is English. In Slovakia, it is common to use global apps, as primary school students learn English properly as their second language. Our chapter efforts were to broaden the selection of more apps that can be utilized for teaching about biodiversity at schools. However, effective implementation will require collaboration between educators, scientists, and technology developers to ensure that CS projects align with educational goals and provide meaningful learning experiences.

Acknowledgments This chapter was written within the VEGA project 1/0080/23.

References

1. Cesco S The future challenges of scientific and technical higher education | Request PDF.
2. BROZ (2020) Strata biodiverzity, vymieranie druhov – čo s tým ?
3. Horizon Magazine (2024) Climate change and biodiversity loss should be tackled together | Research and Innovation. https://projects.research-and-innovation.ec.europa.eu/en/horizon-magazine/climate-change-and-biodiversity-loss-should-be-tackled-together. Accessed 31 May 2024
4. Daehler CC (2003) PERFORMANCE COMPARISONS OF CO-OCCURRING NATIVE AND ALIEN INVASIVE PLANTS: Implications for Conservation and Restoration. Annu Rev Ecol Evol Syst 34:183–211
5. Mckinney M, Lockwood J (1999) Biotic Homogenization: A Few Winners Replacing Many Losers in the Next Mass Extinction. Trends Ecol Evol 14:450–453
6. Primack, Richard B, Pavel K, Jana J (2011) Úvod do biologie ochrany přírody. PORTÁL s. r. o.

7. Bodáczová J. (2021) "Čo vás v tej škole učia." Analýza stavu formálneho environmentálneho vzdelávania na Slovensku. Bratislava: Inštitút environmentálnej politiky (2021).
8. Bonney R, Cooper CB, Dickinson J, Kelling S, Phillips T, Rosenberg KV, Shirk J (2009) Citizen Science: A Developing Tool for Expanding Science Knowledge and Scientific Literacy. Bioscience 59:977–984
9. Zhang E, Hundley C, Watson Z, Farah F, Bunnell S, Kristensen T (2023) Learning by doing: A multi-level analysis of the impact of citizen science education. Sci Educ. https://doi.org/10.1002/sce.21810
10. Conrad CC, Hilchey KG (2011) A review of citizen science and community-based environmental monitoring: issues and opportunities. Environ Monit Assess 176:273–291
11. Kotovirta V, Toivanen T, Järvinen M, Lindholm M, Kallio K (2014) Participatory surface algal bloom monitoring in Finland in 2011–2013. Environ Syst Res 3:24
12. Graham E, Henderson S, Schloss A (2011) Using mobile phones to engage citizen scientists in research. EOS Trans. https://doi.org/10.1029/2011EO380002
13. Buytaert W, Zulkafli Z, Grainger S, et al (2014) Citizen science in hydrology and water resources: opportunities for knowledge generation, ecosystem service management, and sustainable development. Front Earth Sci. https://doi.org/10.3389/feart.2014.00026
14. Dawson T, Hambly J, Kelley A, Lees W, Miller S (2020) Coastal heritage, global climate change, public engagement, and citizen science. Proc Natl Acad Sci 117:201912246
15. Freitag A, Meyer R, Whiteman L (2016) Correction: Strategies Employed by Citizen Science Programs to Increase the Credibility of Their Data. Citiz Sci Theory Pract. https://doi.org/10.5334/cstp.91
16. Buesching CD, Newman C, Macdonald DW (2014) How dear are deer volunteers: the efficiency of monitoring deer using teams of volunteers to conduct pellet group counts. Oryx 48:593–601
17. Nov O, Arazy O, Anderson D (2011) Dusting for science: Motivation and participation of digital citizen science volunteers. pp 68–74
18. Chandler M, See L, Buesching CD, Cousins JA, Gillies C, Kays RW, Newman C, Pereira HM, Tiago P (2017) Involving Citizen Scientists in Biodiversity Observation. In: Walters M, Scholes RJ (eds) GEO Handb. Biodivers. Obs. Netw. Springer International Publishing, Cham, pp. 211–237
19. Dickinson J, Zuckerberg B, Bonter D (2010) Citizen Science as an Ecological Research Tool: Challenges and Benefits. Annu Rev Ecol Syst 41:149–172
20. Jordan R, Gray S, Howe D, Brooks W, Ehrenfeld J (2011) Knowledge Gain and Behavioral Change in Citizen-Science Programs. Conserv Biol J Soc Conserv Biol 25:1148–54
21. McKinley D, Miller-Rushing A, Ballard H, et al (2015) Investing in Citizen Science Can Improve Natural Resource Management and Environmental Protection. Issues Ecol. https://doi.org/10.1016/j.biocon.2016.05.015
22. Toomey AH, Domroese MC (2013) Can citizen science lead to positive conservation attitudes and behaviors? Hum Ecol Rev 20:50-62,67
23. Crall AW, Jordan R, Holfelder K, Newman GJ, Graham J, Waller DM (2013) The impacts of an invasive species citizen science training program on participant attitudes, behavior, and science literacy. Public Underst Sci 22:745–764
24. Shirk J, Ballard H, Wilderman C, et al (2012) Public Participation in Scientific Research: A Framework for Deliberate Design. Ecol Soc 17:29–48
25. otvorena veda (2023) Otvorenaveda.cvtisr.sk Otvorenaveda -. https://otvorenaveda.cvtisr.sk/. Accessed 1 Jun 2024
26. Minedu (2023) Štátny vzdelávací program pre základné vzdelávanie | Ministerstvo školstva, výskumu, vývoja a mládeže Slovenskej republiky. https://www.minedu.sk/statny-vzdelavaci-program-pre-zakladne-vzdelavanie/. Accessed 1 Jun 2024
27. Doyle C, David R, Li Y, Luczak-Roesch M, Anderson D, Pierson C (2019) Using the Web for Science in the Classroom: Online Citizen Science Participation in Teaching and Learning. 80

28. Hiller S, Kitsantas A (2014) The Effect of a Horseshoe Crab Citizen Science Program on Middle School Student Science Performance and STEM Career Motivation. Sch Sci Math. https://doi.org/10.1111/ssm.12081
29. Arthur A (2015) The Inclusion of Environmental Education in Formal Education through the Use of Educational Technology, a Study of Project Noah.
30. Schaen R, Zydney J, Angelone L (2023) It Takes a Village. Sci Child 60:45–49
31. Spellman K, Sparrow E, Chase M, Larson A, Kealy K (2018) Connected Climate Change Learning Through Citizen Science: An Assessment of Priorities and Needs of Formal and Informal Educators and Community Members in Alaska. Connect Sci Learn. https://doi.org/10.1080/24758779.2018.12420500
32. Robles MC, Amos HM, Dodson JB, Bouwman J, Rogerson T, Bombosch A, Farmer L, Burdick A, Taylor J, Chambers LH (2020) Clouds around the World: How a Simple Citizen Science Data Challenge Became a Worldwide Success. Bull Am Meteorol Soc 101:E1201–E1213
33. Echeverria A, Ariz I, Moreno J, Peralta de Andrés J, González E (2021) Learning Plant Biodiversity in Nature: The Use of the Citizen–Science Platform iNaturalist as a Collaborative Tool in Secondary Education. Sustainability 13:735
34. Smith H, Allf B, Larson L, Futch S, Lundgren L, Pacifici L, Cooper C (2021) Leveraging Citizen Science in a College Classroom to Build Interest and Efficacy for Science and the Environment. Citiz Sci Theory Pract 6:29
35. Jakab I, Zigová M, Pucherova Z (2019) Modernization of Environmental Education with the Use of Project-Based Learning, Outdoor Education, and Mobile Learning Supported by Information and Communication Technology. pp 223–248
36. Coşkunserçe O (2024) Use of a mobile plant identification application and the out-of-school learning method in biodiversity education. Ecol Evol 14:e10957
37. Jozef Kahan et al. (2020) Indikátory prostredia. Teaching Green
38. Radicchi A (2017) Hush City. A new mobile application to crowdsource and assess “everyday quiet areas” in cities.
39. Guillaume G, Aumond P, Bocher E, Can A, Ecotière D, Fortin N, Foy C, Gauvreau B, Petit G, Picaut J (2022) NoiseCapture smartphone application as pedagogical support for education and public awareness. J Acoust Soc Am 151:3255–3265

Utilizing the UTAUT2 Model to Elucidate Student Perceptions of Chatbot Acceptance as a Communication Medium in Higher Education

Marta Slepankova, Katerina Kostolanyova, Tatiana Havlaskova, and Tomas Javorcik

1 Introduction

In recent years, educational institutions and the process of education itself have undergone numerous changes, primarily due to the emergence of new technologies. The rapid development of Information and Communication Technologies (ICT) and the rise of the internet have continuously transformed the methods of teaching and learning. These new technologies not only influence how teachers impart knowledge and communicate with students but also reshape the traditional classroom setup, as organizations increasingly rely on ICT for efficient daily operations. Among these emerging technologies, artificial intelligence (AI) is gaining popularity, leading to a shift from the traditional teacher-centered model to a student-centered educational system. This new model is characterized by a strong emphasis on personalization [1].

The utilization of AI has gained significance in various domains; however, its application in educational communication is still in its infancy. Despite several scientific experiments and the existence of educational AI applications for commercial use, this technology is not yet widely employed in the education sector [2]. Although effective communication between teachers and students is a crucial educational activity, especially in online education, AI solutions such as chatbots have not been extensively adopted within educational institutions. The term "communication" may vary in some instances; nevertheless, researchers recognize the communication between students and teachers as fundamental to effective education. The advancements in communication technologies have enabled nearly

M. Slepankova (✉) · K. Kostolanyova · T. Havlaskova · T. Javorcik
University of Ostrava, Ostrava, Czech Republic
e-mail: marta.slepankova.s01@osu.cz; katerina.kostolanyova@osu.cz; tatiana.havlaskova@osu.cz; tomas.javorcik@osu.cz

M. Turčáni (ed.), *15th International Scientific Conference on Distance Learning in Applied Informatics*, EAI/Springer Innovations in Communication and Computing,
https://doi.org/10.1007/978-3-031-81261-3_19

instantaneous communication among people. The increased number of students in schools is reflected in the heightened communication between students and educational institutions (communication initiated by students toward teachers).

Although several studies have been conducted to determine the acceptance of products with AI elements, little attention has been paid to exploring the primary factors influencing the acceptability of communication with AI elements from the perspective of students. Additionally, the existing literature lacks a comprehensive model encompassing the essential factors for the acceptance of AI chatbots among university students. For instance, Almahri, Bell, and Merhi [3] focused on researching the acceptance of AI chatbot technology, investigating students' motivation to use chatbots based on the Unified Theory of Acceptance and Use of Technology 2 (UTAUT2). The results revealed that Performance Expectancy (PE), Effort Expectancy (EE), and Habit (HA) are the primary predictors of students' intention to use chatbots. The quantitative research was conducted by collecting data from 431 students at Brunel University in London. However, it is important to note that the results of the study could not be generalized due to the data being derived solely from a specific department within the university. The aim of this qualitative research was to identify key factors influencing how university students perceive communication with AI elements.

2 Theoretical Background

2.1 AI Chatbots in the Educational Sector

Communication between students and the university is an important educational activity, especially in online education. The term communication can vary in some cases; however, researchers see the importance of communication between students and teachers. It might highly influence the students' success [4, 5]. However, the increased number of students makes adequate guidance difficult for teachers; thus, the positive interaction between teacher and student is lowered [6]. It has been proven that chatbots can: (a) repeat the course lessons, (b) gather feedback for a course to improve learning and teaching, (c) answer frequently asked questions about a specific subject, or (d) support administrative tasks [7].

Although the term AI chatbot, or conversational agent, encompasses a wide range of systems with various capabilities and purposes, most researchers refer to AI-powered communication applications, where a human engages in dialogue with a machine [8]. Despite the continuous expansion of AI utilization, AI chatbots are seldom employed and are overshadowed by the more prevalent rule-based chatbots. In rule-based systems, once a specific keyword is identified in the text, the chatbot provides a predefined response. In contrast, AI chatbots utilize machine learning and natural language processing (NLP) techniques, enabling them to simulate human conversation through certain messaging platforms [9]. In other words, AI chatbots employ algorithms to learn and comprehend unstructured information and improve their understanding through interactions with users.

Currently, AI chatbots are being used in various public and private institutions for services and tasks such as calendar management, scheduling appointments, reading, writing, and sending emails [10]. However, chatbots have the potential to assist both students and teachers in providing useful information [8]. Moreover, they can be utilized in situations where the teaching staff faces significant bureaucratic and communication burdens [11]. Numerous studies have already focused on practical solutions to support communication between students and educational institutions. For instance, researchers like Ranoliya et al. [12] examined time consumption in human-to-human communication. The increasing number of students directly impacts the waiting time for responses. To address this, scientists proposed employing AI chatbots as a communication medium due to their time efficiency. The interaction between students and chatbots consists of initiation, response, and feedback elements [13]. According to research findings, chatbots could assist students by providing material recaps and functioning as online tutors, effectively reducing the workload for teachers [11]. Furthermore, users could approach chatbots with queries related to promotion terms, program information, and academic fields [8].

2.2 Acceptance of Technology

The acceptance of a technology by its consumers is "vital" and a key factor for the further development of any new solution. For organizations to effectively integrate technology into production, it is essential to recognize consumers' needs and attitudes toward the solution, especially in the early stages. Therefore, conducting research on the factors influencing consumers' decision to adopt or reject the technology is beneficial during the development phase.

Various studies have been conducted to forecast the adoption and acceptance of technological solutions. Models like the Technology Acceptance Model (TAM) [14] and The Unified Theory of Acceptance and Use of Technology (UTAUT) have been frequently employed. UTAUT, proposed by Venkatesh et al. [15], consists of eight models, with TAM being one of the included models. This theory aids researchers in evaluating the likelihood of success for new technology and understanding the factors influencing its acceptance. Additionally, according to Venkatesh et al. [15], UTAUT highlights methods to mitigate the impact of skeptics and less inclined users to adopt the new technology, such as through marketing or training interventions.

The UTAUT model comprises four variables: PE, EE, Facilitating Conditions (FC), and Social Influence (SI). A distinction can be made between UTAUT and UTAUT2, introduced by Venkatesh, Thong, and Xin [16]. UTAUT mainly focuses on the organizational context, aiming to predict whether employees will accept and use the technology. On the contrary, UTAUT2 has been proposed to understand consumer behavior. Both models include age, gender, and experience as moderators.

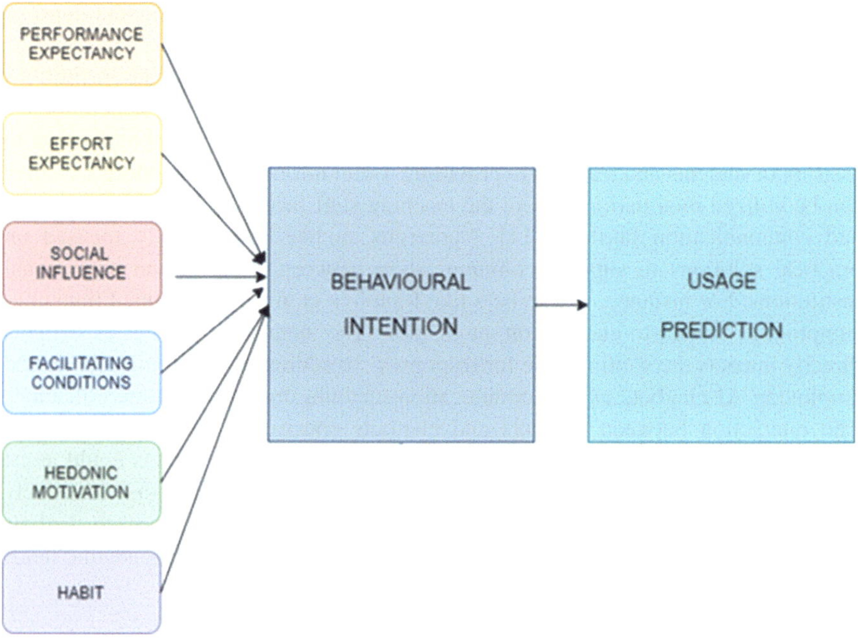

Fig. 1 UTAUT2 model by Almahri, Bell, and Merhi (2020)

Additionally, they share predictive variables such as PE, EE, SI, and FC. UTAUT2 extends the model by incorporating three additional factors: Hedonic Motivation (HM), Price Value (PV), and HA, aiming to explain consumer behavior.

3 Research Model

The research utilized the UTAUT2 model employed in the quantitative study by Almahri, Bell, and Merhi [3]. In their study, the authors excluded age, gender, and experience as moderators from the model. However, they suggested that future work should focus on modifying the predictive model by incorporating additional constructs. As a result, this study used the same predictive model but aimed to identify a new construct that could complement the current model (Fig. 1).

According to researchers Gharrah and Aljaafreh [17] and Huang and Kao [18], the seven factors are further described as follows: PE can be explained as an individual's belief in how the technology will function, support specific tasks, and enhance work performance. In other words, it measures the extent to which new technologies can bring benefits to consumers in performing activities. EE can be described as the extent of effort expended by a person using the system to perform a specific task. SI can be explained as the importance of others' opinions on the use of technology. It reflects the degree to which the surrounding environment believes that the consumer should utilize the technology. FC can be described as the user's confidence in the organization and technical infrastructure in which the new technology is incorporated. It measures the extent to which an individual believes in technical and organizational support. In other words, it is the trust that the existing infrastructure will be able to support the new technology. HM represents the comfort, stimulation, or pleasure a person experiences while using the new technology. It is a motivation based on consumer satisfaction. HA is a routine activity in which a person automatically performs a given task. PV indicates the relationship between the benefits and the financial costs of using the technology. However, this factor was removed from the model as it was deemed irrelevant. It is assumed that any university would provide such a service to students for free. These mentioned factors lead to a situation where the consumer decides and formulates a conscious decision to use or not to use the technology in the future. This phase is called behavioral intention. The five mentioned constructs are predictors of acceptability in this study.

4 Research Methodology

4.1 Participants and Data Collection

The aim of the constructivist qualitative research was to explore and capture various perspectives and beliefs of the participants. Although the participants' gender, academic year, study program, and socioeconomic status might have varied, this merely reflects the real-life diversity of students. In total, the research included 20 student participants. Primary data collection was conducted through voluntary questionnaires and interviews in March 2021. The questionnaire survey consisted of ten open-ended questions to assess participants' perceptions regarding the potential use of AI chatbots in higher education. Participants were later referred to as respondents (R1–R17). The average response time for each participant was 10 min. Three semi-structured interviews were conducted in March 2021, via online video conferencing (Zoom), with each interview lasting approximately 45 min.

5 Findings

This study employed inductive data analysis, specifically content analysis, using the NVivo software program. Data saturation was achieved with participant number 16, where no new information, coding, or themes emerged from the data of the remaining participants. Four primary themes (Rapid Support, Limited Environmental Opportunities, Non-judgment, and Simplicity Kills) along with three subthemes

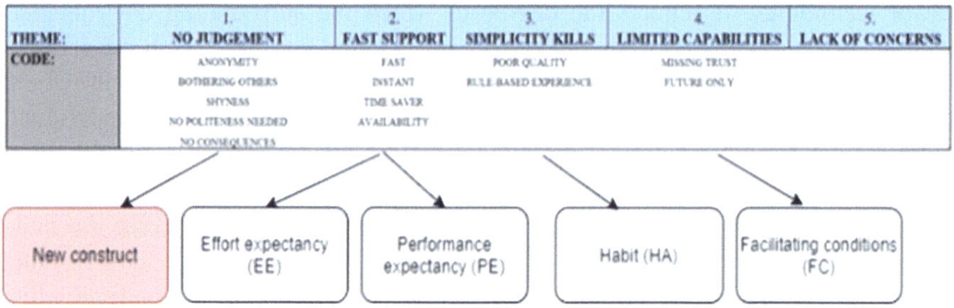

Fig. 2 Identified themes and codes compiled with UTAUT2

and 12 codes were identified during the content analysis. Once the final thematic categories were established, they were incorporated into the UTAUT2 model (Fig. 2).

No Judgment—A total of 20 participants provided 39 relevant responses related to this theme. Some participants expressed negative feelings associated with communicating with educators and waiting for responses. They mentioned feelings of shyness in asking questions publicly and a preference for anonymous questioning. Concerns were raised about how teachers would perceive their questions. Some participants felt nervous or uncomfortable while waiting for a response. The study suggests that students fear being judged for their inquiries while seeking attention from their educators. Figuratively speaking, students want to obtain information promptly, but the respondent should have no right to express opinions or have any subjective attitude toward the questioner. This theme is further divided into three subthemes. The first subtheme (Ideal State) is based on responses indicating that students would prefer to conceal their identities for certain communication with teachers for various reasons. The second subtheme (Emotions) is based on what some students feel during communication with educators. The third subtheme (Less Formality) is based on students' responses suggesting that certain aspects of communication could be skipped/eliminated. For instance, students mentioned feeling obligated to be more polite than usual when communicating with educators. This behavior could be described as walking on thin ice, in contrast to communication with a chatbot that allows students to get to the point without any decorum or excessive formality.

Fast Support—A total of 33 responses were associated with this theme. Most participants emphasized the expected speed of response and 24/7 availability. On the contrary, waiting for a response was perceived as a very unpleasant aspect of communication. None of the responses showed patience.

Simplicity Kills—Within this theme, 29 related responses were found, indicating that participants do not distinguish between AI chatbots and rule-based chatbots. Participants mainly mentioned a limited vocabulary, lack of understanding, and comprehension, leading to the assumption that consumers have more experience with rule-based chatbots than with AI chatbots, which is misleading.

Limited Capabilities—A total of 12 responses were associated with this theme. Participants mainly mentioned the lack of capabilities of AI technology and the shortage of technical resources on the part of the university.

6 Discussion

When the primary themes were compared with the constructs of the UTAUT2 model, three of the primary themes corresponded to the constructs, and one primary theme pointed to a completely new construct in the predictive model. The construct "Non-Judgmental Expectancy" (NE) is defined in this study as the extent to which an individual perceives that using the system frees them from personal judgments and biases of others. Specifically, it is the degree to which students perceive that using AI chatbots as an unbiased communication medium allows them not to worry about any prejudices from educators. AI chatbots can enable students to obtain information without the fear of offending anyone or being judged by others. These findings are consistent with previous research conducted by Schrodt, Whitt, and Truman [19], which suggests that teachers have rewarding and punishing power over students.

As machine learning enables the AI chatbot to learn from datasets and experiences and form opinions accordingly, data must be carefully interpreted, and the results must be evaluated through the actual deployment in production [20]. It is essential to ensure that the algorithm does not use personal information to harm individuals and that this technology does not become biased based on biased data. For this reason, it can be argued that this new construct of "Non-Judgmental Expectancy" within the UTAUT2 model is highly suitable for researching AI technology.

The integration of "Non-Judgmental Expectancy" into the UTAUT2 model complements and expands the theory to encompass the psychological and emotional attitudes of individuals toward self-assessment. This aspect has not been addressed in the model for other technologies (without AI) as they do not exhibit bias. In contrast to the quantitative study by Almahri, Bell, and Merhi [3], this qualitative study suggests different results, even though it uses the same research model. The authors of the previous study stated that PE, EE, and HA are the main predictors of a student's acceptance of an AI chatbot. However, this qualitative study suggests that PE, EE, and the new construct of NE are the main predictors of the intention to use AI chatbots.

Students prefer AI chatbots primarily because of their ability to provide information quickly. This aligns with the findings of Linnes and Metcalf [21], who argue that this generation often expects immediate responses. By using unbiased technology as a communication medium to obtain information, biases, punishment, or stigma associated with asking questions can be eliminated. Therefore, if students perceive that using AI chatbots to communicate with the university significantly contributes to their emotional and social well-being, they may be positively inclined to adopt this technology. However, to achieve this state, the AI chatbot system must be unbiased (using unbiased data).

7 Conclusion

The development of communication technologies has accelerated the communication process and enabled instant communication among people, including students and teachers. Temporary experiments related to the adoption of AI chatbots in education have so far focused only on a single location. This study aimed to fill this gap. The research focused on the factors of acceptability of AI chatbots. The UTAUT2 theory was selected for the research to understand students' intentions to use AI chatbots as a communication medium between them and teachers.

The conclusions of the study indicate that the newly proposed construct, "Non-Judgmental Expectations" (NE), along with the constructs of "Performance Expectancy" and "Effort Expectancy," are the main predictors of acceptability. These constructs can positively influence the acceptability of AI chatbots. NE can positively impact students' well-being and mental health by eliminating negative emotions associated with communication, such as judgment or offense based on impartiality. Furthermore, this construct is a suitable addition to the predictive model UTAUT2, making it highly suitable for addressing features of machine learning.

It is essential that the implementation of the solution is based on unbiased data, ensuring that the technology itself remains impartial. The study suggests a direct impact of technology attributes on individuals' willingness to use this technology.

8 Future Research

The newly proposed construct could be further explored in different domains to enhance its applicability. For instance, it could be investigated in UTAUT2 research involving other solutions that utilize AI elements. By studying its effects in various contexts, researchers could gain a deeper understanding of its value and its potential impact on user acceptance and behavior in different technological applications. This cross-domain exploration would contribute to the broader knowledge of how the "Non-Judgmental Expectations" construct interacts with AI-based technologies and its relevance in shaping user perceptions and attitudes.

References

1. Sandu, N., and Gide, E.: Adoption of AI-Chatbots to enhance student learning experience in higher education in India. In: 18th International Conference on Information Technology Based Higher Education and Training (ITHET), pp. 26–27. IEEE, New York (2019).
2. Jaakola, H., Henno, H., Lahti, A., Järvinen, J-P., and Mäkelä, J.: Artificial Intelligence and Education. In: MIPRO, 43rd International Convention on Information, Communication and Electronic Technology, pp. 28–02. IEEE: New York (2020).
3. Almahri, F.A.J., Bell, D. and Merhi, M.: Understanding Student Acceptance and Use of Chatbots in the United Kingdom Universities: A Structural Equation Modelling Approach. In: 6th International Conference on Information Management (ICIM). IEEE: New York (2020).
4. Bajracharya, D.: Impact of Teacher-Student Communication on "High-Risk Droupout" Students. In: The International Institute for Science, Technology and Education (IISTE) 18 (5), pp. 88–100, (2015).
5. Krahe, C., Escamilla-Fajardo, P., and López-Carril, S.: The influence of teacher-student communication on the importance of physical education. In: Physical Education and Sport 19, (2021).
6. Dimitriadis, G.: Evolution in Education: Chatbots. In: Homo Virtualis, 3 (1), pp. 47–54, (2020).
7. Hien, H. T., Cuong, P.-N., Nam, L. N. H., Nhung, H. L. T. K., and Thang, L. D.: Intelligent assistant in higher-education environments: the FIT-EBot, a chatbot for administrative and learning support. In: SoICT', 9th International Symposium on Information and Communication Technology. ACM: New York (2018).
8. Mekni, M., Baani, Z., and Sulieman, D.: A Smart Virtual Assistant for Students. In: Proceedings of the 3rd International Conference on Applications of Intelligent Systems, 15, pp. 1–6, (2020).
9. Schuetzler, R. M., Grimes, G. M., and Giboney, J. S.: The impact of chatbot conversational skill on engagement and perceived humanness. In: Journal of management information systems, 37 (3), pp. 875–900, (2020).
10. Dale, R.: The return of the chatbots. In: Natural Language Engineering, 22 (5), pp. 811–817, (2016).
11. Lee, L.-K., Fung, Y.-C., Pun, Y.-W., Wong, K.-K, Yu, M. T.-Y., and Wu, N.-I.: Using a Multiplatform Chatbot as an Online Tutor in a University Course. In: IEEE, 6th International Symposium on Educational Technology (ISET), 21 (2), pp. 11–26, (2017).
12. Ranoliya, B. R., Raghuwanshi, N., and Singh, S.: Chatbot for university related FAQs. In: International Conference on Advances in Computing, Communications and Informatics (ICACCI). IEEE: New York (2017).
13. Fleming, M., Riveros, P., Reidsema, C., and Achilles, N.: Streamlining student course requests using chatbots. In: Proceeding 29th Australasian Association for Engineering Education Conference, pp. 207–211, (2018).
14. Davis, F. D.: Perceived usefulness, perceived ease of use, and user acceptance of information technology. In: MIS quarterly, pp. 319–340, (1989).
15. Venkatesh, V., Morris, V., Davis, G., and Davis, F.: User Acceptance of Information Technology: Toward a Unified View. In: Management Information Systems Quarterly, 27 (3), pp. 425–478, (2003).
16. Venkatesh, V., Thong, J., and Xin, X.: Consumer Acceptance and Use of Information Technology: Extending the Unified Theory of Acceptance and Use of Technology. In: MIS Quarterly, 36 (1), pp. 157–78, (2012).
17. Gharrah, A., and Aljaafreh, A.: Why students use social networks for education: Extension of UTAUT2. In: Journal of Technology and Science Education, 11, pp. 53, (2021).
18. Huang, C.-Y., and Kao, Y.-S.: UTAUT2 Based Predictions of Factors Influencing the Technology Acceptance of Phablets by DNP. In: Mathematical Problems in Engineering, pp. 1–23, (2015).

19. Schrodt, P., Whitt, P. L., Truman, P. D.: Reconsidering the measurement of teacher power use in the college classroom. In: Communication Education, 56 (3), pp. 308–332, (2007).
20. Stachowicz-Stanusch, A., and Amann, W.: Artificial Intelligence at universities in Poland. In: Organizacja I Zarzadzanie: kwartalnik naukovy, [e-journal] 2, pp. 65–82, (2018).
21. Linnes, C., and Metcalf, B.: iGeneration And Their Acceptance of Technology. In: International Journal of Management & Information Systems, 21 (2), pp. 11–26, (2017).

Digital and Reading Competency of Secondary School Pupils and Their Comprehension of Everyday and Educational Texts

Dana Vicherková, Martin Kolář, Veronika Murinová, and Josef Malach

1 Introduction

Digital and reading literacy are essential to functional literacy and should be cultivated through lifelong learning in formal and informal education settings. The significance of digital literacy in the job market was emphasized in the Thematic Report (TR) of the Czech School Inspectorate (CSI) [3] in light of the growing interest in ITC technologies. In 2024, the Czech Republic will hold public expert discussions on proposed revisions to national curricula, Framework Educational Programs (FEPs), and School Education Programs (SEPs) for primary education. The Long-Term Plan for Education and Development of the Education System of the Czech Republic 2023–2027 was established to address the digital transformation of society [13]. At the European level, the Digital Education Action Plan (2021–2027) outlines priorities and goals for digital education, focusing on the development of digital skills throughout life, adapting to changing labor market needs, mitigating digital exclusion risks, and enhancing coordination among public, private, and non-profit sectors in digital education [5]. The Czech Republic emphasizes providing schools with modern information and digital technologies and developing pupils' digital competencies necessary for life in the twenty-first century. "Digitally literate citizens fully utilise the potential of available technologies for their personal development, improving quality of life, and contributing to the long-term prosperity and competitiveness of the Czech Republic in the information and digital economy" [5]. The Ministry of Education and Culture of the Czech Republic (2021) has set eight sub-goals for the planned development of digital skills, covering safe and healthy

D. Vicherková (✉) · M. Kolář · V. Murinová · J. Malach
Department of Education and Adult Education, University of Ostrava, Ostrava, Czech Republic
e-mail: dana.vicherkova@osu.cz; L20487@student.osu.cz; D23704@student.osu.cz; josef.malach@osu.cz

M. Turčáni (ed.), *15th International Scientific Conference on Distance Learning in Applied Informatics*, EAI/Springer Innovations in Communication and Computing,
https://doi.org/10.1007/978-3-031-81261-3_20

living in a digital environment, digital transformation, modernizing educational content, cross-sectional development of digital skills, using digital technology and introducing innovations in teaching, sustainable development of school digital infrastructure, and creating school plans connected to digital education.

2 Digital Competency in the Twenty-First Century

According to Růžičková et al. [18], digital literacy can be defined as "a set of digital competencies (knowledge, skills, attitudes, values) that an individual needs to safely, confidently, critically, and creatively use digital technologies at work, in learning, in free time, and in their involvement in social life." The Czech Republic is gradually revising the FEP across all levels of schools while maintaining a two-level system of curriculum documents. In addition to the FRP, each school has its SEP. Significant trends include revisions of the Information and Communication Technology (ICT) in the FEPs at individual levels of schools: primary schools (2023/2024) [17], grammar schools, and secondary vocational schools (2025). As part of the revision process, digital competency is "the ability to navigate the digital environment, handle technologies, and have an overview of what affects everything around us. To use them safely, confidently, critically, and creatively at work, while learning, in free time, and when participating in society and civic life are key skills these days." Digital technologies are being integrated into the content of individual subjects. The cross-cutting topic "Human and the Digital World" in the secondary school environment also focuses on developing digital competencies. The new trends in digital education include upskilling, which offers strategies for learning new skills for one's job, and reskilling, with a critical focus on new skills for another job, for example, in a different field. The school environment also addresses the issues of artificial intelligence, cybersecurity, and the influence of digital technologies on health, humanity, society, childhood, and school, emphasizing digital well-being. The current trend in the development of Czech education is responding to changes in the development of society with further formally and informally managed digital education, for example, by the Ministry of Education, Youth and Sport, as well as the National Pedagogical Institute. The aim is to help Czech schools successfully introduce new technologies into teaching and to focus, with a varied offer of courses and consultations in the field of digital literacy, on educating pupils and teachers in digital competencies and working with different types of texts, or the creation of educational texts with multimedia interactive elements with interdisciplinary overlap.

3 Reading Competency in the Twenty-First Century

Reading is a crucial part of education and literacy. Even in today's digital world, we still gain knowledge through reading. There are various approaches to teaching reading, such as analytic, synthetic, global, and analytic-synthetic. The main aspects of reading acquisition include reading technique and comprehension. There are two essential phases during reading: decoding and comprehension, and they interact. Repeated reading enhances our reading skills, understanding of the information we read, and vocabulary. When we read, we visually perceive the shape of letters based on visual and spatial perception. Reading with comprehension is an essential stage of reader maturity, as it connects the elements of the text with the reader's knowledge on multiple levels. Factors affecting text comprehension include external and internal aspects such as knowledge of morphology, syntax, decoding automation, cognitive reading maturity, reading experience, motivation, joy of knowledge, reading environment, and reading pattern. In everyday and educational settings, it is vital to strengthen cognitive processes and seek opportunities to develop logical thinking, memory, attention, and vocabulary. Vicherková [21] states that "the reader's diverse relationships with the text influence cognitive reflection and regulated understanding from a metacognitive perspective." It is crucial to prioritize the care of sight and hearing in various aspects of life, such as family, leisure, school, and work. Encouraging a positive attitude toward reading diverse texts and promoting the enjoyment and pursuit of knowledge through reading is crucial. Zápotočná [22] emphasizes the need to focus on strategies to improve learning efficacy through reading comprehension in secondary education. Mareš [12] highlights the significance of self-regulated learning through reading, stating that a high level of self-regulation involves activities that individuals engage in independently because they find them interesting or important. Working with texts and understanding them consists of searching for information, understanding, assessment, and reasoning. International PISA research (2000, 2009, 2018, 2022) focuses on the reading literacy of 15-year-old pupils and three key areas when working with texts: (1) searching for information, (2) understanding, and (3) assessment and reasoning. According to the PISA research [16], reading literacy is defined as "the ability to understand a text, think about it, judge it, deal with it, and use it to achieve personal goals, develop knowledge, and actively participate in society" [16]. The research findings of PISA [16] indicate that the reading literacy level among Czech 15-year-old pupils remains comparable to 2018. Given the rapid development of artificial intelligence tools encountered by pupils, the ability to understand, think critically, and assess text is increasingly important. Pupils with inadequate reading literacy may experience disadvantages in meeting their study or work responsibilities and everyday social interactions and engagement [16].

Digital and reading literacy in Society 4.0 are essential for developing key communication and digital skills as part of lifelong learning. The CSI research (TR) [4] focused on assessing the reading literacy level of second-year secondary school pupils, the conditions for developing reading literacy in schools, reading activities and texts in teaching, and pupils' behavior and attitudes toward reading. According to the key results of the research [4], it is evident that 21% of second-year secondary school pupils had an inadequate level of reading literacy, meaning that "the tested group of secondary school pupils was not able to solve even the simplest test items of the verified topics" [4]. Pupils who found the texts they read in class interesting and thematically relevant and had frequent experience working with authentic texts performed significantly better in the reading literacy test. A noteworthy finding is that less than a fifth of secondary school pupils described the teaching texts as entirely uninteresting. Another finding, as per ČSI TR [4], is that secondary school pupils are minimally or weakly engaged in reading online, and 30% of them stated that they hardly read or listen to books.

3.1 Typology of Texts and Text for Teaching

A text is a product of human thought and serves various functions in everyday life, entertainment, education, and work. From a professional perspective, texts can be categorized based on their style, syntax, language, phonetics, morphology, and vocabulary. They also have pedagogical, psychological, social, ethical, and philosophical dimensions. Different readers will interpret texts differently based on their level of maturity. Through the lens of semiotics, Hoflerová ([7], p. 6) defines a text as "a structure composed of signs intended for communication between people. A sign is an entity related to various degrees of similarity, relationship, vividness, and relevance to both concrete phenomena (such as people, objects, and nature) and abstract phenomena (like thoughts, voluntary activity, events, and states)."

In stylistics, the functions of communicative utterances, functional styles, and stylistic factors play an essential role in understanding the content and form of text communication. The function of texts, such as informative, educational, entertaining, aesthetic, or critical, is crucial in understanding the message. Čechová et al. ([1], p. 98) classified texts based on their functional purpose into six categories: (1) simply informative, (2) journalistic, (3) professional, (4) administrative, (5) artistic, and (6) rhetorical. The aesthetic function is the critical function in artistic texts. According to Kostečka [10], professional-style texts can be grouped into four categories based on the degree of expertise and the specific communication goal: "scientific, educational, popular-educational, essayistic." While printed text is considered traditional, the electronic form of texting, as defined by Karlík, Nekula, and Pleskalová [9] as "a non-linear way of organising text during computer processing of optical and audio information," continues to enhance its functionality through quantitative growth and qualitative communication meaning. Kostečka [10] suggests that "the mixing and blending of elements of different functional styles and stylistic procedures is quite common today, corresponding to the complexity of our (post)modern world."

In the context of digital communication, direct face-to-face interaction is absent. Čmejrková and Hoffmannová [2] highlight "conversationalisation, marketisation, and technologisation of public discourse" in digital communication.

According to Štěpáník et al. [20], teaching should focus on expanding language awareness rather than just imparting factual knowledge. It should help pupils gain insight into language use in communication contexts, understand language as a flexible and ambiguous phenomenon, and encourage them to acquire knowledge and use it to create new knowledge. In practical terms, this approach involves moving away from a prescriptive view of language teaching.

4 Research Methodology

In a quantitative questionnaire survey, we focused on eight secondary vocational schools in three regions of the Czech Republic. The research took place from January to April 2024, and we gathered data from 594 respondents. The semi-structured questionnaire, containing **28** semi-open items grouped into four categories A–D, was used:

A. Comprehension and reflection of read information (7)
B. Text of an everyday nature (6)
C. Electronic texting and reader interest (8)
D. Reading and digital strategies, competencies (7)

4.1 Research Objectives

The research aimed to identify the characteristics and forms of text that influence reading comprehension and digital literacy. The specific goals of the research were to determine:

- The most frequently read types of texts by pupils
- The amount of time pupils spend reading texts daily
- Pupils' self-perception of their literacy levels
- Pupils' evaluation of electronic/hybrid text
- The devices pupils use to read electronic texts
- The types of texts that enhance pupils' reading competency
- The types of texts that enhance pupils' digital competency
- The preferred thematic content for pupils
- The pupils' preference for printed texts
- The types of learning texts pupils prefer
- The types of texts pupils consider to be complex or innovative

The research problem was broken down into the following research questions:

1. Is there a connection between the most frequent reading of texts that pupils consider difficult and the time spent reading texts daily? (A1/B1)
2. Is there a connection between the pupil's perception of their literacy level and their daily reading time? (A2/B1)
3. Is there a connection between the most frequent reading of texts that pupils consider difficult and the pupils' perception of their level of literacy? (A1/A2)
4. Is there a connection between the pupil's opinion of the hybrid text and the device they use to read electronic texts? (C3/C5)

5 Results

The following chapter presents the formulation of four hypotheses based on the research objectives and questions and their verification results.

5.1 Descriptive Data

The data provides detailed insights into how different content and text formats influence reading comprehension, digital literacy, and competency in various contexts. The majority of 391 (65.82%) respondents reported reading simple texts, while 203 (34.18%) said they read demanding or complex texts. When asked about the time spent reading texts (e.g., books, newspapers, and online articles), 345 (58.08%) respondents read texts online for less than 30 min a day, while 249 (41.92%) read online texts for more than 30 min a day. It is worth noting that 399 respondents (67.17%) consider hybrid texts innovative for readers, while 195 (32.83%) find them difficult. Additionally, 300 (60.48%) respondents read electronic texts due to the immediate availability of information, while 196 (39.52%) do so because of convenience. The survey results show that 180 respondents (30.30%) prefer reading about current social events, while 146 (24.58%) are interested in fantasy reading. Additionally, 54 (9.09%) respondents confirmed their interest in an artistic topic, 60 (10.10%) preferred a historical topic, and 34 (5.72%) preferred technically oriented reading. Notably, 33 (5.56%) respondents prefer a culturally and scientifically oriented topic. The majority of 507 (85.35%) respondents perceive the importance of regular reading for developing knowledge, and 69 (11.62%) respondents consider it necessary to read daily for fun. Only 18 (3.03%) respondents answered that reading daily for other purposes is unimportant. The survey results indicate that 412 out of 594 respondents (69.36%) use a smartphone to read electronic texts, while only a fifth (21.55%) reported using a tablet or laptop for the same purpose. Additionally, 54 respondents (9.09%) used an e-reader to read electronic texts. Notably, almost two-thirds of the 377 respondents (63.47%) consider themselves literate, while more than a third of the 217 respondents (36.53%) do not. Moreover, the research highlights that two-thirds of the 406 respondents (68.35%) enjoy

Table 1 Detected and expected frequencies to H1

Q A1	Q B1—less than 30 min	Q B1—more than 30 min	Line totals
Simple texts	249 (227.10)	142 (163.90)	391
Complex texts	96 (117.90)	107 (85.10)	203
Column totals	345	249	594

Pearson´s chi-square = 14.7474374 degree of freedom = 1 significance p=0.0001229

Table 2 Detected and expected frequencies to H2

Q A2	Q B1—less than 30 min	Q B1—more than 30 min	Line totals
Yes	196 (218.96)	181 (158.04)	377
No	149 (126.04)	68 (90.96)	217
Column totals	345	249	594

Pearson´s chi-square = 15.727487 degree of freedom = 1 significance p = 7.31534E-05

Table 3 Detected and expected frequencies to H3

Q A1	Q A2—yes	Q A2—no	Line totals
Simple texts	233 (248.16)	158 (142.84)	391
Complex texts	144 (128.84)	59 (74.16)	203
Column totals	377	217	594

Pearson´s chi-square = 7.41788 degree of freedom = 1 significance p = 0.0064579

Table 4 Detected and expected frequencies to H4

Q C3	otázka C5—smartphone	otázka C5—e-reader	otázka C5—tablet, notebook	Line totals
I consider them innovative	263 (276.75)	43 (36.27)	93 (85.98)	399
I consider them demanding	149 (135.25)	11 (17.73)	35 (42.02)	195
Column totals	412	54	128	594

Pearson´s chi-square = 7.62686296 degree of freedom = 2 significance p = 0.02207231

reading printed books, and a third of the 188 respondents (31.65%) regularly listen to audiobooks.

5.2 *Relational Research Results*

Four hypotheses were verified Using Pearson's chi-square test for the contingency table. Tables 1, 2, 3, and 4 present the source data and corresponding statistical results.

In the questionnaire, pupils answered questions A1 (regarding which texts they read most often on the Internet) and B1 (about how much time they spend reading texts daily).

Hypothesis 1, stating that pupils who frequently read complex texts on the Internet spend over 30 min a day reading, compared to those who primarily read simple texts, **has been confirmed** (Table 1). A statistically significant relationship exists between the texts' complexity and the daily reading time. Pupils answered questions about literacy (question A2) and daily reading time (question B1).

Hypothesis 2, which states that pupils who perceive themselves as literate spend more than 30 min reading text on the Internet daily compared to pupils who spend less than 30 min reading, **has been confirmed** (Table 2). The results show a statistically significant relationship between the pupil's self-perception of literacy and the time they spend reading texts on the Internet daily. In the survey, pupils responded to question A1, which asked about the complexity of texts they typically read, and question A2, which inquired about their self-perception of literacy.

Hypothesis 3, which suggests that pupils who read complex texts daily consider themselves more literate than pupils who read simple texts daily, **has been confirmed** (Table 3). There is a statistically significant relationship between the answers to both questions regarding the complexity of the texts read and whether they consider themselves literate. In the questionnaire, pupils answered question C3 about their opinion on hybrid texts and question C5 about the type of device they use for reading electronic texts.

Hypothesis 4, which suggests pupils who find hybrid texts innovative for reading, use technological devices for reading electronic texts more frequently than pupils who find hybrid texts too demanding for reading, **has been confirmed** (Table 4). A statistically significant relationship exists between the responses to both questions (attitude toward hybrid texts and the type of device used for reading electronic texts).

A quantitative questionnaire study involving 594 Czech secondary school pupils revealed that the content and format of a text impact the reading skills development of secondary school pupils. The primary goal is to enhance reading comprehension and foster reading motivation, curiosity, creativity, imagination, and the reader's thirst for knowledge and entertainment. Additionally, pupils appreciate the opportunity to explore the unknown through stories, secrets, technical and technological innovations, and the experience of beauty and art by integrating sound and technical effects when reading electronic or hybrid texts, listening to audiobooks, and engaging with texts that help develop technical thinking. Interestingly, most pupils view printed books as traditional texts, which are more prevalent among secondary school readers than electronic books. Regarding availability, flexibility, innovation, and complexity of educational materials, pupils prefer electronic and hybrid texts for reading but spend less time reading them than printed materials. Additionally, 180 (30.30%) of the pupils prioritize current social events as the main thematic content for reading, followed by fantasy and fiction for 146 (24.58%) pupils. Only 33 (5.56%) pupils show interest in scientific topics and culture. Regarding non-artistic texts, pupils predominantly read academic and professional texts electronically during their studies and in their free time. Administrative texts are the least popular among pupils, although they recognize their importance for everyday life. The research indicates that most respondents (412 or 69.36%) use smartphones to read electronic texts.

6 Discussion and Conclusion

The results of the questionnaire-based research indicate that 76.77% of the 456 respondents believe that using methods to develop reading strategies and critical thinking is adequate for understanding the content of the information being read. The research also found that pupils do not face formal barriers to reading electronic texts and use various technical devices. Specifically, 69.36% of the 412 respondents use smartphones, 21.55% use tablets or laptops, and only 54 (9.09%) use e-readers to read electronic texts. The research also established a statistically significant relationship for four proposed hypotheses.

- H1 **has been confirmed** (Table 1). There is a statistically significant relationship between the complexity of the texts read and the time spent reading the texts daily.
- H2 **has been confirmed** (Table 2). A statistically significant relationship exists between whether individuals consider themselves literate and the time spent reading texts daily.
- H3 **has been confirmed** (Table 3). There is a statistically significant relationship between the texts' complexity and whether individuals consider themselves literate.
- H4 **has been confirmed** (Table 4). A statistically significant relationship exists between an individual's opinion on hybrid texts and the type of device used for reading electronic texts.

The research conducted by Kalmárová et al. [8] and Mazáčová et al. [15] focused on the use of digitized cultural heritage in education. Marinič and Pěkný [14] investigated the advantages and disadvantages of distance learning and the effects of connectivism. They also looked into the popularity of hybrid learning through a survey of 162 secondary school pupils. The results indicated that pupils prefer face-to-face education over distance learning. In line with this, it was found that secondary school pupils appreciate education that incorporates digital technologies relevant to everyday life and educational practice. In 2018, Swedish academics Spante et al. [19] conducted a systematic review focusing on the definitions of "digital competency" and "digital literacy" in higher education research. The goal was to understand how these terms have evolved and how different research approaches in various countries have intersected. The study analyzed 107 publications from three databases (Web of Science, Scopus, and Education Resources) from 1997 to 2017. Of these publications, 28 related to digital skills and 79 to digital literacy. The research findings revealed that the definitions of these terms vary based on whether they are politically oriented, research-oriented, or focused on technical or social aspects. Finally, Kubíková [11] focused on text comprehension and reading skills among younger pupils, while Vicherková [21] addressed the reading strategies of 15-year-old pupils. The research conducted by Hašková et al. [6] examined the efficacy of various information presentation methods within the curriculum context. This case study involved 24 pupils from secondary vocational schools in Slovakia. The findings indicated that video content did not emerge as the most influential factor for fostering motivation and enhancing information retention. Instead, the research highlighted the significance of textual materials in effectively conveying pertinent information. The study addressed the challenges of communicating technical information relevant to control unit programming.

Acknowledgments The paper was created with the financial support of the SGS project at the Faculty of Education of the University of Ostrava. The authors thank for the support provided to SGS17/PDF/2024 Understanding of (e)non-artistic texts among students of secondary vocational schools with a technical orientation in the Moravian-Silesian and Zlin regions.

References

1. Čechová, M., Krčmová, M., & Minářová, E.: *Současná stylistika*. Praha: Nakladatelství Lidové noviny (2008). [Čechová, M., Krčmová, M. and Minářová, E.: Contemporary stylistics. Prague: Lidové noviny Publishing House (2008).]
2. Čmejrková, S., & Hoffmannová, J. (ed.): *Mluvená čeština: hledání funkčního rozpětí*. Praha: Academia (2011). [Čmejrková, S., & Hoffmannová, J. (ed.): Spoken Czech: the search for a functional span. Prague: Academia (2011).]
3. ČŠI Tematická zpráva 2019: *Rozvoj informační gramotnosti na středních školách ve školním roce 2018/2019*. Dostupné z: https://www.csicr.cz/Csicr/media/Prilohy/PDF_el._publikace/Tematick%c3%a9%20zpr%c3%a1vy/TZ_informacni-gramotnost_2018-2019.pdf, [cit. 2024-05-30] [CSI Thematic report 2019: Development of information literacy in secondary schools in the 2018/2019 school year. Available from: https://www.csicr.cz/Csicr/media/Prilohy/PDF_el._publikace/Tematick%c3%a9%20zpr%c3%a1vy/TZ_informacni-gramotnost_2018-2019.pdf, [quot. 2024-05-30]]
4. ČŠI Tematická zpráva 2024: *Čtenářská gramotnost na základních a středních školách ve školním roce 2022/2023*. Dostupné z: https://www.csicr.cz/CSICR/media/Prilohy/2024_p%c5%99%c3%adlohy/Dokumenty/TZ_Ctenarska_gramotnost_final_12032024.pdf, [cit. 2024-05-30] [CSI Thematic Report 2024: Reading literacy in primary and secondary schools in the 2022/2023 school year. Available from: https://www.csicr.cz/CSICR/media/Prilohy/2024_p%c5%99%c3%adlohy/Dokumenty/TZ_Ctenarska_gramotnost_final_12032024.pdf, [quot. 2024-05-30]]
5. Digitální Česko:. Digitální vzdělávání. Strategický rámec pro rozvoj digitálního vzdělávání. Dostupné z https://vlada.gov.cz/assets/ppov/rvis/zapisy_rvis/DiV_2023.pdf [cit. 2024-05-10]

[Digital Czech Republic:. Digital education. A strategic framework for the development of digital education. Available from https://vlada.gov.cz/assets/ppov/rvis/zapisy_rvis/DiV_2023.pdf [quot. 2024-05-10]]

6. Hašková, H., Kuna, P., & Palaj, M. Information Transfer in Education: Case Study Based on Teaching Lowcost Control Unit Programming in K. Kumar (Ed.) Engineering Pedagogy Towards Outcome-Based Education, Taylor & Francis Group (2022). https://doi.org/10.1201/9781003083160.
7. Höflerová, E.: Elektronický učební text. VŠB – technická univerzita Ostrava (2011). ISBN 978-80-248-2472-7. [Höflerová, E.: Electronic learning text. VŠB – Ostrava University of Technology (2011). ISBN 978-80-248-2472-7.]
8. Kalmárová, K., Černý, M., Mazáčová, P., & Martonová, M.: Možnosti využití digitalizovaného kulturního dědictví ve vzdělávání. *ProInflow,* 13(1) (2021). Dostupné z: https://doi.org/10.5817/ProIn2021-1-3 [cit. 2024-05-15]. [Kalmárová, K., Černý, M., Mazáčová, P., & Martonová, M.: Possibilities of using digitised cultural heritage in education. ProInflow, 13(1) (2021). Available from: https://doi.org/10.5817/ProIn2021-1-3 [quot. 2024-05-15].]
9. Karlík, P., Nekula, M., & Pleskalová, J. (eds.): *Encyklopedický slovník češtiny.* Praha: Nakladatelství Lidové noviny, 604 (2002). ISBN 80-7106-484-X. [Karlík, P., Nekula, M., & Pleskalová, J. (eds.): Encyclopedic dictionary of the Czech language. Prague: Nakladatelství Lidové noviny, 604 (2002). ISBN 80-7106-484-X.]
10. Kostečka, J.: *Český jazyk pro 3. ročník gymnázií.* Praha: SPN – pedagogické nakladatelství (2002). ISBN 80-7235-191-5. [Kostečka, J.: Czech language for the 3rd year of grammar schools. Prague: SPN – pedagogical publishing house (2002). ISBN 80-7235-191-5.]
11. Kubíková, A.: *Porozumění textu a žáků mladšího školního věku z hlediska hlasitého čtení a poslechu* (Disertační práce). Praha: Univerzita Karlova, Pedagogická fakulta. Vedoucí práce: Špačková, K. (2020) [Kubíková, A.: Comprehension of text and pupils of younger school age in terms of reading aloud and listening (Dissertation). Prague: Charles University, Faculty of Education. Supervisor: Špačková, K. (2020)]
12. Mareš, J.: *Pedagogická psychologie.* Praha: Portál (2013). ISBN 978-80-262-0174-8. [Mareš, J.: Educational psychology. Prague: Portal (2013). ISBN 978-80-262-0174-8.]
13. Mareš, J., Beran, J., Třešťíková, L. Horáčková, D. a kol autorů MŠMT: *Dlouhodobý záměr vzdělávání a rozvoje vzdělávací soustavy České republiky na období 2023-2027.* Praha: MŠMT (2023). ISBN 978-80-87601-57-0. [Mareš, J., Beran, J., Třešťíková, L. Horáčková, D. and other authors of the Ministry of Education, Youth and Sport: Long-term plan for education and development of the educational system of the Czech Republic for the period 2023–2027. Prague: Ministry of Education, Youth and Sport (2023). ISBN 978-80-87601-57-0.]
14. Marinič, P., & Pěkný, J.: The Advantages and Disadvantages of Distance Teaching and Learning from the Perspective of Secondary School Pupils. *Ad Alta: Journal of Interdisciplinary Research,* 13(1), 189–194 (2023). https://doi.org/10.33543/j.1301.189194
15. Mazačová, P., Černý, M. Kalmárová, K., Martonová, M., Vokřál, J., & Štěpánek, J.: *Digitalizované kulturní dědictví jako prostředek vzdělávání: sbírka metodik.* 1. vyd. Brno: Flow (2022). ISBN 978-80-88123-33-0. [Mazačová, P., Černý, M. Kalmárová, K., Martonová, M., Vokřál, J., & Štěpánek, J.: Digitised cultural heritage as a means of education: a collection of methodologies. 1st ed. Brno: Flow (2022). ISBN 978-80-88123-33-0.]
16. Mezinárodní šetření PISA: *Národní zpráva.* Praha: Česká školní inspekce (2022). Dostupné z: https://www.csicr.cz/CSICR/media/Prilohy/2023_p%c5%99%c3%adlohy/Mezin%c3%a1rodn%c3%ad%20%c5%a1et%c5%99en%c3%ad/PISA_2022_e-verze-9.pdf [cit 2024-05-28]. [International PISA survey: National report. Prague: Czech School Inspectorate (2022). Available from: https://www.csicr.cz/CSICR/media/Prilohy/2023_p%c5%99%c3%adlohy/Mezin%c3%a1rodn%c3%ad%20%c5%a1et%c5%99en%c3%ad/PISA_2022_e-verze-9.pdf [cit. 2024-05-28]

17. Národní pedagogický institute: Revize rámcových vzdělávacích programů 2023/24. Dostupné z https://revize.edu.cz/digitalni-kompetence#co-to-je. [cit. 2024-05-25] [National Pedagogical Institute: Revision of framework education programs 2023/24. Available from https://revize.edu.cz/digitalni-kompetente#co-to-je. [quot. 2024-05-25]]
18. Růžičková, D., Fanfulová, E., Maněnová, M., Podrázká, M., Rambousková, J., Berki, J., Janata, D., Lána, M., Olbrich, L., Roubal, P., Vyvial, A., & Hawiger D.: *Digitální gramotnost v uzlových bodech vzdělávání*, verze 07 (2018). Dostupné z https://docplayer.cz/105497169-Digitalni-gramotnost-v-uzlovych-bodech-vzdelavani.html, [cit. 2024-05-20] [Růžičková, D., Fanfulová, E., Maněnová, M., Podrázká, M., Rambousková, J., Berki, J., Janata, D., Lána, M., Olbrich, L., Roubal, P., Vyvial, A., & Hawiger, D.: Digital literacy at the nodal points of education. Available from https://docplayer.cz/105497169-Digitalni-gramotnost-v-uzlovych-bodech-vzdelavani.html, version 07 (2018) [quot. 2024-05-20]]
19. Spante, M., Hashemi, S. S., Lundin, M., Algers, A., & Wang, S.: Digital competency and digital literacy in higher education research: Systematic review of concept use. *Cogent Education,* 5(1) (2018). https://doi.org/10.1080/2331186X.2018.1519143
20. Štěpáník, S. a kol.: Školní výpravy do krajiny češtiny. Nakladatelství Fraus (2020). ISBN 978-80-7489-595-1. [Štěpáník, S. et al.: School trips to the Czech countryside. Fraus Publishing House (2020). ISBN 978-80-7489-595-1]
21. Vicherková, D.: Pohled na úroveň čtenářských strategií patnáctiletých žáků v Moravskoslezském kraji. Ostrava: Ostravská univerzita, Pedagogická fakulta (2018). [Vicherková, D.: A view of the reading strategies of fifteen-year-old pupils in the Moravian-Silesian Region. 1st ed. Ostrava: University of Ostrava, Faculty of Education (2018).
22. Zápotočná, O.: Metacognitive Processes in Reading, Learning and Education. Bratislava: TYPI Universitatis Tyrnaviensis – Veda, Ed. SAV (2013). ISBN 978-80-8082-742-7.

Application of Arduino Microcomputer in the Creation of Aeronautical Data Recorder

Martin Vozár and Nikolas Ludas

1 Introduction

Flight recording device considers several types of flight recorders which are an extremely important tool for identifying the cause of an aircraft accident. The device is usually represented by two individual recorders: a Flight Data Recorder (FDR) and a Cockpit Voice Recorder (CVR) [1]. According to ICAO Annex 6—Attachment D [2], all aircraft with a maximum take-off mass of more than 5700 kg must be equipped with an FDR regardless of the date of the certificate of airworthiness [1]. According to Annex 6 as in the sentence below, the FDR must record the parameters required to accurately determine the flight path, speed, position, engine power, configuration, and operation of the airplane. The reason for installing a CVR device in an aircraft follows the fact that human error has been documented as a primary factor in more than 80% of aircraft accidents. This mostly refers to errors and mistakes in the activities of the flight crew. Recorded sounds such as cockpit noise, audio warnings, and certain actions such as steering movements and turning on/off instruments, give investigators key information in the reconstruction of the whole event [1]. CVR recorders must record on the flight recorder all communications over the data link from and to the aircraft. The minimum length of the recording must be the same as for the sound recorder in the cabin and must be relative to the recorded sound in the cabin [1].

In our chapter, we will discuss the design and implementation of our own types of FDR and CVR devices based on the Arduino microcomputer. Since, in real practice, there is no possibility of connecting to the aircraft instrumentation due to safety in

M. Vozár (✉) · N. Ludas
Constantine the Philosopher University in Nitra, Nitra, Slovakia
e-mail: mvozar@ukf.sk; nicolas.ludas@student.ukf.sk

M. Turčáni (ed.), *15th International Scientific Conference on Distance Learning in Applied Informatics*, EAI/Springer Innovations in Communication and Computing,
https://doi.org/10.1007/978-3-031-81261-3_21

air traffic, we have limited ourselves to the use of custom sensors and the writing of selected parameters.

2 Related Work

There are various devices supporting the Internet of Things (IoT) such as Arduino, Raspberry PI, and other microelectronic devices. IoT is able to use the Internet and wireless technology to create an environment of remote-control system to create a domestic automated system at home/in the office to monitor/control [3]. Arduino is a popular microcontroller for student projects or amateur-built projects, which is programmable in v C/C++ via a Universal Serial Bus (USB) port [4]. There are several solutions for how to program the Arduino and how to visualize the data from the connected sensors. For our project, we used the Arduino IoT Cloud platform, which is designed to simplify the creation of such projects. Self-evidently, we did not avoid adding our own source code for the setup and calibration of the individual sensors used, which we write about in the next section of this post.

Arduino has been used in "prototypes" of CubeSats to demonstrate such functions as weather forecasting [5] or attitude determination [6]. There is a significant advantage in Arduino being popular and open source, as there are a large number of resources (e.g., examples and tutorials) available on the Internet [4]. Lay [4] states in his paper that the parts used for his Arduino microcontroller-based project are not built for a space mission and have lower performance and lower reliability. This same idea can be carried over to our project of create-by-flight recorders. And for this very reason, it is necessary to perform more verification tests for projects like this. As well as these tests, the procurement of the individual components is not too expensive.

There have been attempts at various flight recorder projects in the recent past. Ankan Ashish along with S. B. Chougule (2012) [7] focused on the development of an intelligent system to record real-time flight data using wireless technology. Their device recorded data on engine temperature, fuel level, speed, and geographical position. The technology they used was wireless transmission of data to a ground computer via an XBEE-RF module that was capable of communicating over long distances. The received GPS coordinates were processed on the ground by a microcontroller and after processing were sent via the Internet to the server. The data from the server was then used by the device, which displayed the real-time position of the aircraft on a digital map on a monitor. The focus of this project was on the two main objectives of the system. The first was to provide sufficient data in the short term to analyze air disasters and the second, in the long term, was to try to contribute to a reduction in the number of air disasters.

Another solution was presented by Meivel and Maguteeswaran (2016) [8], who addressed the design of a flight data transmitter for black box detection in aviation accidents. Due to signal failure in aircraft crash, parameters such as direction, acceleration, vibration, temperature, abandoned, or lost in the destroyed aircraft are recorded. Therefore, the authors developed an embedded system that is independent of GPS. In this work, an Arduino microcontroller was used to collect data from all sensors and process them according to the program. The authors used EEPROM memory to store the data and flash it to a PC using the Zigbee module in the 2.4 Industrial, Scientific and Medical (ISM) band. The data was further received at the base station. Thus, the proposed system uses the Arduino functions to send the uploaded flight data in Excel format within 40 s without using the Internet network.

Another and more recent approach has been implemented by Garcia-Barrientos, Torres-Uresti, et al. (2022) [9]. They dealt with the problem of the application and use of a microcomputer in the creation of an event recorder. In contrast to the previous projects mentioned, this case involves the presentation of the design and implementation of an automotive black box system using a Raspberry Pi microcomputer and an IoT module. This system is created using a Raspberry Pi microcomputer and various sensors including GPS, camera module, sound module, alcohol sensor module, and signals from the electronic control unit. The data is stored on a secure digital card and also in the cloud using a Waveshare SIM7600G-H 4G module. The results obtained show that this embedded system can acquire and process image, sound, GPS data, alcohol concentration, speed, temperature, and so on. Finally, a graphical user interface is developed to process the stored files. This system is similar to the black box in airplanes, which records all the flight information in the black box using a specific algorithm. This makes the recorded flight data accessible to investigating authorities if needed.

3 Implementation of Recorders

For our project, we chose the Arduino UNO microcomputer, which can be connected to a variety of sensors for our recorders. It has an 8-bit microcontroller with 32 kB of flash memory. The device supports a wider voltage range with 24 V tolerance, which facilitates seamless integration with servo motors and other sensors using a single power supply. The processor speed ranges from 8 to 400 MHz, the average Arduino board operates at 16 MHz, and it is optimized for a variety of applications. However, users are limited to integrated development environment (IDE) platforms, which simplifies code development but limits advanced external programming functionality [10, 11]. The Arduino board includes built-in Human Interface Devices (HID) support, which allows it to simulate mouse or keyboard activity when connected to a computer via USB, thus allowing easy transmission of keystrokes and mouse movements.

With the ESP32-S3 module, the Arduino UNO R4 adds WiFi® and Bluetooth® connectivity, allowing makers to wirelessly connect their projects and use the Arduino IoT Cloud for remote monitoring and control. And that is the main reason we chose this particular type for the FDR. It features a Qwiic I2C connector, making it easy to connect to the vast Qwiic ecosystem, and adapter cables ensure compatibility with sensors based on other connectors. Finally, it includes a runtime error diagnostic function that captures failures and provides detailed explanations and tips for troubleshooting the code causing the problem. For the audio recorder, we reached for the Arduino UNO R3, which won't need a WiFi module.

For programming, we used the Arduino IDE, which works on the Java platform. It contains a code editor and includes mechanisms for compiling and uploading programs to the Arduino board. It also includes a text console, a message area, and a toolbar with buttons for common functions and a hierarchy of operational menus. An open-source advantage is the use of free software libraries, while the Arduino IDE comes with a set of standard libraries for commonly used functions. These functions provide basic communication and support for the most commonly used hardware accessories such as LCD displays and servo motors [11].

3.1 Characteristics of the Sensors Used for CVR

Real-Time Clock (RTC)

Time is a very important aspect of flight data recording. This must be maintained even after the device has been disconnected from power. We have therefore taken advantage of this integrated feature of the Arduino UNO R4 WiFi board. It was therefore necessary to supplement our assembly with a 3.3 V coin cell battery on the VRTC pin of the Arduino. We placed the battery in the coin battery holder. Specifically, we used a CR2032 battery type, which has a larger kappa-capacity endurance, specifically 225 mAh. To work with the RTC, the RTC.h library was used and we set the current time and date using its functions.

For the audio recorder, we used an Arduino UNO R3, which does not have pins for a battery, so we supplemented this with a DS1302 RTC module. This is used to maintain the real date and time after the Arduino board is disconnected from the power supply. To connect it, we needed to connect five wires, namely, power supply VCC to 5 V, GND to ground, clock pin (CLK) to Arduino digital pin number 5, data pin (DAT) to pin 4, and reset pin (RST) to pin 2. The Rtc by Makuna library was used to work with this module.

Storage of Recorded Data

We used a module with an SD card slot to store the recorded data. We chose this module because it is a standard module that is easy to plug in and works well with Arduino when using an SD card. The module supports both reading and writing on the SD card. It works with both 3.3 and 5 V voltages. Working with the SD card requires fast communication with the Arduino for writing and reading data and therefore the module is used with the Serial Peripheral Interface (SPI). To use this type of interface, a standard wiring with four pins plus two pins for power/ground is provided. We used a 5 V pin to power the module, and we connected the CS (Chip Select) pin to pin 10, the SCK (Serial Clock) pin to pin 13, the MISO (Master In Slave Out) pin to pin 12, and the MOSI (Master Out Slave In) pin to pin 11. The SD.h and SPI.h libraries were used to work with this module.

Sound Recording

For sound recording, we used a MAX9814 microphone module with an adaptive amplifier. The audio signal is captured in the range of 20–20 kHz, and thanks to the MAX9814 amplifier designed specifically for audio applications, we get a clean signal at the output. The output of the amplifier is a maximum of 2 V at 1.25 V DC bias. This means that the module can be connected to an AD converter with a 3.3 V input.

Once connected, we recorded simple code that measured the input at A0 using the analogRead function during the first test. We saw the data output on the Serial Monitor. We found that the readings changed occasionally without changing the ambient sound, which made us judge that these measured readings were not accurate. We modified the code by adding a sampleWindow variable, which we set to 50, meaning that the sample window width was 50 ms. This value is sufficient to measure sound levels with frequencies as low as 20 Hz. Next, we added an algorithm to filter the data, which can contain values from 0 to 1023, and we set limits. Finally, for better clarity, the measured values were converted to values in volts. For the different sound measurements, we obtained a maximum value of 2.43 V for the loudest sound.

To record the audio, we initially needed to import the TMRpcm library, which has all the different settings for working with the microphone module and for recording audio. Once configured, we programmed the code to ensure constant sound recording for 1 min. In this way, the recorded sound will be stored in 1-min intervals in .wav files on the SD card. To minimize file sizes and maximize the length of recorded time, we experimented with the frequency of recording. In the end, we optimized the recording to a frequency of 8000 Hz, where the sound was of sufficient quality but with occasional recording dropouts. To eliminate dropouts and achieve better sound quality, we set the buffSize in the pcmConfig library to 128. The throw-not can be increased but at the expense of the microcontroller's RAM. By using SD cards with higher data write speeds, we were able to increase the frequency to 16,000 Hz. With a card size of 32 GB, this storage space gave us 563.75 h of recording time at the above quality, several times the requirements for such a device.

3.2 *Sound Recorder Implementation*

Block diagrams provide insight into the layout of the modules, show the data transfers between all the modules and components involved, and also show their interconnections and layouts.

After testing the microphone module, we connected the MAX9814 microphone module and the SD card slot to the Arduino UNO R3 and followed the block diagram in Fig. 1. We implemented the wiring diagrams in Fritzing software. Figure

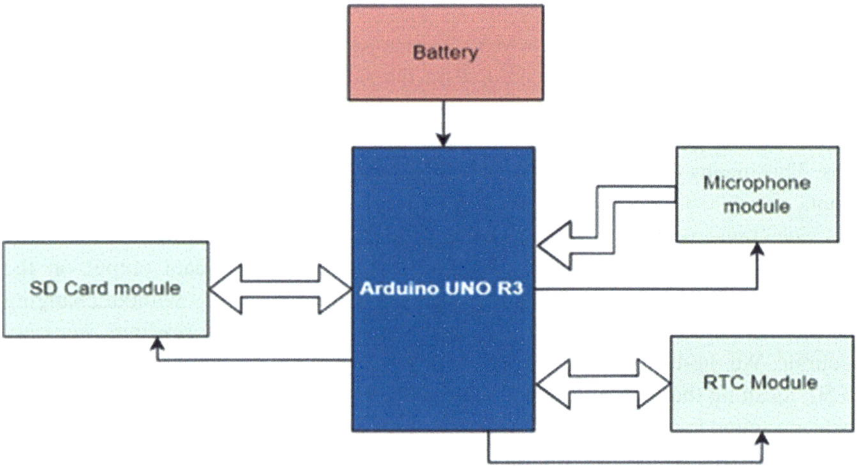

Fig. 1 Block diagram of Arduino UNO R3 for audio recording

2 shows the circuit diagram of the components for the cockpit audio recording device working with Arduino UNO 3.

To power both devices, we used 9 V batteries with a capacity of approximately 550 mAh, which we connected to the Arduino board using the Power Jack. We also wired a simple switch to turn the devices on and off. To indicate the functionality of the device, we connected an Light-Emitting Diode (LED). In the circuit, it indicates by lighting up that sound is being recorded.

3.3 Characteristics of the Sensors Used for the FDR

Sensors for Pressure Measurement

For recording the air pressure, we opted for a sensor type MS5611-01BA03 with a barometric pressure sensor from MEAS Switzerland. The pressure measurement range is from 10 to 1200 mbar with an absolute accuracy of ±1.5 mbar over a pressure range from 450 to 1100 mbar. This barometric pressure sensor is optimized for altimeters and variometers with an altitude resolution of 0.012 mbar, which is approximately 10 cm above sea level. One of the outstanding features of the MS5611-01BA module is the built-in temperature sensor. This sensor plays a key role in compensating for changes in the environment by calibrating its readings based on temperature changes. This provides very accurate data used in the calculation of the over-sea altitude. This calibration process increases the accuracy and stability of the measurements and provides users with highly accurate data. The temperature sensor measures temperature from −40 to +85 °C with an accuracy of ±0.8 °C. The module also includes a precision 3.3 V MIC5205 on-voltage regulator

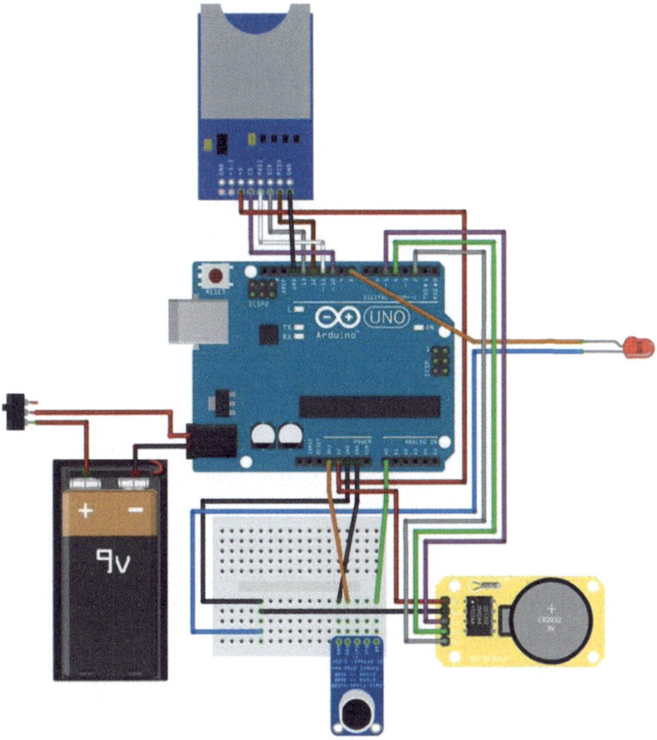

Fig. 2 Cab sound recording equipment wiring diagram

and voltage level converter, so it can be used with a 3.3 or 5 V connection on a microcontroller without worry. The MS5611 consumes less than 1.4 mA during measurements and less than 0.15 μA during standby. It can be connected to the microcontroller using either the SPI or IC2 (InterIntegrated Circuit) bus interfaces.

We used an Arduino UNO R3 microcontroller for the first wiring, testing all the modules separately and calibrating them. We took this step to prevent potential damage to the Arduino UNO R4 + WiFi in case the purchased modules contained a possible bug or if there was some mistake when wiring them. After the hardware wiring, it was necessary to install the MS5611 library with the I2C interface. Using the readPreasure() function, we obtained the measured pressure value in Pascals, which was used to calculate the relative and absolute altitude provided by the getAltitude() function getAltitude().

Sensors for GPS Measurement

The sensors used to record GPS coordinates were the NEO-M8 module from the manufacturer u-blox, which is built on the high-performance GNSS ublox M8 core. These modules use simultaneous reception of up to three GNSS systems (GPS/Galileo along with BeiDou or GLONASS), recognize multiple constellations simultaneously, and provide excellent positioning accuracy even in scenarios involving urban canyons or locations with weak signals. The NEOM8 series also supports message integrity protection, geofencing, and spoofing detection with configurable interface settings that are easily adapted to customer applications. For our project, we used a replica of the NEO-M8M module to meet the requirements of our facility. It operates at 1.575 GHz, achieves time to first fix from a cold start in 26 s, provides horizontal position accuracy of 2.5 m, and requires an operating supply voltage of 1.65–3.6 V. It supports multiple interface types including I2C, SPI, UART, and USB. The module has a built-in micro USB port for powering and sending AT commands and a port for the antenna. There is a ceramic patch antenna mounted on the back of the module [12].

Before the first connection, the module had to be configured using the u-center software supplied by the module manufacturer. After configuration, we tested the GPS signal outdoors. Because it is a replica, it was necessary to connect an external antenna with a frequency of 1575.42 MHz to the module, since the built-in antenna in the module is not powerful enough. By using the antenna we reduced the detection of satellites during cold start from the original about 20 to 1–3 min. The module lights the built-in diode red when searching for signals. When the exact position is found, the diode starts blinking. To successfully determine the position, we need to have a stable signal with at least four satellites.

Our module was usually able to establish communication with a maximum of five satellites, and of these, all were of the GPS type that is of US origin. Russian satellites of the GLONASS (Globalnaya Navigazionnaya Sputnikovaya Sistema) type, the European Union's Galileo satellites, and China's BeiDou satellites could not be intercepted by the module. Figure 3 shows a listing of the sensor data on the console on the left and a graphical representation of the satellites and GPS position on the map on the right.

After testing and calibrating the module, we connected it to the Arduino. The data from this module will be used to display the position of the aircraft and also to calculate its velocity with respect to its movement towards the ground. We needed to use the SoftwareSerial.h library to work with the module because the Arduino UNO needs a software serial port emulator. In our case, we set these ports to 2 and 3 (since the module was connected to the Arduino on these ports) as follows: SoftwareSerial gpsSerial(2,3). The gpsSerial.read() function provides us with receiving the message from the GPS module. We then had the received data written out to the Serial monitor in raw format. The data that was output to us were NMEA_GLL (Geographic Position—Latitude Longitude), NMEA_RMC (Recommended Minimum Specific GNSS Sentence), NMEA_VTG (Course Over Ground and Ground Speed), NMEA_GGA (GPS Fix Data), NMEA_GSA (GNSS DOPS and Active Satellites), and NMEA_GSV (GNSS Satellites in View). We converted the latitude and longitude values from these data using a separate custom code.

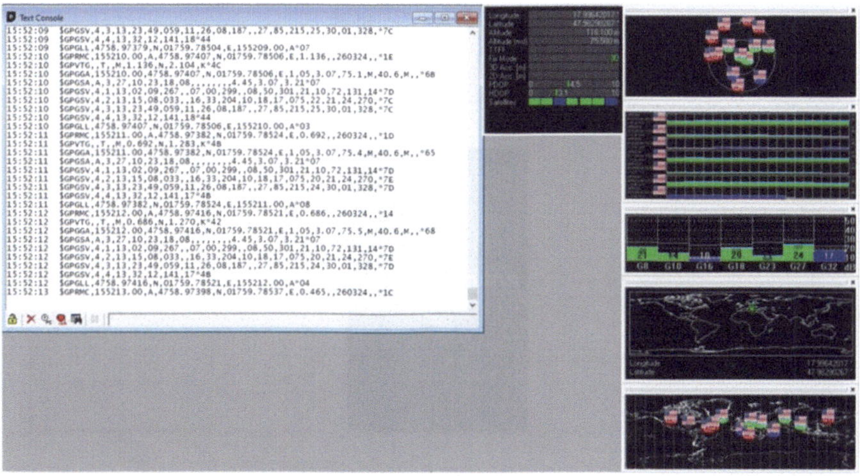

Fig. 3 View listing and positioning in u-center

3.4 Wiring, Setting Up, and Programming the FDR

Figure 4 shows the wiring of the Arduino UNO R4 WiFi with the MS5611 module, GPS module, SD card module, and connection to Arduino Cloud with a dashboard for visualization.

Similar to the audio recorder, in this circuit, the LED indicates by flashing that the Arduino is connected to the WiFi network. Similarly, a 9 V 550 mAh battery was used for this device, connected to the Arduino board using a power jack. The device was also completed with a simple switch to turn the device on and off. To indicate the functionality of the device, we connected an LED. Due to the lack of power and ground pins on the Arduino boards, we used a smaller contact box (breadboard) to ground and connect the individual modules and components used (Fig. 5).

4 Results

To keep track of the logged parameters, we created a dashboard to visualize the data. We used the Arduino IoT Cloud platform, which offers an integrated solution to connect Arduino-based devices to the Internet and manage them remotely. Arduino IoT Cloud provides a web-based interface where users can register, manage, and monitor their IoT devices. It uses secure protocols such as Message Queuing Telemetry Transport [13] to transfer data between the devices and the cloud server. We provided functionality signaling using an attached LED to indicate a successful WiFi connection and a built-in LED to indicate the functionality of writing and reading to and from the SD card. The GPS module already has a built-in LED that

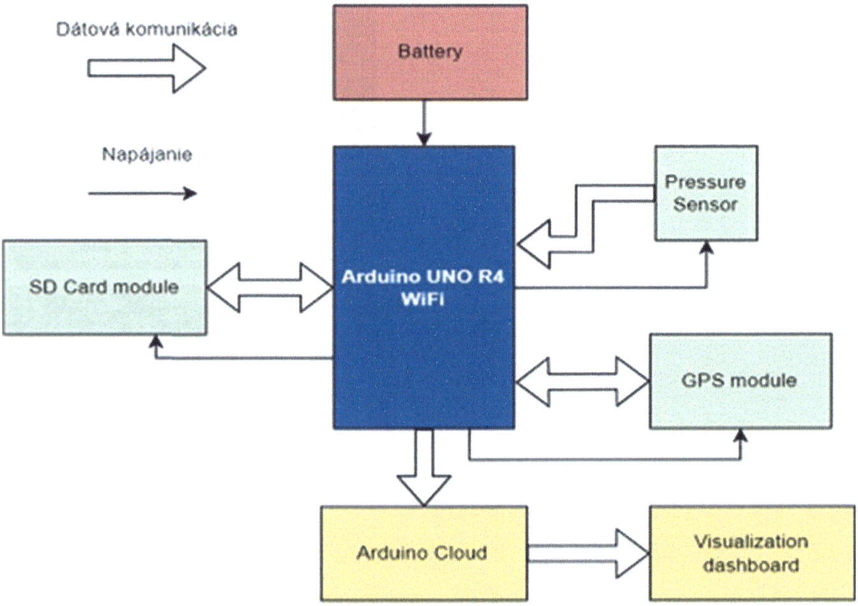

Fig. 4 Block diagram of Arduino UNO R4 WiFi for sensor data recording

flashes every second if the GPS signal is stable. This signaling has made it easier to analyze in the event of a failure of one of the modules. We write the sensor data to a text file in the form:

2024/4/13 16:41:34, 48.00, 18.15, 22.84, 133.80, 8.43, 38.14, 1010.00, 0.05, 1

In order, the data is: date time, latitude, longitude, absolute altitude Barometer, absolute altitude GPS, relative altitude, temperature, pressure, speed, and gps fix quality. We chose this format because of the ease of access for further data processing.

We have deployed all the necessary components on the dashboard and the sensor data is sent to them via cloud variables. These components include GPS signal status, speed in knots, conversion of speed to km/h by multiplying speed in knots by 1.852, temperature in degrees Celsius, atmospheric pressure in hPa unit, relative altitude, absolute altitude from GPS module, absolute altitude from the MS5611 module in meters, and also Google map to plot the exact location. Figure 6 shows the mobile version of the IoT Remote application.

To verify the correct measurement of the altitude, we compared the measured data from the GPS sensor and the barometric sensor. The recorded data is shown in Fig. 7.

The recorded pressure altitude should be between 30 and 200 m and 100 and 700 ft., respectively, according to Aircraft Operation Regulation L6 [2]. By comparing the data from the two sensors mentioned above and also by comparing

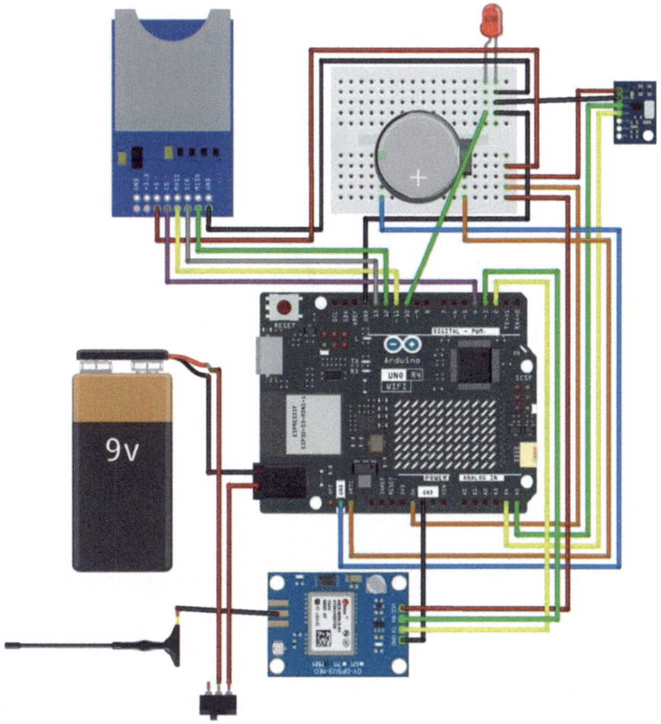

Fig. 5 Wiring diagram of sensors to Arduino UNO R4 WiFi

the data in real time during the flight, we did not observe any deviations that were outside the above interval.

In a similar manner, we verified the recording of the aircraft position from the recorded geographic coordinates using the NEO-M8 GPS module. Figure 8 shows the recorded flight paths from the two devices. On the left is the route recorded by our device working with Arduino UNO 4 WiFi. On the right is the flight path recorded by SkyDemon's aeronautical navigation.

The GPS module data is also used to calculate the speed, which we compared during the flight with the navigation device working with the SkyDemon air navigation in the aircraft [14, 15]. The recorded differences between these speeds were a maximum of 2 kt (3.7 km/h), which translates to only 1.3%. The recorded speed in the original flight recorders must meet an accuracy of 5%, so, in our case, we achieved excellent accuracy and met the accuracy condition for the recorded speed.

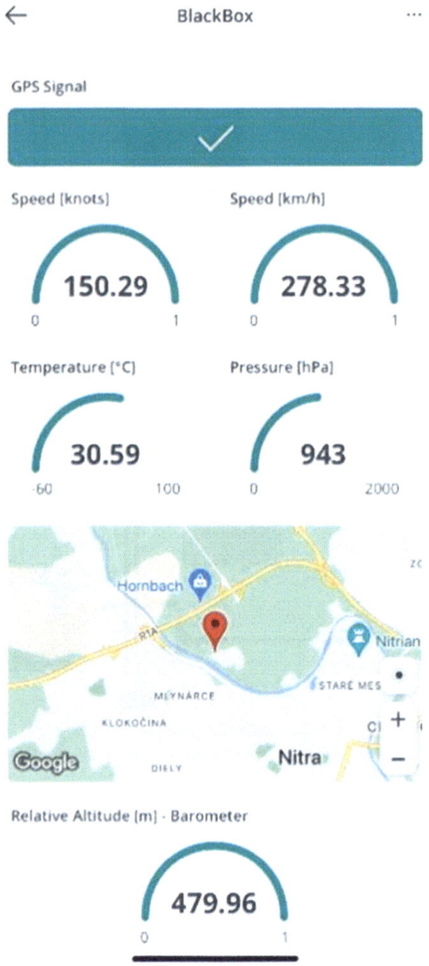

Fig. 6 IoT Remote app displaying in-flight updates

5 Conclusion

We have created recorders for the category of aircraft with a maximum flight weight of less than 5700 kg. These will be able to be used mainly in the flight school that teaches future pilots. The CVR device can be used to play back pilot communications with air traffic controllers. In this way, both the student pilot and the instructor can replay the recording and point out errors in the communication from which the student can learn. The second FDR device can be used within the flight school to visualize and compare the flown route with the planned route, to examine the adherence to prescribed speeds or prescribed altitudes. By adding a SIM card to the FDR device and connecting to the mobile internet, it is possible to

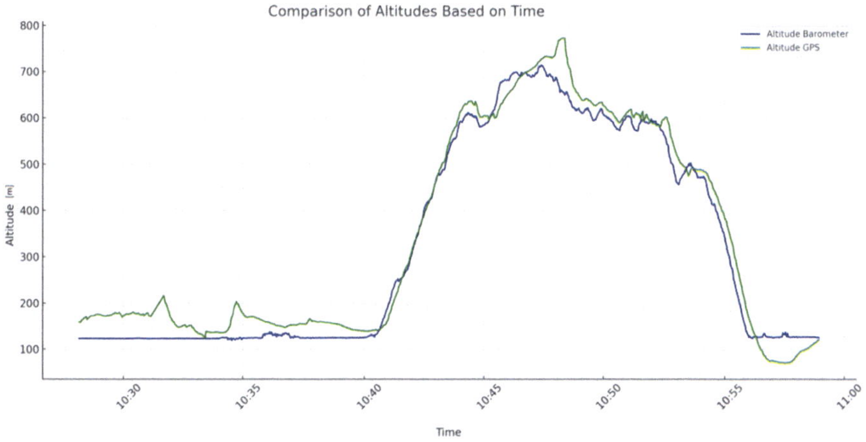

Fig. 7 Comparison of recorded altitudes obtained from the GPS module and the barometer module

Fig. 8 Visualization of recorded data of flown routes by blue lines. On the left is the route recorded by our device, and on the right is the route recorded by air navigation

monitor the recorded data remotely via the internet. This appears to be a convenient check of the aircraft position by the instructor.

Acknowledgments This work was supported by the Scientific Grant Agency of the Ministry of Education of the Slovak Republic (ME SR) and the Slovak Academy of Sciences (SAS) under Contract No. VEGA 1/0385/23.

References

1. Vidović, A., Franjić, A., Štimac, I., Ozmec Ban, M.: The Importance of Flight Recorders in the Aircraft Accident Investigation, Transportation Research Procedia, Volume 64, 2022, Pages 183-190, ISSN 2352-1465, https://doi.org/10.1016/j.trpro.2022.09.022.
2. Predpis L6, Ministerstvo dopravy, pôšt a telekomunikácií SR, http://letectvo.nsat.sk/wp-content/uploads/sites/2/2014/07/L-6-Prev%C3%A1dzka-lietadiel-II.-%C4%8Das%C5%A5-V%C5%A1eobecn%C3%A9-letectvo-let%C3%BAny1.pdf
3. Koprda, Š., Balogh, Z., Magdin, M. Developing System from Low-Cost Devices to Build a Security and Fire System as a Part of IoT. In: Huang, DS., Huang, ZK., Hussain, A. (eds) Intelligent Computing Methodologies. ICIC 2019. Lecture Notes in Computer Science(), vol 11645, 2019, Springer, Cham. https://doi.org/10.1007/978-3-030-26766-7_14
4. S. Lay, K. S., Li, L., Okutsu, M.: High altitude balloon testing of Arduino and environmental sensors for CubeSat prototype, HardwareX, Volume 12, 2022, e00329, ISSN 2468-0672, https://doi.org/10.1016/j.ohx.2022.e00329.
5. M.R. Laskar, R. Bhattacharjee, M. Sau Giri, P. Bhattacharya: Weather forecasting using Arduino based cube-sat, Procedia Comput. Sci. 89 (2016) 320–323, https://doi.org/10.1016/j.procs.2016.06.078.
6. M. Clarke, J. Guo, B. Sanders, A. Gasiewski, Affordable and Accessible Attitude Control Validation Test Methods for CubeSats, in: Paper presented at the 67th International Astronautical Congress (IAC), Paper IAC-16-B4.2.8, Guadalajara, Mexico, September 26–30, 2016.
7. ASHISH, A., CHOUGULE, S. B. 2012. Wireless Flight Data Recorder (FDR) for airplanes. Research Gate. [online] 2012. [cit. 2024-06-01]. <https://www.researchgate.net/publication/272108873_Wireless_Flight_Data_Reorer_FDR_for_airplanes>
8. MEIVEL, S., MAGUTEESWARAN, R. 2017. Design of flight data transmitter for black-box detection at airplane crash. Research Gate. [online] 2017. [cit. 2024-06-01]. https://www.researchgate.net/publication/326893076_Design_of_flight_data_transmitter_for_black-box_detection_at_airplane_crash
9. GARCIA-BARRIENTOS, A., TORRES-URESTI, D., CASTILLO-SORIA, F. R., PINEDA-RICO, U., HOYO-MONTAÑO, J. A., PEREZ-CORTES, O., ORDAZOLIVER, P. : Design and Implementation of a Car's Black Box System Using a Raspberry Pi and a 4G Module. MDPI Journal. [online] 2022. [cit. 2024-06-01]. <https://www.mdpi.com/2076-3417/12/11/5730?type=check_update&version=1>
10. SZCZEPANIAK, A. 2023. Raspberry Pi or Arduino—when to choose which? Leo Rover. [online] 2023. [cit. 2024-05-20]. <https://www.leorover.tech/post/raspberry-pi-or-arduino-when-to-choosewhich#:~:text=Raspberry%20Pi%20has%20a%20superb,good%20for%20developing%20software%20applications>
11. EARL, B., BARELA, A. 2013. All About Arduino Libraries. Adafruit. [online] 2013. [cit. 2024-01-20].<https://learn.adafruit.com/adafruit-allabout-arduino-libraries-install-use/arduino-libraries>
12. U-BLOX. 2021. NEO-M8 u-blox M8 concurrent GNSS modules. U-blox modul Data sheet. [online] 2021. [cit. 2024-05-23]. <https://content.ublox.com/sites/default/files/NEO-M8-FW3_DataSheet_UBX-15031086.pdf >
13. ANAND, S. 2024. How does Arduino IoT Cloud work? LinkedIn. [online] 2024. [cit. 2024-03-20]. <https://www.linkedin.com/pulse/how-doesarduino-iot-cloud-work-saurabh-anand-k7goc/>
14. Uvaliyeva, Indira, et al. "The Development of the Staking-Ensemble of Methods for Analyzing Academic Data." *Acta Polytechnica Hungarica* 19.11 (2022): 7–25.
15. Kovari, Attila, and Monika Rajcsanyi-Molnar. "Mathability and creative problem solving in the MaTech math competition." *Acta Polytechnica Hungarica* 17.2 (2020): 147–161.

Artificial Intelligence as a Key Element of the Personal Digital Environment of University Students

Viola Vrbová, Dominik Frolík, and Lucie Rohlíková

1 Introduction

We live in a digital age, and this greatly affects the methods of education. We are surrounded by technology, and most likely use it ourselves, even if we do not even know it. The use of digital devices and tools is a common part of the educational process, which is gradually changing, due to the digital age [1]. Teachers can include digital elements in teaching in different ways—it can be the use of digital devices and tools during face-to-face teaching, when creating e-learning courses that are available to their students at any time, they can practice blended learning, and so on.

University students can also purposefully support their knowledge acquired in the formal system by using various digital tools; they do not have to, and should not, just be mere recipients of information. Students belonging to the generation of digital natives [2] are increasingly immersed in their own digital world, which consists of all digital elements, devices, applications, and technologies with which they normally come into contact. All this forms their personal digital environment, and if they use these tools in their daily life to learn, it is already a *personal learning environment* (PLE). These tools can include applications adapted for content sharing and subsequent interaction, various content editors, communication platforms, Massive Open Online Courses (MOOC) courses, and much more [3]. Thanks to the appropriate use of digital technologies, students can personalize their learning environment to suit their own individual needs—each user should have a slightly different form of their *PLE* [4]. For example, one tool that can play a key

V. Vrbová (✉) · D. Frolík
Faculty of Science, University of Hradec Králové, Hradec Králové, Czechia
e-mail: violavrb@kvd.zcu.cz

L. Rohlíková
Faculty of Education, University of West Bohemia, Pilsen, Czechia

M. Turčáni (ed.), *15th International Scientific Conference on Distance Learning in Applied Informatics*, EAI/Springer Innovations in Communication and Computing,
https://doi.org/10.1007/978-3-031-81261-3_22

role in students' digital learning environment is artificial intelligence (AI) tools. AI tools are increasingly appearing in the educational process, as they have become more easily available and become more widely known since the introduction of ChatGPT to the public in October 2022. AI has a huge potential to transform the form of higher education, while one of the reasons may be a better way of personalizing the learning process through the use of a personal AI assistant, which would allow adaptive teaching based on the interaction of the student with the study material, and the adaptation of learning procedures to his/her specific needs and preferences [5]. We must not forget that properly used AI tools can lead to a better provision of individual feedback reflecting the strengths and weaknesses, as well as the specific educational needs, of the individual [6].

AI has already been involved in the learning process. It found its place in language teaching because its use can have an impact on the higher quality of students' language skills—AI tools can help them improve their ability to write in a foreign language and interpret difficult words if they use suitable applications or websites [7]. An example of a popular application that stands outside the formal education system is Duolingo, which used AI to create personalized lessons before [8], but now, thanks to the connection with modern language models, they can offer an even more sophisticated learning experience [9]. AI tools are part of some already commonly used applications in education, from grammar checks to methods of searching in search engines [10].

Artificial intelligence in education is gaining increasing importance. With the advent of generative AI comes new ways to use AI tools in teaching and in your own studies. For example, teachers can use the ChatGPT application to help them create a course structure or even just an individual lecture, and they can also recommend tools to their students; students can use it as a helper to solve problem tasks or to get individual feedback [11]. Providing the right individual feedback and using AI tools to evaluate your own learning progress will lead to faster personalization of the learning environment [12].

Related to all of this are the ethical aspects of using AI to support the study and the attitudes that different groups take toward AI. AI brings with it various opportunities and challenges for the traditional model of education. Therefore, we consider it important to explore the ways in which AI tools are used by college students, who will then become teachers themselves, and most likely pass on their skills and experience.

The opinions of pre-service teachers have been addressed by various studies, for example, Ayuso-del Puerto and Gutierrez-Esteban [13] investigated the opinions of future kindergarten teachers and concluded in their research that AI enriches the learning environment in the context of higher education and also awakens the students' interest in using technology for their future teaching practice. In Ghana [14], insight into student confidence in AI tools from pre-service teachers was gained through quantitative data collection. A significant finding is that 77% of 231 university students actively use AI in their studies, while another 70% of students perceive these AI tools as the reason for their better academic results because they support their education [14]. A qualitative research project [15] was carried out in

Turkey, in which they focused on the feelings that pre-service teachers feel regarding AI. They revealed that fear and anxiety about AI dominate among students, but also enthusiasm. At the time of the Turkish study (2019), the tools of generative AI models such as Chat GPT, Lamma, Bard, and so on, were not yet publicly available.

The question of what perspective university students take on generative AI and its impact on higher education was addressed by a qualitative study [16]. Students' ideas about higher education have changed as a result of AI, as students themselves design new learning outcomes related to AI literacy and ways to directly learn and work with AI tools. At the same time, they also mention adaptability related to the use of different approaches that AI enables as a key output. A study carried out at the University of Petra [17] among IT students, where 82% of the 122 students use AI during learning, mainly to solve programming problems or to express their ideas, and is devoted to ascertaining their opinion on the use of AI during the learning process. A study from the University of Poland [18] shows that out of 231 respondents, 72% of students expressed that they directly use the ChatGPT tool in their learning process to search for information, while another 65% use it to solve complex problems. At the same time, they use it extensively as a source of inspiration for the topics of study papers. They also mention that ChatGPT can serve as an assistant, thanks to which they can understand complex concepts (42%), and can provide support during their own study projects (39%).

2 Methodology

The research carried out as part of this study is an initial insight into the personal educational environment of pre-service teachers at the Faculty of Education of the University of West Bohemia in Pilsen, with a targeted focus on AI tools. The aim was to investigate how some pre-service teachers perceive generative AI tools and how they use them in their own studies at university. The research carried out is qualitative, and elements of grounded theory [19] were used in it. Thanks to the qualitative nature, we can better understand the ways in which AI tools are used by individual students.

The main research question for the conducted survey was: *How do pre-service teachers use AI tools in their educational environment?* This question is also related to finding out their opinion on AI tools not only as a study aid. If they fear or distrust AI tools, they will not use them, and vice versa.

Qualitative data was obtained through semi-structured interviews. A total of six interviews were conducted with student teachers. The selection of respondents consisted of approaching students who showed a general interest in AI tools during classes in the winter semester. During the first interview with a student who we knew was definitely using AI tools, we asked him/her for recommendations of classmates who could participate in the survey. We contacted them further, some agreed to participate in the survey, and we obtained further contacts from them. We therefore used the snowball method to obtain additional respondents [20]. A total of six interviews were finally conducted with pre-service teachers, as not all of those interviewed agreed to participate in the survey. All of them met the condition that they independently use some AI tools, which was also reflected in the questions asked. Five respondents are studying informatics as one of the two compulsory approvals, the sixth respondent shows his own personal interest in informatics but does not study it formally. The age range of respondents was 23–25 years.

Semi-structured interviews were conducted at the Faculty of Education. Respondents had their own devices on which they self-reported working with AI tools: three of the respondents had their own laptops, two of the respondents had their own tablets, and all of them used smartphones. At the beginning of the interview, the students were introduced to the course and subsequent methods of data processing and use in accordance with the ethical principles of conducting qualitative research. The interviews were subsequently recorded and transcribed into text so that the data could be processed in the statistical software ATLAS.ti. The shortest interview lasted 18 minutes, while the longest was 26 minutes. The obtained data was then subjected to open coding, developed within the framework of the analytical apparatus of grounded theory [19], where we divided the individual parts into meaning units, and assigned them codes that described their meaning. We first formulated the codes with regard to the research question, then according to the specific statements of our respondents, so we also established some "in vivo" codes.

The students were asked several framing questions, the most important for our inquiry being the following: *Can you show me specifically how you use AI tools?* and *When you use AI tools, how does it make you feel?* In the course of the interview, additional questions responding to what the respondents communicated were also asked. Thanks to this, we were able to see how they work with AI tools.

3 Results

In this chapter, we present the results of a survey conducted among students of the follow-up teaching program at the Faculty of Pedagogy of the University of West Bohemia in Pilsen. All of the semi-structured interviews took place during the second half of December 2023 and the beginning of January 2024. All of the respondents who voluntarily participated in the research use AI tools voluntarily, either in their personal life or in connection with their studies.

One of the ways that students use generative intelligence tools for studying can be described as a search engine: *"I use it as always, when I have a question, I go straight there... Well, I use it both for school and for my personal life... I just go and I ask."* The same respondent who made the previous statement adds that he/she also searches for professional resources through AI tools. Other respondents also confirm this way of using it to search for resources: "*... so I use it mainly for the creation of thesis papers, for example. Exactly for those term papers, not that I would give it the text to generate it for me, but I would simply look for sources through it so that I wouldn't have to just look at several different books, different* *publications, and the like."* and *"... it was rather hard even for Google to find what I needed. So the generative AI helped me quite a bit with that. It found me two good sources there, from which I actually drew my entire Bachelor's degree."* However, the search for sources does not always turn out as expected, as one of the respondents proves by reading the answer that the text AI tool gave him: *"'I'm sorry, but I cannot provide specific citations from the literature, because I don't have access to actual records or specific sources.' So I didn't have any of that."*

In all of the obtained statements that were subjected to open coding, we revealed codes related to text processing. Generative AI tools were most often used in connection with the creation of an outline or framework for some work, determining the structure and stylization of already inserted text. Respondents gave examples where one of the tools suggested the goals and structure of the planned lesson after entering the topic (*"In the beginning, you enter the topic of the lesson, if the children already had some basics about the topic as before, or what and for which class you want it. And then it spits it out for you based on the objective of the lesson, and now you can either like approving that you want the goals that it thought up, or you can even change the goals of that lesson."*), further, an example of use to propose points for the assignment of a semester paper, and the use of an AI chatbot in such a way that *"it actually formulates my own thoughts,"* which the respondent describes to him in individual, often disjointed terms. Another example can be the following statement: *"I want to formulate something, some really big sentence—and I just have a problem with Czech in general, so I'll throw it to it so that it can remake it in some academic style."*

Essential for the independent learning process is the use of an AI chatbot or other AI tool as a study partner. Some tools (e.g., Exam Cram and PrepSup) allow you to upload a file that you then edit into different flashcards, or directly offer a function where *"... it automatically asks questions that the file contains and you answer it in the chat and it evaluates whether your answer is right or wrong."* This tool aroused interest in the respondent himself/herself to use the tool effectively: *"I'm going to, like, when I have some kind of test, like, where there's, like, more content that I have to learn by heart, so that it kind of helps me with that."*Regarding the detected direct references to the names of individual AI tools, all of the respondents mentioned in their statements that they had used ChatGPT at some point. It was the most frequently inflected tool. However, not all of the respondents were satisfied with this tool—three respondents said that they use another AI tool, in this case, Microsoft Copilot, which they also called Bing Chat. The reasons why these respondents use Microsoft Copilot more are varied. One of the respondents claims that, in his opinion, Copilot better lists the sources of the generated information: *"... in any case, with the Bing chat, when it worked, what I really liked about it was that it was mostly well-sourced."* The same respondent highlights another Bing chat option—setting how accurate or creative the user wants the outputs. In his/her opinion, *"sourcing"* also depends on this: *"It is therefore the most important thing to tell Bing that you want really accurate results ..."*.

In his/her statement, one of the respondents quite forcefully justified why he/she does not use ChatGPT directly, following the free availability of the application: *"I wouldn't use ChatGPT anymore because it's not worth it, because it's Microsoft Copilot, which ... they kind of bought chat and OpenAI, and it actually uses chatGPT four for 30 messages for free, so, like, I have no reason to use normal chat at all, unless the person is looking for 200 messages."* The answers of the respondent, who rather focuses on the generation of image content, which is also free in the Microsoft Copilot application, thanks to the use of the GPT 4 model, are related to this opinion about free-of-charge; ChatGPT was only used in early 2023 for text questions: *"It was something about programming, so that it could advise me on something here, which actually, as I remember, that one question really helped me, that it simply wrote something that maybe was clear, but I probably didn't use something like that. So it was great, but then I've pretty much stopped using it here for that since then."* The last respondent who mentioned tools from Microsoft, although he/she does not use Microsoft Copilot at the expense of ChatGPT, suggested Microsoft as a possible candidate for the creation of an AI tool that combines moving image generation, voice generation, and automatic presentation creation into one, which, in his/her words, *"would be like a completely heavy bomb. Like, for example, for that Microsoft, if they could do it like that, because actually they ... now their complete innovation was actually the designers, what they did, so they actually try ...".*

It follows from all the statements that the respondents consider AI tools to be a good helper to complement study matters. The most identified code in the entire investigation was the code for *streamlining activities*, under which, among other things, statements talking about making work easier and saving time are hidden. Respondents often mentioned this streamlining in connection not only with their studies but also with their work, whether it was teaching or working in the private sector, in the field of programming *("I have the feeling that when I use it specifically, so it saves me an awful lot of time ...," "Of course I can, yes, write the conclusion myself, but this is much more practical for me. It speeds up my time.", "I'll just take the text and possibly edit it as I need, if there's anything there.")*

All of the statements also resonated with the need to control the outputs that students get from not only generative AI, which sometimes contradicts the facilitation of work: *"Well, I thought it was a terrible facilitation, until I found out that sometimes it spits out the wrong answers and I have to check it anyway."* The necessity of checking is emphasized by the following statement: *"Of course, it's always a good idea to read it after, because sometimes it can throw something in there that shouldn't be there."* According to another respondent, it is advisable to be able to formulate your requirements, which is also why he/she personally takes the following approach to entering input: *"I feel that it's like some subordinate of mine who actually creates on the basis of what I command him/her to do. Or I try to guide him/her in some way. And the more precise the task I actually create, the more I try to get the result that I would expect."*

Of all the opinions, the students are satisfied with the tools (we dare to highlight the very clear statement *"It's great, it's free money,"* because time means money in this respondent's opinion), but one of the respondents directly mentioned fear: *"... at first I was afraid of it, but one must adapt."* However, according to the statement, this fear has gradually disappeared. An interesting opinion, from which slight concerns are evident, is a statement about the use of very complex AI tools (combining sound, image, and text, for example, with the aim of creating educational material) that could replace teachers: *"It's hard to say, but it would make the job easier, but how can artificial intelligence do everything for humans? I don't know."*

4 Discussion

All of the survey participants independently mentioned the ChatGPT AI chatbot, which has become the main pioneer of generative AI. They gave examples of how they (don't) use ChatGPT, and compared it to another AI tool, Microsoft Copilot/Bing Chat. One respondent was very appreciative that Microsoft Copilot, in his opinion, lists good resources. This may be related to one of the characteristics that Copilot is distinguished by: unlike, for example, ChatGPT, it does not suffer from hallucinations, that is, it does not invent resources just to satisfy the user. A comparison of these two AI chatbots was made in this study [21], where they mention, among other things, precisely the hallucinations that ChatGPT suffers from, whereas Bing Chat from Microsoft does not. However, respondents are unanimously aware of the need to check, verify, and adapt all outputs from various AI tools to their own individual needs.

Respondents use AI tools as an assistant that helps them find some information and formulate their ideas, which can make their work more efficient [22]. Thanks to the above, they can act as their *"learning mate"* [22], which greatly supports the implementation of AI technologies in the educational process. Some tools even test students and check the correctness of their answers. This immediate feedback on their answers can encourage them and can be greatly appreciated by the students themselves [23].

In the study [15], some of the respondents estimated that AI could benefit education and information retrieval, which is in line with our respondents' views. Elements of AI have already begun to be integrated into various research applications (e.g., scite.ai and the Web of Science Research Assistant) because it can actually make searching for relevant information more efficient.

Contradictory to the same study are the findings that participants in our survey are largely unafraid of AI. However, these results cannot be generalized to a larger scale, as we only had a small sample of students who actively use AI tools. However, it is understandable that in the study [15], students were afraid of AI because, at the time, it was not as accessible to everyone as it is now.

Thanks to the obtained results, we specifically found out some of the tools with which the students work. In the interviews, they themselves even compared them, which proves that they really do think about the use of AI tools, and verify the results. These tools are easy to abuse and generate various outputs, and students can easily generate without thinking, for example, an entire term paper, raising concerns about proper ethical use [24]. According to our respondents, with the outputs of AI, they always work further and develop them according to other sources, which may be the right way to integrate AI tools into the regular educational process. The tools can function as the previously mentioned "learning mate" or assistant, with whom the student will mainly talk about the possible form of the study output. However, it is necessary to talk to students about these ethical aspects of using AI tools and instill the right principles in them. In some studies, the students themselves directly mention that it is necessary to train on the correct use of AI tools, in order to avoid their misuse or abuse [17].

5 Conclusion

The data we obtained shows that survey participants share a rather positive on about the use of AI. But most importantly for us, we looked directly into the ways in which some students use AI tools. We identified common features, as all of the respondents use generative AI to generate the outline or structure of not only coursework. A significant place is occupied by the use of AI to search for resources, and it is also gratifying to find out when students use AI as a study "partner" that tests them. All of the above-mentioned activities inherently belong to the educational process. As yet, it cannot be stated that generative AI inherently belongs to the personal digital environment of university students. However, judging by the ways in which students are using AI to streamline learning activities, it could become a key element over the course of time.

The conducted research has its limits. Only a small group of students took part in the interviews, so the obtained results cannot be globally applied to all of the teaching students at the Faculty of Education. It is not possible to generalize the results we obtained, but this insight into the students' digital environment is a good springboard for further research, which is almost certainly necessary, due to the general increased interest in the use of AI. However, this insight into the digital environment of students is a good springboard for further research, which is almost certainly necessary, due to the general increased interest in the use of AI. and ways of using it are constantly evolving, and further developments will also be dynamic in the area of using AI within the personal digital environment of students. That is why we plan to repeat this investigation in the future, on a different and larger group of students.

References

1. Bates, A.W.: Teaching in a Digital Age: Guidelines for Designing Teaching and Learning, Vancouver BC: Tony Bates Associates Ltd. (2015). ISBN: 978-0-9952692-0-0.
2. Prensky, M.: Digital Natives, Digital Immigrants Part 1. On the Horizon, vol. 9, no. 5, pp. 1-6 (2001).

3. Attwell, G.: Personal Learning Environments-the future of eLearning? In: ELearning Papers, vol. 2, no. 1 (2007). ISSN 1887-1542.
4. Zounek, J., Juhaňák, L., Staudková, H. and J. Poláček: E-learning: učení (se) s digitálními technologiemi: kniha s online podporou, 2., aktualizované vydání. (E-learning: learning with digital technologies: a book with online support, 2nd, updated edition) Prague: Wolters Kluwer (2021).
5. Drozdowski, M. J.: 5 Ways Artificial Intelligence Will Transform Higher Education (2023). https://www.bestcolleges.com/news/analysis/5-ways-ai-will-transform-higher-education/. Accessed 25 April 2024.
6. Toyokawa, Y., Horikoshi, I., Majumdar, R. et al.: Challenges and opportunities of AI in inclusive education: a case study of data-enhanced active reading in Japan. Smart Learn. Environ. 10, 67 (2023). https://doi.org/10.1186/s40561-023-00286-2
7. Dewi, H. K., Putri, R. E., Rahim, N. A., Wardani, T. I., & Pandin, M. G. R., Dr., M. Si., M.Phi., M.Psi., Psi.: THE USE OF AI (ARTIFICIAL INTELLIGENCE) IN ENGLISH LEARNING AMONG UNIVERSITY STUDENT: CASE STUDY IN ENGLISH DEPARTMENT, UNIVERSITAS AIRLANGGA. In: SocArXiv (2021). https://doi.org/10.31235/osf.io/x3qr6
8. Marr, B.: The Amazing Ways Duolingo Is Using Artificial Intelligence To Deliver Free Language Learning. Forbes. Accessed April 26, 2024, available at https://www.forbes.com/sites/bernardmarr/2020/10/16/the-amazing-ways-duolingo-is-using-artificial-intelligence-to-deliver-free-language-learning/
9. Marr, B.: The Amazing Ways Duolingo Is Using AI And GPT-4. Forbes. Accessed April 28, 2024, from https://www.forbes.com/sites/bernardmarr/2023/04/28/the-amazing-ways-duolingo-is-using-ai-and-gpt-4/
10. Maghribi, L.: Artificial intelligence is already everywhere, we need to adapt, The National, 20 May 2021, (2021). https://www.thenationalnews.com/world/artificial-intelligence-is- already-everywhere-we-need-to-adapt-1.1225794, accessed April 26 2024
11. Qadir, J.: Engineering Education in the Era of ChatGPT: Promise and Pitfalls of Generative AI for Education. In: 2023 IEEE Global Engineering Education Conference (EDUCON), Kuwait, Kuwait, pp. 1-9, (2023). https://doi.org/10.1109/EDUCON54358.2023.10125121.
12. Tapalova, O., Zhiyenbayeva, N.: Artificial Intelligence in Education: AIEd for Personalised Learning Pathways. In: Electronic Journal of e-Learning, vol. 20, no. 5, pp. 639-653, (2022). https://doi.org/10.34190/ejel.20.5.2597.
13. Ayuso-del Puerto, D., Gutierrez-Esteban, P.: Artificial Intelligence as an Educational Resource during Preservice Teacher Training. In: Ried-Revista Iberoamericana de Educacion a Distancia, vol 25, no. 2, pp. 347–362 (2022). Available at: https://revistas.uned.es/index.php/ried/article/view/32332/25162
14. Butakor, P. K.: Exploring Pre-service Teachers´ Beliefs about the Role of Artificial Intelligence in Higher Education in Ghana. In: International Journal of Innovative Technologies in Social Science, vol. 3, no. 39 (2023). https://doi.org/10.31435/rsglobal_ijitss/30092023/8057
15. Haseski, H. I.: What Do Turkish Pre-Service Teachers Think About Artificial Intelligence? In: International Journal of Computer Science Education in Schools, vol. 3, no. 2, pp. 3–23 (2019). https://doi.org/10.21585/ijcses.v3i2.55.
16. Chiu, T. K. F.: Future research recommendations for transforming higher education with generative AI. In: Computers and Education: Artificial Intelligence, vol. 6. (2024). ISSN 2666920X. Available at: https://doi.org/10.1016/j.caeai.2023.100197.
17. El-Khalili, N., Al-Nashashibi, M., Al-E'mari, S.: The Impact of Generative Artificial Intelligence on Learning: A Case Study at the University of Petra. In: 2024 2nd International Conference on Cyber Resilience (ICCR). IEEE, pp. 1-6 (2024). ISBN 979-8-3503-9496-2. Available at: https://doi.org/10.1109/ICCR61006.2024.10532824.
18. Niedbał, R., Sokołowski, A., Wrzalik, A.: Students' Use of the Artificial Intelligence Language Model in their Learning Process. In: Procedia Computer Science, vol. 225, pp. 3059–3066 (2023). ISSN 18770509. Available at: https://doi.org/10.1016/j.procs.2023.10.299.
19. Glaser, B., Strauss, A.: The Discovery of Grounded Theory: Strategies for Qualitative Research. Mill Valley, CA: Sociology Press (1967).

20. Shaheen, M., Pradhan, S., Ranajee: Sampling in Qualitative Research. In: Qualitative Techniques for Workplace Data Analysis, pp. 25–51 (2019). https://doi.org/10.4018/978-1-5225-5366-3.ch002.
21. Rudolph, J., Tan, Sh., Tan. S.: War of the chatbots: Bard, Bing Chat, ChatGPT, Ernie and beyond. The new AI gold rush and its impact on higher education. In: Journal of Applied Learning & Teaching, vol. 6, no. 1, pp. 364–389 (2023). https://doi.org/10.37074/jalt.2023.6.1.23.
22. Kim, J., Cho, Y. H.: My teammate is AI: understanding students' perceptions of student-AI collaboration in drawing tasks. In: Asia Pacific Journal of Education, pp. 1–15 (2023). ISSN 0218-8791. https://doi.org/10.1080/02188791.2023.2286206
23. Pesovski, I, Santos, R., Henriques, R., Trajkovik, R.: Generative AI for Customizable Learning Experiences. In: Sustainability, vol. 16, no. 7 (2024). https://doi.org/10.3390/su16073034.
24. Hosseini, M., Resnik, D. B., Holmes, K.: The ethics of disclosing the use of artificial intelligence tools in writing scholarly manuscripts. In: Research Ethics, vol. 19, no. 4, pp. 449–465 (2023). https://doi.org/10.1177/17470161231180449

Science and Engineering Project Using the DALL-E Tool: Exploring Primary School Students' Communication Competence: A Pilot Study

Dejan Zemljak and Maja Kerneža

1 Introduction

The fusion of artificial intelligence (AI) with primary education is not only transformative—it is redefining the boundaries of learning and creativity. DALL-E, a cutting-edge generative AI model, exemplifies this shift by enabling students to visualize and design with unprecedented freedom and precision. This pilot study focuses on the ability to articulate science and engineering project concepts and overcome technical challenges. The research assesses the broader pedagogical implications of AI, including its role in fostering essential communicative and technical skills.

Technological advances have profoundly impacted the acquisition of knowledge [11]. Recent research supports the idea that contemporary learning modalities resonate better with certain student demographics than traditional educational approaches [11]. A 2020 report from the United Nations International Children's Emergency Fund (UNICEF) highlights the significant benefits of integrating digital technologies, rather than saying that they make educational experiences more engaging and accessible. Similarly, Moltudal et al. [6] emphasize how adaptive learning technologies address the real-time needs of students, improving instructional strategies and classroom dynamics. UNICEF [13] analysis also demonstrates the reliability of the effectiveness of educational technologies and suggests that their strategic integration—taking into account regional and contextual nuances—is

D. Zemljak
Faculty of Natural Sciences and Mathematics, University of Maribor, Maribor, Slovenia
e-mail: dejan.zemljak1@um.si

M. Kerneža (✉)
Faculty of Education, University of Maribor, Maribor, Slovenia
e-mail: maja.kerneza1@um.si

M. Turčáni (ed.), *15th International Scientific Conference on Distance Learning in Applied Informatics*, EAI/Springer Innovations in Communication and Computing,
https://doi.org/10.1007/978-3-031-81261-3_23

crucial to optimizing learning outcomes. Taken together, these findings underscore the transformative potential of AI tools such as DALL-E in primary education and highlight their central role in promoting digital literacy and creating a dynamic, customized learning environment.

Building on this technological foundation, the concept of the communication competence process is a central point. First children learn to understand simple words and sentences. McCroskey's [5] examination of communication competence offers a nuanced perspective and acknowledges that it is difficult to define such a broad construct consistently across different domains. He suggests that communication competence should be understood as the sufficient ability to convey ideas through speaking or writing. This definition captures the essence of communication competence, which is not only about performing communicative behaviors, but also understanding and strategically selecting them.

Between the ages of 6 and 11 years, children's communication skills undergo critical development. This stage is characterized by a rapid expansion of vocabulary and an increasingly sophisticated understanding of grammatical structures, which improves their ability to communicate in a nuanced way [1, 10]. In addition, children refine their pragmatic skills to adapt to different social contexts [12] and develop complex narrative skills that are important for both academic success and social interaction [8]. During this time, their ability to effectively interpret and utilize verbal and nonverbal cues in their growing social networks also improves [7].

Considering that effective communication skills are cultivated in an enriching learning environment that fosters the development of relevant behavioral and thinking skills, it is crucial in today's education (e.g., [14]) to foster these skills in novel environments. The importance of communication skills in technical and scientific disciplines is emphasized at different levels of education and in different contexts. Studies emphasize the need to integrate effective communication training into STEM (science, technology, engineering and mathematics) and engineering curricula to improve both educational outcomes and future employability (Wu et al., [15]). Research in undergraduate settings reveals that establishing a foundation of communication skills is critical to the effective transfer of scientific knowledge [9]. Communication skills are also crucial in the context of science and engineering projects, especially when advanced tools such as AI are used in education. Effective communication is important clearly articulating science and engineering project concepts and refining these ideas with others. It allows students to express their technical thoughts and justify their science and engineering project decisions, leading to a better understanding and application of their ideas in practical scenarios.

Research Aim and Goals

The aim of this study is to explore how the level of communication competence among primary school students affects their interactions with AI, specifically using the DALL-E tool, in the context of science and engineering projects. The goals include assessing the initial communication competence levels of students, examining how varying communication skills influence interactions with DALL-E during the science and engineering project process, evaluating the impact of these competencies on the quality and innovativeness of the science and engineering project, and identifying educational strategies to enhance the effective use of AI tools in educational settings.

2 Research Methodology

2.1 General Background

The pilot study aims to explore the integration of the DALL-E tool in the educational context, focusing on science and engineering projects and their impact on the communication competences of elementary school students. This study addresses two primary educational outcomes: the enhancement of science and engineering skills and the development of oral communication skills. By allowing students to use DALL-E to conceptualize science and engineering projects, this research seeks to understand how technological tools can augment traditional learning methods and influence student competencies in both science and engineering projects and language use. Given the focus on how students' abilities might influence AI interactions, the research question stated is as follows:

How does the level of communication competence among primary school students influence their interaction with the AI tool DALL-E in the context of science and engineering projects?

2.2 Participants

The study involves four participants, two from the first educational period (6 (first grade, male) and 8 years (third grade, female) old) and two from the second educational period (10 (fourth grade, female) and 12 years (sixth grade, male)) of elementary education. The participants are from a rural background, all attending the same primary school and coming from similar socio-economic conditions. These age groups were chosen to provide insights into the developmental differences in handling technology-based science and engineering project tasks and communication abilities at different stages of cognitive and linguistic growth. This decision was also informed by studies finding that younger students exhibit a deficit in computer usage skills [2, 3], allowing for a comparative analysis across different educational levels.

2.3 Instrument and Procedures

Each participant attended an introductory session to familiarize themselves with the DALL-E interface and its capabilities. The introductory session was scheduled for each participant individually to ensure personalized attention and an environment conducive to learning without distractions. The session lasted approximately 1 h, which was sufficient to cover all essential aspects of the tool while keeping the children engaged and attentive. The primary goal of the session was to ensure that each participant was comfortable with the basic functions of the DALL-E interface and understood how to interact with the tool to generate images based on verbal or text-based inputs. This included tool introduction (a concise introduction to DALL-E, highlighting its capabilities, the demonstration of example outputs to enhance understanding and stimulate interest), technical walkthrough (step-by-step demonstration on how to operate DALL-E tool, including entering descriptions and interpreting results), hands-on practice (the use of DALL-E by entering simple prompts to generate various images, helping them gain familiarity and confidence with the technology), questions, and answers (questions and answers allowed students to address any uncertainties or concerns regarding the tool's functionalities).

During the instruction phase, the teacher assisted a 6-year-old and an 8-year-old with writing down instructions. The assistance was provided to the 6-year-old due to their emerging literacy skills, and to the 8-year-old, who, although able to read and write without difficulty, experienced slowness in writing due to unfamiliarity with the keyboard layout. In these cases, the teachers assumed the role of the scribe based on the children's instructions during the experiment. For all students, the teacher acted as a facilitator, providing guidance on using the tool while deliberately refraining from influencing the students' creative processes. The teacher also ensured that each student understood the technical aspects of the tool and could troubleshoot basic issues.

After becoming familiar with DALL-E, the next day, each participant was given the challenge to independently conceptualize and sketch an eco-friendly playground using the tool. This task was designed to simulate a real-world science and engineering project scenario, encouraging creativity and application of the learned skills. The task was carried out individually by each student to ensure that the outcomes reflected their personal understanding and ability to use DALL-E without external influences. Teacher intervention was minimal and strategically needed, focusing exclusively on the use of the tool itself. The primary focus during these interactions was on assessing how effectively the students could articulate their science and engineering project ideas and their proficiency in utilizing the DALL-E tool to put their ideas into practice.

The videos, products, and communication of the child with AI were reviewed and evaluated by two experts, one in the field of didactics of technology and engineering and the other in the field of didactics of the mother tongue. Their observations centered on students' communication competence in the context of the subject of technology and engineering.

2.4 Data Analysis

Data collected from this pilot study includes video recordings of the session, expert evaluations, and transcripts of interactions between children and the DALL-E tool. The analysis focuses on students' ability to articulate and respond to technological concepts. Specific observational points included:

- *Clarity of expression*: Note how clearly the child communicates ideas.
- *Responsiveness*: Observe how the child responds to the AI's outputs and makes adjustments.
- *Use of technical terms*: Check for the correct and situation-appropriate use of simple technical terms.
- *Interactive communication*: Note the interactions that reflect understanding and instruction.
- *Adaptability*: Assess how well the student adapts communication based on AI feedback.
- *Vocabulary*: Evaluate the appropriate use of more complex technical vocabulary.
- *Precision in communication*: Observe precision and accuracy in conveying complex ideas.
- *Effectiveness in persuasion*: Assess how convincingly the students present science and engineering project ideas.

The qualitative data from the verbal interactions and science and engineering project outputs were coded and categorized to identify patterns and draw conclusions about the effectiveness of using DALL-E in enhancing technical and communicative competencies. This pilot study provides preliminary insights that can inform larger-scale studies into the integration of AI tools in educational settings.

3 Research Results

This chapter outlines the findings from the exploration of the DALL-E tool in primary education, assessing communication competence in science and engineering project development.

The observational assessment of the 6-year-old participant (Table 1) interacting with the DALL-E tool shows typical communication competencies for his age. His expressions were simple and directly related to visible outcomes, indicating a focus on immediate and tangible aspects of the technology rather than abstract concepts. The participant's responses to the AI outputs were largely emotional, such

Table 1 Descriptive assessment of a 6-year-old student's communication assessment

Age	Example
6	*Clarity of expression*: The participant communicated ideas simply, focusing on immediate reactions to the DALL-E outputs. His expressions were direct but limited to describing observable aspects rather than conceptual details. *Responsiveness*: The participant showed spontaneous reactions to the DALL-E's often expressing delight or confusion without articulating specific reasons. His responses were based more on emotional cues than analytical observations. *Use of technical terms*: Technical vocabulary was minimal and typically repeated from instructions without a demonstrated understanding of the meaning. *Interactive communication*: Participant's interaction with DALL-E involved basic commands. He demonstrated a straightforward understanding of the task, following simple steps and responding to direct outcomes. *Adaptability*: Adaptation to feedback was basic, showing initial signs of recognizing cause and effect, although his adjustments were not always logically structured. *Vocabulary*: The participant used a limited range of vocabulary, suited for his age, including simple terms related to the task. *Precision in communication*: Communication was generally broad, with general descriptions used more frequently than specific directives. *Effectiveness in persuasion*: Attempts to persuade were characterized by assertive or emotional appeals rather than structured elements.

as showing delight or confusion, without deep analytical reasoning, which aligns with the expected cognitive development at this age. His use of technical terms was minimal, often echoing instructions, suggesting a nascent stage of understanding. Communication involved basic commands and showed an emerging awareness of cause and effect, though adjustments lacked logical structuring. Vocabulary was limited and appropriate for his age, with broad communication that often lacked specificity. His attempts at persuasion relied on emotional appeals rather than structured arguments, typical of early childhood where logical reasoning is still developing. These findings highlight the developmental stages of young children interacting with advanced technology and underscore the need for tailored educational strategies to enhance learning and interaction skills effectively.

For the younger student, the level of communication competence primarily influenced his ability to interact with DALL-E in straightforward and immediate ways. His basic command usage and limited vocabulary focused on direct and observable outcomes rather than engaging with more complex or abstract science and engineering project aspects. This suggests that at this developmental stage, interactions with AI in science and engineering project tasks might be more effective when the tool provides clear, visual, and immediate feedback that aligns with their cognitive and communicative capabilities. It should also be pointed out that it would have been impossible for a 6-year-old student to complete the task without the teacher writing down the child's instructions and basic and advanced computer use.

The 8-year-old participant (Table 2) demonstrated moderate clarity in expressing what she was doing, indicating an enhanced understanding of detailed activities compared to younger children. She responded to DALL-E outputs by adjusting her

Table 2 Descriptive assessment of an 8-year-old student's communication assessment

Age	Example
8	*Clarity of expression*: The participant was able to describe what she was doing with moderate clarity, using language that reflected a better grasp of detailed activities. *Responsiveness*: The participant adjusted her actions based on the DALL-E's outputs and showed developing problem-solving skills by trying different commands when initial attempts did not yield the expected results. *Use of technical terms*: The participant used technical terms more accurately, with occasional minor errors in the application. *Interactive communication*: The participant engaged more actively with the DALL-E, utilizing it to achieve specific goals. Her use of the AI showed emerging strategic thinking. *Adaptability*: The participant began to show meaningful adaptability, making changes based on feedback, although she occasionally required guidance to optimize her approach. *Vocabulary*: The participant's vocabulary included a mix of general and technical terms, used mostly correctly, with better contextual appropriateness. *Precision in communication*: The participant provided more exact instructions and could explain her actions with improved precision. *Effectiveness in persuasion*: The participant's persuasive efforts were based on simple logical connections, though these were not always fully developed.

actions and experimenting with different commands, reflecting developing problem-solving skills. Her use of technical terms was generally accurate, with occasional errors showing her growing familiarity with relevant vocabulary. In interactions, she engaged actively with DALL-E, aiming to achieve specific goals and showing initial strategic thinking. Her adaptability was evident as she made changes based on feedback, although she sometimes needed guidance to optimize her approach. Her vocabulary was a mix of general and technical terms, mostly used correctly and with contextual appropriateness. She provided precise instructions and could explain her actions clearly, although her persuasive efforts, based on simple logical connections, were not always fully developed.

This student showed a more developed communication competence, which allowed for slightly more complex interactions with DALL-E. His ability to use a mix of general and technical terms and to adjust actions based on AI feedback indicates a transitional competence level where students begin to apply problem-solving skills in their interactions. This age group benefits from AI tools that support explorative learning and provide opportunities to experiment with different commands, which in turn helps them refine their understanding of technical concepts and the science and engineering project process. Like the 6-year-old, this student also needed some help from the teacher with advanced computer use, and the teacher also wrote down the instructions for the DALL-E tool for this student, as she was not familiar enough with the layout of the letters on the keyboard and the child's motivation started to drop quickly after only a few minutes of initial attempts.

Table 3 Descriptive assessment of a 10-year-old student's communication assessment

Age	Example
10	*Clarity of expression*: The participant articulated ideas clearly, explaining both the actions she was taking and the reasoning behind them. *Responsiveness*: The participant showed a thoughtful response to DALL-E's feedback, making informed adjustments that reflected an understanding of the science and engineering project process. *Use of technical terms*: The participant demonstrated a good command of relevant technical vocabulary, applying terms accurately and appropriately. *Interactive communication*: The participant's communication with DALL-E was effective and purposeful. She gave clear instructions and asked relevant questions that indicated a deeper understanding of the task. *Adaptability*: The participant exhibited strong adaptability, using feedback effectively to refine her approach and improve her science and engineering project outcomes. *Vocabulary*: The participant's use of vocabulary was robust, including appropriate technical terms relevant to the task, and used accurately. *Precision in communication*: The participant communicated with high precision, providing detailed and specific instructions and feedback. *Effectiveness in persuasion*: The participant was capable of making persuasive arguments supported by logical reasoning and evidence.

The 10-year-old participant (Table 3) articulated her ideas with clear expression, detailing both the actions taken and the reasoning behind them, showcasing a solid understanding of the task at hand. She responded thoughtfully to DALL-E's feedback, making informed adjustments that indicated a deep comprehension of the science and engineering project process. Her command of technical vocabulary was strong, with accurate and appropriate application of terms. Her communication with DALL-E was effective and purposeful; she issued clear instructions and posed relevant questions that demonstrated a thorough grasp of the technology and task. This participant showed strong adaptability, effectively using feedback to refine her approach and enhance science and engineering project outcomes. Her vocabulary was robust and included technical terms relevant to the task, used accurately. She communicated with high precision, providing detailed instructions and feedback. Her persuasive abilities were well-developed and supported by logical reasoning and evidence.

This student exhibited a further enhanced communication competence, which allows for more meaningful and effective interactions with DALL-E. She was capable of articulating clear and detailed explanations of their science and engineering project actions and the reasoning behind them, demonstrating a solid understanding of both the task and the technology. Student's advanced vocabulary and ability to adjust based on sophisticated AI feedback suggest that integrating AI tools that challenge their cognitive and communicative skills at this age, can significantly enhance their learning and science and engineering project capabilities.

Table 4 Descriptive assessment of a 12-year-old student's communication assessment

Age	Example
12	*Clarity of expression*: The participant articulated detailed processes and abstract concepts with high clarity, adapting his language to suit the context and audience effectively. *Responsiveness*: The participant's responsiveness to DALL-E outputs was nuanced, capable of making precise adjustments based on sophisticated feedback interpretation. *Use of technical terms*: The participant utilized advanced technical vocabulary, demonstrating a deep understanding of the subject matter. *Interactive communication*: The participant's interaction with DALL-E was mature, managing complex science and engineering project tasks effectively and using advanced communication strategies to maximize AI utility. *Adaptability*: The participant showed advanced adaptability, strategically using feedback to optimize science and engineering project processes and outcomes. *Vocabulary*: The participant's vocabulary was advanced and specialized, effectively used in discussing complex ideas and technical details. *Precision in communication*: The participant exhibited exceptional precision in conveying intricate instructions and complex ideas, ensuring clarity and accuracy. *Effectiveness in persuasion*: The participant's persuasive efforts were sophisticated, using both emotional and logical appeals to convincingly support his science and engineering project choices.

The 12-year-old participant (Table 4) exhibited a high level of clarity in expressing detailed processes and abstract concepts, effectively adapting his language to suit the context and audience. His responsiveness to DALL-E outputs was nuanced, showing the ability to make precise adjustments based on sophisticated feedback interpretations. He used advanced technical vocabulary, demonstrating a deep understanding of the subject matter. His interactions with DALL-E were mature and efficient, managing complex science and engineering project tasks effectively and employing advanced communication strategies to maximize the tool's utility. His adaptability was advanced; he strategically used feedback to optimize science and engineering project processes and outcomes. His vocabulary was both advanced and specialized, used effectively in discussing complex ideas and technical details. He showed exceptional precision in communication, ensuring clarity and accuracy in conveying intricate instructions and complex ideas. His persuasive efforts were sophisticated, utilizing both emotional and logical appeals to convincingly support his science and engineering project choices.

The 12-year-old student displayed the highest level of communication competence among the groups, engaged in sophisticated interactions with DALL-E. His ability to articulate detailed processes and abstract concepts and to use advanced technical vocabulary effectively enabled him to maximize the utility of AI in complex science and engineering project tasks. The student showed an advanced adaptability and strategic use of AI feedback, indicating that at this stage, students can fully exploit AI capabilities to enhance their learning outcomes in technical education.

4 Discussion

This study revealed that the level of communication competence significantly influences how students interact with AI tools like DALL-E in the context of science and engineering project tasks. Younger students tend to engage with AI in more basic, command-following ways, focusing on direct and observable outcomes, while older students engage in more complex and strategic interactions. A notable finding is that students with higher communication competence not only navigate AI tools more effectively but also use these interactions to further enhance their communication skills, which in turn influences their overall academic performance and social capabilities [4].

The findings of this study align with existing literature that underscores the importance of communication skills in educational settings, where strong communication skills are linked to better academic outcomes and social navigation [4]. However, our study extends this understanding by demonstrating how these competencies influence and are influenced by interactions with advanced technological tools like AI. The strengths of the study lie in its detailed observational assessment and the direct comparison across age groups, offering insights into developmental stages in AI interactions. A limitation, however, is the rapid pace of technological advancements that outstrip changes in educational curricula, with significant gaps since the last updates in Slovenian educational standards (2011 for technology and 2018 for the Slovenian language). An unexpected finding was the extent to which younger students used AI feedback to incrementally adjust their communication strategies, even within their limited scope. This study also confirmed, in comparison with the Kordigel Aberšek and Kerneža [2] and Legvart et al. [3], that younger students are hindered in their computer use by their computer usage competence and their not (yet) fully developed writing and reading abilities.

The purpose of this study was to explore how students' communication competence affects their interaction with the AI tool DALL-E during science and engineering project tasks. The significance of this study lies in its potential to inform educational strategies that utilize AI tools to cater to diverse educational needs, adapting to different cognitive and communicative development levels. As students mature, their communicative interactions with AI become increasingly complex and strategic, suggesting a need for versatile and adaptable AI tools to effectively support educational growth [4]. Future research should explore how educational curricula can be rapidly updated to keep pace with technological advancements and further study how AI can be used to enhance not just technical but also communicative competencies in a broader range of educational contexts.

Acknowledgments This research was founded by the Slovenian Research Agency (ARRS), grant number P5-0433 for the research program Digital Restructuring of Deficit Occupations for Society 5.0 (Industry 4.0).

References

1. Hoff, E.: Language Development. Cengage Learning, Belmont (2013).
2. Kordigel Aberšek, M., Kerneža, M.: Primary school students' functional literacy in digital learning environments—how to teach. In: M. Turčani, Z. Balogh, M. Munk, M. Magdin, L. Benko, J. Francisti (eds.), 14th International Scientific Conference on Distance Learning and Applied Informatics (DIVAI 2022), pp. 174–185. Wolters Kluwer, Bratislava (2022).
3. Legvart, P., Kordigel Aberšek, M., Kerneža, M.: Primary school students' natural science digital literacy competence in digital learning Environments. In: V. Lamanauskas (ed.), *Science and Technology Education: Developing a global perspective: Proceedings of the 4th International Baltic Symposium on Science and Technology Education (BalticSTE2021)*, Šiauliai, 21–22 June 2021, pp. 105–114. Scientia Socialis, Siauliai (2021).
4. Mashburn, A. J., Pianta, R. C.: Social relationships and school readiness. Early Education and Development, 17(1), 151–176 (2006).
5. McCroskey, J. C.: Communication competence: The elusive construct. In: R. N. Bostrom (ed.), Competence in Communication: A multidisciplinary approach, pp. 259–268. Sage, Beverly Hills (1984).
6. Moltudal, S. E., Hovdenak, S. S., Aadland, H.: Adaptive learning technology in primary education: Implications for professional teacher knowledge and classroom management. Frontiers in Education, 7, 830536 (2020). https://doi.org/10.3389/feduc.2022.830536
7. Parker, J. G., Asher, S. R.: Friendship and friendship quality in middle childhood: Links with peer group acceptance and feelings of loneliness and social dissatisfaction. Developmental Psychology, 29(4), 611-621 (1993). https://doi.org/10.1037/0012-1649.29.4.611
8. Peterson, C., McCabe, A.: Developmental psycholinguistics: Three ways of looking at a child's narrative. Plenum. New York (1983).
9. Shivni, R., Cline, C., Newport, M., Yuan, S., Bergan-Roller, H. E.: Establishing a baseline of science communication skills in an undergraduate environmental science course. International Journal of STEM Education, 8, 47 (2021). https://doi.org/10.1186/s40594-021-00304-0
10. Snow, C., Burns, M. S., Griffin, P.: Preventing reading difficulties in young children. National Academy Press, Washington (1998).
11. Szymkowiak, A., Melović, B., Dabić, M., Jeganathan, K., Singh Kundi, G.: Information technology and Gen Z: The role of teachers, the internet, and technology in the education of young people. Technology in Society, 65, 101565 (2021). https://doi.org/10.1016/j.techsoc.2021.101565
12. Taguchi, N.: Pragmatic development as a dynamic, complex process: General patterns and case histories. The Modern Language Journal, 96(4), 605–627 (2012). https://doi.org/10.1111/j.1540-4781.2011.01246.x
13. UNICEF: Effectiveness of digital learning solutions to improve educational outcomes: A review of the evidence (2020), https://www.unicef.org/media/103246/file/Effectiveness%20of%20digital%20learning%20solutions%20to%20improve%20educational%20outcomes%20-%20A%20review%20of%20the%20evidence.pdf, last assessed 2024/5/11.
14. Weng, X., Cui, Z., Ng, O. L., Jong, M. S. Y., Chiu, F. K. F. Characterizing students' 4C skills development during problem-based digital making. Journal of Science Education and Technology, 31(3), 372–385 (2022). https://doi.org/10.1007/s10956-022-09961-4
15. Wu, Y, Xu, L., Philbin, S. P.: Evaluating role of the communication skills for engineering students on employability according to the outcome-based education (OBE) theory. Sustainability, 15(12), 9711 (2023). http://dx.doi.org/10.3390/su15129711

Rethinking the Role of Intelligent Systems in Education

Dejan Zemljak, Urška Martinc, and Maja Kerneža

1 Introduction

The development and research of intelligent systems (from here ISs) in education have been ongoing for longer than it might initially appear. Particularly in recent years, the use of artificial intelligence (from here AI) and humanoid robots (from here HR) has experienced a rapid increase, both in general and within the educational sector. These modern tools promise numerous benefits; however, there remains the question of whether AI and HR genuinely contribute to enhancing the pedagogical work of teachers or if they potentially detract from essential teaching responsibilities.

Currently, the field of AI and HR in education is highly relevant for research. The rapid integration of these technologies in educational settings presents a complex problem: while they offer numerous potential advantages such as new possibilities in the areas of curriculum development, motivation of students, interest arousal, and soft skills development [1], as well as stimulating student interest and developing their creativity, social skills, and digital literacy [2], they also raise significant concerns about their impact on the traditional roles and responsibilities of teachers. The core problem lies in determining whether these ISs serve as valuable aids

D. Zemljak
Faculty of Natural Sciences and Mathematics, University of Maribor, Maribor, Slovenia
e-mail: dejan.zemljak1@um.si

U. Martinc
University of Maribor, Maribor, Slovenia
e-mail: urska.martinc1@um.si

M. Kerneža (✉)
Faculty of Education, University of Maribor, Maribor, Slovenia
e-mail: maja.kerneza1@um.si

M. Turčáni (ed.), *15th International Scientific Conference on Distance Learning in Applied Informatics*, EAI/Springer Innovations in Communication and Computing,
https://doi.org/10.1007/978-3-031-81261-3_24

that enhance the teaching process or if they risk undermining the essential human elements of education, such as personalized instruction and emotional support. The new era of education, marked by the effective use of teaching robots, can help develop essential skills like logical thinking, creativity, and problem-solving, which are crucial for the future [3]. Overall, Nagy et al. [4] suggest that effective human–robot interaction is essential for the successful implementation of the NAO robot and that further research is needed to better understand the dynamic of human–robot interaction and its impact on the effectiveness of the robot. However, amidst the deluge of information and results, which sometimes get lost in the plethora, there is a renewed need for a thorough reconsideration of the role of IS in education. It is crucial to understand that ISs not only automate administrative tasks but can also transform pedagogical methods and enable personalized learning tailored to individual students. This transformation can lead to a fundamental shift in how education is delivered, potentially affecting the quality and nature of teacher–student interactions.

Moreover, specific pedagogical benefits and challenges of using AI and HR need to be emphasized. The ethical issues involved with integrating these technologies into the educational system also warrant careful consideration. This systematic overview study examines the utilization of ISs in education and their impact on the pedagogical role of teachers. The primary research problem addressed in this study is to determine whether and how these technologies serve as aids to teachers or if they replace tasks that should remain within the domain of human teaching. The research process focuses on understanding which tasks can be effectively handled by ISs to support teachers while emphasizing the preservation of the teacher's essential role.

Key research questions addressed in this study include the following:

- How do ISs currently affect the pedagogical work of teachers?
- What are the advantages and disadvantages of using ISs in the classroom?
- Which tasks should ISs take over to best support teachers?

This study contributes to a broader understanding and discussion about the role of ISs in education, highlighting the dilemmas and challenges that modern educational systems face with the integration of new technologies. The importance of this topic is particularly emphasized in light of the recently adopted legislation by the European Union, which in May 2024 approved the *Artificial Intelligence Act*. This legislation is the first of its kind in the world and establishes a comprehensive regulatory framework for the use of AI, including provisions related to the use of AI in education. The European Union aims to ensure that the use of these technologies is safe, ethical, and oriented towards the benefit of users, including both teachers and students. This research aims to contribute to a better understanding of how ISs can best support teachers in their work and thus improve the quality of education.

2 Theoretical Framework

AI and robots have been part of the educational landscape for over three decades [5–7]. Interest in AI and other ISs has particularly surged with the launch of the IS ChatGPT at the end of 2022. To properly contextualize this research, it is essential to define the area of interest. AI includes systems that mimic human intelligence to perform tasks such as speech recognition, decision-making, and pattern recognition [8]. The European Commission [9] defines AI systems as "software that is developed with one or more of the techniques and approaches listed in Annex I and can, for a given set of human-defined objectives, generate outputs such as content, predictions, recommendations, or decisions influencing the environments they interact with." The mentioned annex lists techniques and approaches such as machine learning, logic- and knowledge-based approaches, statistical approaches, Bayesian estimation, search, and optimization methods.

United Nations Educational, Scientific and Cultural Organization (UNESCO) [10], in its document *Artificial intelligence in education: challenges and opportunities for sustainable development*, examines the role of AI in education, emphasizing its potential to enhance teaching and learning practices and advance progress towards sustainable development goals. The main highlights include the following:

- *Personalized Learning*: AI enables the customization of educational content to the needs and abilities of individual students, leading to a more efficient and inclusive learning environment.
- *Data Analysis*: AI facilitates the collection and analysis of large amounts of data, improving the monitoring of student progress, identifying their needs, and providing targeted support.
- *Automation of Administrative Tasks*: AI can alleviate teachers of routine tasks, allowing them to focus more on teaching and interacting with students.
- *Ethics and Regulation*: UNESCO stresses the need for discussions on ethical issues and the regulation of AI in education to ensure that its use aligns with human rights and societal values.

AI can also be integrated into robots, presenting another innovation that can enhance the learning experience for students [11], as they can adapt to students' capabilities [12]. HRs are designed to mimic human form and behavior, facilitating more natural interaction with people. These machines are powered by AI, providing them with the capacity to perform complex tasks with minimal or no supervision [13]. In educational settings, they improve learning motivation and student engagement, help students develop various skills, including problem-solving, and enhance the overall learning experience while fostering integration between students and teachers [14].

Due to varying approaches to learning and technology adoption in educational environments, it is important to highlight learning theories that offer multidimensional frameworks for analyzing how students acquire, process, and use information and how educational methods and technologies can be adjusted to optimize learning outcomes.

Although some teachers are cautious about robots [15], a survey conducted in 14 countries showed that educators in higher education are open to innovations like educational robots, which can improve students' creativity and self-expression [16]. Constructivism emphasizes that learning is an active process in which students build knowledge through experiences and interactions with their environment [17], and the development of children's intellectual abilities is largely dependent on the guidance, direction, and explanation provided by teachers [18]. AI and HR can support constructivist learning by enabling simulations, interactive environments, and personalized learning paths. For instance, they can adjust content and delivery methods based on the progress and needs of individual students, promoting deeper understanding and independent problem-solving [19]. Social learning theory highlights that people learn by observing others, whereas AI and HR can be seen as models in the learning process, demonstrating ideal behaviors, strategies, and solutions in various situations. HRs, interacting with students, can demonstrate social and cognitive skills, aiding learning through observation and interaction. Sweller's cognitive load theory (CLT) posits that working memory is limited, and excessive cognitive load can hinder learning. To optimize learning, instructional materials and tasks should be designed to minimize unnecessary cognitive load and focus on key information [20]. AI systems can contribute to this by automating administrative tasks, using guided examples, and interactive learning tools, reducing the cognitive load on students.

Integrating learning theories into the use of AI and HR in education can lead to more tailored, effective, and inclusive learning experiences, contributing to better learning outcomes and holistic student development. As indicated by the aforementioned (though not exhaustive) learning theories, the individualization and personalization of educational content, traditionally emphasized, are critical aspects facilitated by ISs. Tang [21] identifies the primary issue in traditional educational structures as the reliance on a single teaching method by one teacher, which is likely ineffective for some students, as not all learn the same way. Traditional educational structures refer to classroom settings where a teacher delivers the same instruction to all students simultaneously, without considering the individual learning needs and styles of each student. Previous research supports this, showing that AI can significantly contribute to the individualization of the learning process [22, 23] and enhance learning experiences [24]. This is particularly important amid criticisms of the educational system as being inherently designed to suit everyone [25]. However, over-reliance on contemporary technologies may alter teacher–student interactions, as evidenced by some studies [26, 27].

In conclusion, two facts must be highlighted. First, there is a lack of longitudinal studies [28, 29], which have become more prominent in recent years [30, 31], yet still insufficient to draw concrete conclusions. These studies are crucial for understanding the long-term effects of AI in education and developing strategies that ensure sustainable benefits. Second, modern technologies develop unevenly across the world due to cultural and linguistic diversity, which also affects the effectiveness of AI systems in diverse educational settings [32, 33]. There may be differences in the acceptance and effectiveness of AI due to the specific needs of different cultural groups.

3 Analysis and Discussion

The integration of ISs into education suggests that AI and HR could further transform the role of teachers and/or the interaction dynamics between teachers and students, which are crucial for effective teaching. The teacher's role is a complex construct of paramount importance, as the teacher–student relationship (TSR) is key to developing effective teaching, learning, and assessment. Traditionally, this relationship is structured as a process of communication between the teacher and the student (Fig. 1).

With the advent of modern technologies, the ability for students to learn individually, at their own pace, using new systems, becomes increasingly emphasized. This capability significantly alters learning outcomes [34, 35], particularly through adaptive learning systems that facilitate interaction between ISs and students (Fig. 2). This further modifies the interaction between the student and the knowledge provider. Students may no longer need teachers in the traditional sense but instead have virtual tutors providing one-on-one education customized to their needs [36, 37].

The role of the teacher should not be reduced to simply managing technology or being replaced by it. Teachers must remain central to the educational process, bringing a human element that is indispensable for the holistic development of

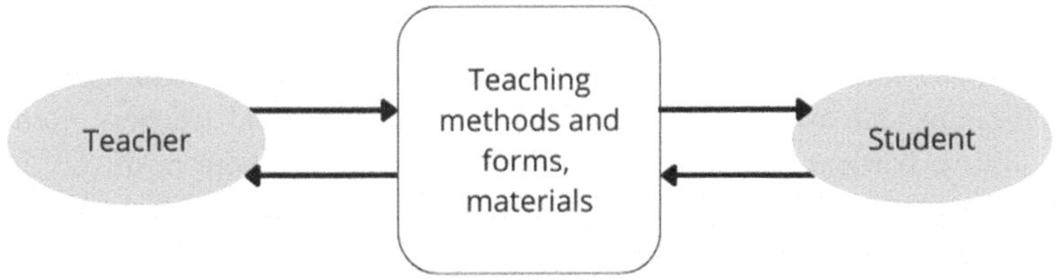

Fig. 1 Teacher–student communication. (Source: Own)

Fig. 2 IS–student communication. (Source: Own)

students. Teacher–student interaction is a crucial factor for both teacher and student well-being, the quality of teaching and learning, and the outcomes of both processes [38–40]. This interaction is already evolving in digital environments [40], and it changes further with the use of IS like AI and HRs. The teacher's role in these new environments should include understanding and using technology to support students rather than replacing them. Maintaining human contact and empathy, which technology currently cannot replicate, significantly contributes to the holistic development of students and their success in the learning process. Concepts such as affective TSRs highlight the emotional bonds that significantly impact various aspects of the educational experience, including academic success, engagement, motivation, and overall student well-being. These relationships are based on emotional support, positive communication, trust, respect, empathy, consistency, and encouragement [41].

Certainly, teachers and IS can collaborate effectively. IS teaching assistants can, for example, free up teachers' time by responding to students' routine, simple inquiries in online discussion forums. With that time, instructors may focus on higher-value tasks. In the context of education, these AI technologies can complete time-consuming administrative activities related to teaching and learning without sacrificing the quality of work [11, 42]. Xiao [40] emphasizes that teachers and students can effectively manage the integration of IS to ensure a well-rounded and meaningful learning experience by identifying and honing their distinct qualities. They argue that the future of education lies in the synergy between human teachers and AI.

Collaboration between teachers and AI is crucial for the future of education. AI assistants can take over routine tasks, allowing teachers to focus on more complex and creative aspects of teaching (Fig. 3). This maintains high educational quality while improving efficiency. Teachers and students must collaborate in integrating AI to leverage the advantages technology offers while retaining the human element essential for successful learning. The synergy between human teachers and IS will ensure a richer and more tailored learning experience that will still contribute to the well-being and overall satisfaction of both students and teachers. Reducing administrative burdens and improving the learning environment positively impact the psychological health and general satisfaction of all stakeholders.

The diagram presented in Fig. 3 illustrates the circular process of interaction and collaboration between teachers, IS, and students, aimed at enhancing the quality of education and learning outcomes. Teachers provide guidance and input to IS, which in turn assist with administrative tasks and supply student progress data. ISs facilitate personalized learning with simple tasks and collect student interaction and feedback, which are then used to improve teaching strategies and outcomes. This collaborative communication results in enhanced teaching quality, improved student outcomes, and reduced administrative workload for teachers. The diagram underscores the comprehensive nature of this process, which includes complex and creative tasks, assessment and evaluation, and holistic development, ensuring that the educational experience is both efficient and effective. Moreover, the integration of IS fosters a more adaptive and responsive educational environment, allowing

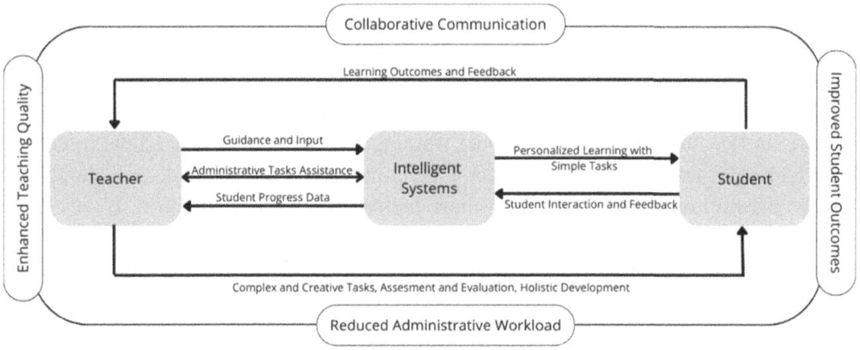

Fig. 3 Teacher–ISs–student communication. (Source: Own)

for real-time adjustments to teaching methods and materials based on continuous feedback. This dynamic interplay not only supports the individualized needs of students but also empowers teachers to focus on high-value educational activities, ultimately contributing to a more engaging and impactful learning experience. In this process, the focus remains on the student and the teacher, while ISs serve as tools for both, enhancing rather than replacing the teacher's role. ISs complement the teacher's work, enabling greater efficiency and flexibility in the learning process. This ensures that the educational environment is not only technologically advanced but also human-centered, which is crucial for the holistic development and success of students.

4 Conclusion

The rapid development and integration of IS such as AI and HR in education have sparked debates about their impact on the pedagogical role of teachers. The core issue lies in understanding whether these technologies enhance teaching or undermine essential teacher functions.

This study examined how ISs are used in education and their impact on teachers' roles. It explored whether these systems assist teachers or replace tasks that should remain within human teaching's domain. The research highlights AI's potential to automate administrative tasks, personalize learning experiences, and transform pedagogical methods. The findings suggest that while AI and HR can significantly enhance personalized learning and efficiency, they should not replace the teacher's role. The integration of these technologies must focus on maintaining the human element in education, ensuring that teachers can still provide emotional support and foster critical thinking and creativity. Additionally, it is essential to recognize that IS should serve as tools that complement the teacher's work rather than substitutes.

Specific pedagogical benefits and challenges of using AI and HR need to be emphasized. Benefits include increased student engagement, improved collaboration during group projects, and enhanced interactive learning environments. Challenges may involve initial resistance from educators, the need for technical support, and the necessity of developing training programs to help teachers integrate these technologies effectively into their teaching methods. Ethical considerations are also paramount when integrating these technologies into education. Issues such as data privacy, potential biases in AI algorithms, and ensuring transparency in AI decision-making processes must be addressed. Establishing ethical guidelines can ensure that the use of AI and HR respects student privacy and promotes fairness and equity.

The integration of IS such as AI and HR in education has the potential to transform the pedagogical role of teachers by automating administrative tasks, personalizing learning experiences, and enhancing pedagogical methods. However, it is crucial to ensure that these technologies complement rather than replace the essential human elements of teaching, such as emotional support and critical thinking. The synergy between human teachers and IS can lead to more tailored and effective learning experience, contributing to the well-being and development of both students and teachers. To achieve this, careful planning, ethical considerations, and continuous professional development for teachers are essential. The EU AI Act provides a regulatory framework to support the ethical and balanced integration of AI in education, promoting trust and transparency while preserving the critical human element of teaching.

In conclusion, the synergy between human teachers and AI is crucial for the future of education. IS can support teachers by taking over routine tasks, allowing teachers to focus on more complex and valuable aspects of teaching. This collaboration can lead to a richer, more tailored learning experience, ultimately contributing to the well-being and overall development of both students and teachers. Maintaining the focus on students and teachers ensures that education remains a deeply human-centered process, with IS enhancing the learning environment rather than detracting from the essential human interactions that characterize effective teaching. To fully realize these benefits, careful planning and consideration of cultural and ethical factors are essential. Moreover, continuous professional development for teachers in the use of these technologies will be vital to ensure they are equipped to integrate IS effectively into their teaching practices.

In this context, the EU AI Act plays a pivotal role by establishing clear guidelines and standards for the ethical and safe deployment of AI technologies. Adhering to this regulatory framework ensures that the integration of IS in education supports, rather than undermines, the fundamental role of teachers. By complying with the EU AI Act, educational institutions can promote a balanced approach that leverages the benefits of AI while preserving the critical human element of teaching, such as emotional support and creative engagement. This alignment fosters trust and transparency, ensuring that the educational environment remains focused on holistic student development and ethical technological advancement.

Acknowledgments This research was founded by the Slovenian Research Agency (ARRS), grant number P5-0433 for the research program Digital Restructuring of Deficit Occupations for Society 5.0 (Industry 4.0).

References

1. Nagy E, Holik I (2023) Educational robots in higher education -findings from an international survey. SACI 2023 – IEEE 17th International Symposium on Applied Computational Intelligence and Informatics, Proceedings 15–19. https://doi.org/10.1109/SACI58269.2023.10158668
2. Nagy E, Holik I (2022) Empirical research on robot-Assisted educational innovations. INES 2022 – 26th IEEE International Conference on Intelligent Engineering Systems 2022, Proceedings 91–95. https://doi.org/10.1109/INES56734.2022.9922652
3. Nagy E (2022) Robots in educational processes. Journal of Applied Multimedia. https://doi.org/10.26648/JAM.2022.1.001
4. Nagy E, Karl É, Molnár G Exploring the Role of Human-Robot Interactions, within the Context of the Effectiveness of a NAO Robot. Acta Polytechnica Hungarica 21:2024–177
5. Chassignol M, Khoroshavin A, Klimova A, Bilyatdinova A (2018) Artificial Intelligence trends in education: a narrative overview. Procedia Comput Sci 136:16–24. https://doi.org/10.1016/J.PROCS.2018.08.233
6. Devedzic V, Devedžić V (2004) Web Intelligence and Artificial Intelligence in Education
7. Peredo R, Canales A, Menchaca A, Peredo I (2011) Intelligent Web-based education system for adaptive learning. Expert Syst Appl 38:14690–14702. https://doi.org/10.1016/J.ESWA.2011.05.013
8. Xu Y, Liu X, Cao X, et al (2021) Artificial intelligence: A powerful paradigm for scientific research. Innovation 2:100179. https://doi.org/10.1016/J.XINN.2021.100179/ASSET/3F13AB5B-F920-46DA-85DE-0376543F7765/MAIN.ASSETS/GR6.JPG
9. European Commission (2021) Proposal for a regulation of the European parliament and of the council. https://eur-lex.europa.eu/legal-content/EN/TXT/HTML/?uri=CELEX%3A52021PC0206. Accessed 3 Sep 2024
10. UNESCO (2019) Challenges and Opportunities for Sustainable Development Education Sector United Nations Educational, Scientific and Cultural Organization
11. Chen L, Chen P, Lin Z (2020) Artificial Intelligence in Education: A Review. IEEE Access 8:75264–75278. https://doi.org/10.1109/ACCESS.2020.2988510
12. Timms MJ (2016) Letting Artificial Intelligence in Education out of the Box: Educational Cobots and Smart Classrooms. Int J Artif Intell Educ 26:701–712. https://doi.org/10.1007/S40593-016-0095-Y/METRICS
13. Newton DP, Newton LD (2019) Humanoid Robots as Teachers and a Proposed Code of Practice. Front Educ (Lausanne) 4:476824. https://doi.org/10.3389/FEDUC.2019.00125/BIBTEX
14. Wang K, Sang GY, Huang LZ, et al (2023) The Effectiveness of Educational Robots in Improving Learning Outcomes: A Meta-Analysis. Sustainability 2023, Vol 15, Page 4637 15:4637. https://doi.org/10.3390/SU15054637
15. Nagy E, Holik I (2022) Some aspects of the use of educational robotics in international and Hungarian contexts. ICCC 2022 – IEEE 10th Jubilee International Conference on Computational Cybernetics and Cyber-Medical Systems, Proceedings 173–178. https://doi.org/10.1109/ICCC202255925.2022.9922898
16. Nagy E, Holik I (2022) Examining negative attitudes towards robots among teachers. IEEE Joint 22nd International Symposium on Computational Intelligence and Informatics and 8th International Conference on Recent Achievements in Mechatronics, Automation, Computer Science and Robotics, CINTI-MACRo 2022 – Proceedings 221–226. https://doi.org/10.1109/CINTI-MACRO57952.2022.10029503

17. Fosnot CT, Perry RS (2005) Constructivism: A Psychological Theory of Learning
18. VYGOTSKY LS (1978) Development of Higher Psychological Processes. Harvard University Press
19. Bandura A (1986) Social foundations of thought and action: A social cognitive theory. Prentice-Hall, Inc, Englewood Cliffs, NJ, US
20. Sweller J (1988) Cognitive Load During Problem Solving: Effects on Learning. Cogn Sci 12:257–285. https://doi.org/10.1207/S15516709COG1202_4
21. Tang Y, Liang J, Hare R, Wang FY (2020) A Personalized Learning System for Parallel Intelligent Education. IEEE Trans Comput Soc Syst 7:352–361. https://doi.org/10.1109/TCSS.2020.2965198
22. Klašnja-Milićević A, Ivanović M (2021) E-learning Personalization Systems and Sustainable Education. Sustainability 2021, Vol 13, Page 6713 13:6713. https://doi.org/10.3390/SU13126713
23. Zavalevskyi Y, Kyrilenko S, Kijan O, et al (2024) The role of AI in individualizing learning and creating personalized programs. Revista Amazonia Investiga 13:200–208. https://doi.org/10.34069/AI/2024.73.01.16
24. Onesi-Ozigagun O, James Ololade Y, Louis Eyo-Udo N, et al (2024) REVOLUTIONIZING EDUCATION THROUGH AI: A COMPREHENSIVE REVIEW OF ENHANCING LEARNING EXPERIENCES. International Journal of Applied Research in Social Sciences 6:589–607. https://doi.org/10.51594/IJARSS.V6I4.1011
25. Bhutoria A (2022) Personalized education and Artificial Intelligence in the United States, China, and India: A systematic review using a Human-In-The-Loop model. Computers and Education: Artificial Intelligence 3:100068. https://doi.org/10.1016/J.CAEAI.2022.100068
26. Kaiyue L, Sun Z, Xu M (2021) Artificial intelligent based video analysis on the teaching interaction patterns in classroom environment. International Journal of Information and Education Technology 11:126–130. https://doi.org/10.18178/IJIET.2021.11.3.1500
27. Zhan Z, Wu Q, Lin Z, Cai J (2021) Smart classroom environments affect teacher-student interaction: Evidence from a behavioural sequence analysis. Australasian Journal of Educational Technology 37:96–109. https://doi.org/10.14742/AJET.6523
28. Hwang S (2022) Examining the Effects of Artificial Intelligence on Elementary Students' Mathematics Achievement: A Meta-Analysis. Sustainability 2022, Vol 14, Page 13185 14:13185. https://doi.org/10.3390/SU142013185
29. Zawacki-Richter O, Marín VI, Bond M, Gouverneur F (2019) Systematic review of research on artificial intelligence applications in higher education—where are the educators? International Journal of Educational Technology in Higher Education 2019 16:1 16:1–27. https://doi.org/10.1186/S41239-019-0171-0
30. Bergdahl J, Latikka R, Celuch M, et al (2023) Self-determination and attitudes toward artificial intelligence: Cross-national and longitudinal perspectives. Telematics and Informatics 82:102013. https://doi.org/10.1016/J.TELE.2023.102013
31. Polyportis A (2023) A longitudinal study on artificial intelligence adoption: understanding the drivers of ChatGPT usage behavior change in higher education. Front Artif Intell 6:1324398. https://doi.org/10.3389/FRAI.2023.1324398/BIBTEX
32. Cave S, Dihal K (2023) Imagining AI: How the World Sees Intelligent Machines. Imagining AI: How the World Sees Intelligent Machines 1–430. https://doi.org/10.1093/OSO/9780192865366.001.0001
33. Eke DO, Wakunuma K, Akintoye S Responsible AI in Africa Challenges and Opportunities
34. Luan H, Tsai C-C (2021) A Review of Using Machine Learning Approaches for Precision Education. Educational Technology & Society 24:250–266
35. Pai KC, Kuo BC, Liao CH, Liu YM (2021) An application of Chinese dialogue-based intelligent tutoring system in remedial instruction for mathematics learning. Educ Psychol (Lond) 41:137–152. https://doi.org/10.1080/01443410.2020.1731427
36. Jonnalagadda A, Rajvir M, Singh S, et al (2022) An Ensemble-Based Machine Learning Model for Emotion and Mental Health Detection. https://doi.org/101142/S0219649222500757 22:. https://doi.org/10.1142/S0219649222500757

37. Basu S, Chakraborty J, Aftabuddin M (2017) Emotion recognition from speech using convolutional neural network with recurrent neural network architecture. Proceedings of the 2nd International Conference on Communication and Electronics Systems, ICCES 2017 2018-January:333–336. https://doi.org/10.1109/CESYS.2017.8321292
38. Brekelmans M, Wubbels T, den Brok P (2002) TEACHER EXPERIENCE AND THE TEACHER–STUDENT RELATIONSHIP IN THE CLASSROOM ENVIRONMENT. Studies in Educational Learning Environments 73–99. https://doi.org/10.1142/9789812777133_0004
39. Hagenauer G, Muehlbacher F, Ivanova M (2023) "It's where learning and teaching begins—is this relationship" — insights on the teacher-student relationship at university from the teachers' perspective. High Educ (Dordr) 85:819–835. https://doi.org/10.1007/S10734-022-00867-Z/FIGURES/1
40. Xiao M, Tian Z, Xu W (2023) Impact of teacher-student interaction on students' classroom well-being under online education environment. Educ Inf Technol (Dordr) 28:14669–14691. https://doi.org/10.1007/S10639-023-11681-0/FIGURES/4
41. Roorda DL, Koomen HMY, Spilt JL, Oort FJ (2011) The Influence of Affective Teacher–Student Relationships on Students' School Engagement and Achievement. http://dx.doi.org/103102/0034654311421793 81:493–529. https://doi.org/10.3102/0034654311421793
42. Felix C V. (2020) The Role of the Teacher and AI in Education. 33:33–48. https://doi.org/10.1108/S2055-364120200000033003

Index

M. Turčáni (ed.), *15th International Scientific Conference on Distance Learning in Applied Informatics*, EAI/Springer Innovations in Communication and Computing,
https://doi.org/10.1007/978-3-031-81261-3

The manufacturer's authorised representative in the EU is Springer
Nature Customer Service Centre GmbH, Europaplatz 3, 69115 Heidelberg,
Germany. If you have any concerns regarding our products, please
contact ProductSafety@springernature.com

Printed and bound by CPI Group (UK) Ltd, Croydon, CR0 4YY
01/10/2026
02225932-0005